Frontiers in Aerospace Science

(Volume 4)

Advanced Control of Flight Vehicle Maneuver and Operation

Edited by

Chuang Liu

Honghua Dai

Xiaokui Yue

&

Yiqing Ma

School of Astronautics,
Northwestern Polytechnical University,
Xi'an 710072, P.R.
China

Frontiers in Aerospace Science

(Volume 4)

Advanced Control of Flight Vehicle Maneuver and Operation

Editors: Chuang Liu, Honghua Dai, Xiaokui Yue and Yiqing Ma

ISSN (Online): 2468-4724

ISSN (Print): 2468-4716

ISBN (Online): 978-981-5050-02-8

ISBN (Print): 978-981-5050-03-5

ISBN (Paperback): 978-981-5050-04-2

First published in 2023.

Bentham Science Publishers Pte. Ltd.
80 Robinson Road #02-00
Singapore 068898
Singapore
Email: subscriptions@benthamscience.net

CONTENTS

PREFACE i

LIST OF CONTRIBUTORS iii

CHAPTER 1 ANTI-DISTURBANCE CONTINUOUS FIXED-TIME CONTROLLER DESIGN FOR AIR-BREATHING HYPERSONIC VEHICLE 1

Yibo Ding, Wenbo Li, Panxing Huang, Jiashun Si and *Fenglin Wang4*

INTRODUCTION 1

FLEXIBLE HYPERSONIC VEHICLE MODEL 2

INPUT/OUTPUT FEEDBACK LINEARIZATION 4

ANTI-DISTURBANCE CONTINUOUS FIXED-TIME CONTROLLER DESIGN 6

Novel Fast Fixed-Time Integral Sliding Surface Design 6

Continuous Fixed-Time Super-Twisting-Like Reaching Law Scheme 14

Uniformly Convergent Observer Design 14

SIMULATION RESULTS 15

Simulations for a Fast Fixed-Time High-Order Regulator 16

Simulations of FAHV using FSMC 17

Simulations of FAHV using ACFTC 20

CONCLUSION 27

CONSENT FOR PUBLICATION 27

CONFLICT OF INTEREST 27

ACKNOWLEDGEMENTS 27

APPENDIX A: DEFINITIONS OF 28

REFERENCES 29

CHAPTER 2 FAST AND PARALLEL ALGORITHMS FOR ORBIT AND ATTITUDE COMPUTATION 31

Xuechuan Wang, Haoyang Feng and *Wei He1*

INTRODUCTION 31

LVIM AND QUASI-LINEARIZATION METHOD 34

Local Variational Iteration Method 34

Quasi-Linearization Method 36

Quasi Linearization of TPBVP 36

Transformation of Linear TPBVP to IVPs 37

PERTURBED ORBIT PROPAGATION 39

COMPARISON OF LVIM WITH MCPI 40

COMPARISON OF LVIM WITH RUNGE-KUTTA 12(10) 46

ERTURBED LAMBERT'S PROBLEM 50

Using LVIM 51

Using Fish-Scale-Growing Method 53

Using QLVIM 59

CONCLUSION 64

CONSENT FOR PUBLICATION 65

CONFLICT OF INTEREST 65

ACKNOWLEDGEMENT 65

REFERENCES 65

CHAPTER 3 ADAPTIVE EVENT-TRIGGERED SLIDING MODE CONTROL FOR SPACECRAFT ATTITUDE TRACKING 67

Jianqiao Zhang

INTRODUCTION 67

PRELIMINARIES 70

Spacecraft Attitude Motion Modeling *Via* Rotation Matrix 70

Attitude Error Dynamics of a Spacecraft 71

Control Objective 74

GEOMETRIC TRACKING CONTROL DESIGN ON SO(3) AND STABILITY

ANALYSIS 74
Fuzzy Logic System 75
FLS-Based Adaptive Event-Triggered SMC Design 76
SIMULAYIONS 82
CONCLUSION 90
CONSENT FOR PUBLICATION 90
CONFLICT OF INTEREST 91
ACKNOWLEDGEMENTS 91
REFERENCES 91

CHAPTER 4 ROBUST FINITE-TIME ADAPTIVE CONTROL ALGORITHM FOR SATELLITE ATTITUDE MANEUVER 95
Li You
INTRODUCTION 95
DYNAMIC AND KINETIC MODEL 97
ATTITUDE STABILIZATION ISSUE 100
Finite-time Controller 100
Adaptive Finite-Time Controller 104
ATTITUDE TRACKING ISSUE 108
SIMULATION 110
Attitude Stabilization Issue 110
Attitude Tracking Issue 117
CONCLUSION 123
CONSENT FOR PUBLICATION 123
CONFLICT OF INTEREST 123
ACKNOWLEDGEMENT 123
REFERENCES 124

CHAPTER 5 ATTITUDE STABILIZATION OF FLEXIBLE SPACECRAFT USING OUTPUT FEEDBACK CONTROLLER 127
Ziyu Yang, Xiaokui Yue and *Chuang Liu*
INTRODUCTION 128
DYNAMICS MODELING 129
PRELIMINARIES 135
CONTROLLER DESIGN 136
SOF $H\infty$ NI Controller Design 136
Decoupling Method Design 138
LMI-based Iterative Algorithm 141
Event-trigger Scheme Design 144
NUMERICAL SIMULATIONS 146
CONCLUSION 152
CONSENT FOR PUBLICATION 152
CONFLICT OF INTEREST 153
ACKNOWLEDGEMENT 153
REFERENCES 153

CHAPTER 6 VIBRATION CONTROL AND ENERGY HARVESTING IN AEROSPACE ENGINEERING USING NONLINEAR ENERGY SINKS 155
Haiqin Li, Ang Li and *Xianren Kong*
INTRODUCTION 155
VIBRATION SUPPRESSION USING AN NES WITH NONLINEAR DAMPING 158
Dynamical Modeling of a System Coupled to an NES 158
Bifurcation Analysis 160
Strongly Modulated Response 163
Vibration Suppression Performance 168
ENERGY HARVESTING BY MEANS OF A VINES 170
Formulation of an Energy Harvesting System with a VINES 170
Non-Smooth Contact Force Modeling 173
Transient Response Evaluation 175
Forced Response Evaluation 178

CONCLUDING REMARKS 181
CONSENT FOR PUBLICATION 182
CONFLICT OF INTEREST 182
ACKNOWLEDGEMENT 182
REFERENCES 182

CHAPTER 7 CONFIGURATION KEEPING TECHNOLOGY OF PARTIAL SPACE ELEVATORS 185
Gefei Shi and Zheng H. Zhu
INTRODUCTION 185
DYNAMICS OF THE PARTIAL SPACE ELEVATOR 188
Dynamic Model 188
Speed Function 190
CONFIGURATION KEEPING CONTROL 191
Control Scheme 191
Disturbance Observer 192
CASE STUDY 203
CONCLUSION 209
CONSENT FOR PUBLICATION 210
CONFLICT OF INTEREST 210
ACKNOWLEDGEMENT 210
REFERENCES 210

CHAPTER 8 ADAPTIVE FIXED-TIME 6-DOF COORDINATED CONTROL OF SPACECRAFT FORMATION FLYINGS 213
Ruixia Liu
INTRODUCTION 213
MODELLING AND PRELIMINARIES 215
6-DOF Dynamic Model 215
Graph Theory 217
Hysteretic Quantizer 218
Preliminaries 219
MULTI-SPACECRAFT NONSINGULAR FIXED-TIME TERMINAL SLIDING MODE 220
DESIGN OF FIXED-TIME ADAPTIVE COORDINATED CONTROL SCHEME 222
ILLUSTRATIVE EXAMPLE 232
CONCLUSION 238
CONSENT FOR PUBLICATION 238
CONFLICT OF INTEREST 238
ACKNOWLEDGEMENTS 238
REFERENCES 238

CHAPTER 9 FTCESO-BASED PRESCRIBED TIME CONTROL FOR SATELLITE CLUSTER RECONSTRUCTION 241
Siyuan Li, Zhaowei Sun, Yang Yang and *Fenglin Wang*
INTRODUCTION 241
DYNAMICS MODELING 243
CONTROLLER DESIGN 246
Disturbance Estimation based on FTCESO 247
Prescribed Time Controller Designing 249
NUMERICAL SIMULATIONS 254
CONCLUSION 260
CONSENT FOR PUBLICATION 260
CONFLICT OF INTEREST 261
ACKNOWLEDGEMENTS 261
REFERENCES 261

SUBJECT INDEX 263

PREFACE

To meet the requirements of future space missions, the high-precision and high-stability control problem of flight vehicles has become increasingly important. However, many uncertain factors, such as environmental and non-environmental disturbances, parameter uncertainty, and other nonlinear perturbations, widely influence the performance of flight vehicle maneuvers and operations. The present generation of flight vehicles should be capable of high-precision pointing and better robustness to external disturbances and various uncertainties. It should be noted that a distributed flight vehicle system may include multiple flight vehicles distributed in one or more orbits according to certain requirements and cooperating to perform space missions, *e.g.*, observation, communication, reconnaissance, and navigation. It should be mentioned that formation flying and satellite clusters both belong to distributed satellite systems. Consequently, multiple complex disturbances will have a considerable influence on the stability of the flight vehicle leading to degradation in the dynamics and control performance of the system and even instability, which pose a huge challenge for control system designers, and advanced control approaches are required to improve robust performance and control accuracy in maneuver and operation to solve these problems.

In order to understand the behavior of flight vehicle maneuvers and operation properly, it is significant to investigate the advanced controller designs. To this end, Chapter 1 discusses air-breathing hypersonic vehicle, Chapter 2 discusses fast and parallel algorithms for orbit and attitude computation, Chapters 3 and 4 discuss rigid spacecraft, Chapter 5 discusses flexible spacecraft, Chapter 6 discusses vibration control using nonlinear energy sink, Chapter 7 discusses partial space elevators, and Chapters 8 and 9 discuss spacecraft formation flying and satellite cluster, respectively. In particular, Chapter 1 investigates anti-disturbance continuous fixed-time controller design for an air-breathing hypersonic vehicle, where a fast fixed-time integral sliding surface, a continuous fixed-time super-twisting-like reaching law and a uniformly convergent observer are combined. As control efficiency is very important in practice, to efficiently solve nonlinear differential equations in aerospace engineering, fast and parallel algorithms can be a good choice, and this results in the writing of Chapter 2, where a simple adaptive local variational iteration method is developed. Chapter 3 provides detailed derivations of adaptive event-triggered sliding mode controller used for attitude tracking, where the communication burden is decreased significantly. Chapter 4 investigates an adaptive finite-time controller for satellite attitude maneuver, where the singularity problem is dealt with based on the properties of Euler rotations. Chapter 5 develops an output feedback controller for attitude stabilization and vibration suppression of flexible spacecraft using negative imaginary and H_∞ theories. To further investigate the vibration control and energy harvesting properties in aerospace engineering, a nonlinear energy sink approach is developed to deal with the influence of rich and complex dynamic environments in Chapter 6. Chapter 7 describes the mathematical model of partial space elevators, and a configure-keeping technology for stable cargo transportation is investigated. Furthermore, Chapters 8 and 9 discuss the distributed flight vehicle system, where adaptive fixed-time 6-DOF coordinated control for spacecraft formation flying and prescribed time control for satellite cluster reconstruction are investigated, respectively.

This book will be helpful to scientists and engineers who are interested in working on the development of flight vehicle maneuvers and operations. Researchers studying control science and engineering and advanced undergraduate and graduate students and professionals involved in the flight vehicle control field will also benefit from the information given in this book. This book covers a wide range of topics in flight vehicle maneuver and operation, *e.g.*, hypersonic vehicle, orbit and attitude computation, single spacecraft, flexible vibration, space elevators, spacecraft formation flying, satellite cluster, *et al.*

The book has a broad scope and helps students and researchers in universities, industries, and national and commercial laboratories to learn the fundamentals and in-depth knowledge regarding thermal modeling and developments in solar thermal systems in the past few years. It is a research-oriented book in which different researchers have contributed in the form of different chapters. I hope that the book will provide sufficient knowledge regarding solar systems and will not discourage the readers. This book can be used as a reference tool for teaching the solar energy and thermal modeling of solar thermal systems to the students and research fellows in universities and research organizations

Chuang Liu
School of Astronautics
Northwestern Polytechnical University
Xi'an 710072
China

List of Contributors

Ang Li	Research Center of Satellite Technology, Harbin Institute of Technology, Harbin 150001, China
Chuang Liu	Northwestern Polytechnical University, Xi'an 710072, China
Fenglin Wang	Beijing Institute of Tracking and Communication Technology, Beijing 100124, China
Gefei Shi	School of Aeronautics and Astronautics, Sun Yat-sen University, Guangzhou, 510275, P.R. China
Haiqin Li	Department of Mechanics, Tianjin University, Tianjin 300350, China
Haoyang Feng	Northwestern Polytechnical University, Xi'an 710072, China
Jianqiao Zhang	Shanghai Institute of Satellite Engineering, Shanghai, 201109, China
Jiashun Si	Beijing Research Institute of Automation for Machinery Industry CO., LTD., Beijing 100124, China
Li You	Xidian University, Xi'an 710126, China
Panxing Huang	Beijing Institute of Control Engineering China Academy of Space Technology, Beijing 100094, China
Ruixia Liu	School of Automation, Xi'an University of Posts & Telecommunications, Xi'an 710121, China
Siyuan Li	Research Center of Satellite Technology, Harbin Institute of Technology, Harbin 150001, China
Wei He	Northwestern Polytechnical University, Xi'an 710072, China
Wenbo Li	Beijing Institute of Control Engineering, China Academy of Space Technology, Beijing 100094, China
Xianren Kong	Research Center of Satellite Technology, Harbin Institute of Technology, Harbin 150001, China
Xiaokui Yue	Northwestern Polytechnical University, Xi'an 710072, China
Xuechuan Wang	Northwestern Polytechnical University, Xi'an 710072, China
Yang Yang	Xi'an Satellite Control Center, Xi'an 710043, China
Yibo Ding	Northwestern Polytechnical University, Xi'an 710072, China
Zhaowei Sun	Research Center of Satellite Technology, Harbin Institute of Technology, Harbin 150001, China
Zheng H. Zhu	Department of Mechanical Engineering, York University, Toronto, Ontario, Canada
Ziyu Yang	Northwestern Polytechnical University, Xi'an 710072, China

CHAPTER 1

Anti-Disturbance Continuous Fixed-Time Controller Design for Air-breathing Hypersonic Vehicle

Yibo Ding[1,*], **Wenbo Li**[2], **Panxing Huang**[2], **Jiashun Si**[3] and **Fenglin Wang**[4]

[1]*School of Astronautics, Northwestern Polytechnical University, Xi'an 710072, China*

[2]*Beijing Institute of Control Engineering, China Academy of Space Technology, Beijing 100094, China*

[3]*Beijing Research Institute of Automation for Machinery Industry CO., LTD., Beijing 100124, China*

[4]*Beijing Institute of Tracking and Communication Technology, Beijing 100124, China*

Abstract: Anti-disturbance, continuous fixed-time controller is designed for faulted air-breathing hypersonic vehicle, including a fast fixed-time integral sliding surface (FFIS), a continuous fixed-time super-twisting-like reaching law (CFSTL) and a uniformly convergent observer. Firstly, the model of a hypersonic vehicle is established. Secondly, an FFIS is designed based on a newly presented fast fixed-time high-order regulator (FFTR). Then, a CFSTL is applied to drive the sliding mode vector and its derivative to achieve fixed-time convergence. Finally, lumped disturbances are estimated by a uniformly convergent observer.

Keywords: Air-breathing hypersonic vehicle, Disturbance observer, Fixed-time control, Sliding mode control.

INTRODUCTION

The air-breathing hypersonic vehicle has recently received great attention [1, 2]. It is necessary to investigate the control algorithm with highly fault-tolerant ability for flexible air-breathing hypersonic vehicle (FAHV). There have been various control algorithms developed for fault tolerant control of hypersonic flight vehicles such as back-stepping control [3, 4], neural networks control, fuzzy control [5], sliding mode control [6, 7] *etc.* Considering the advantages of finite-time convergence laws, such as faster convergence speed and higher precision, sliding mode control is strongly recommended on account of its characteristics of finite-

[*]**Corresponding author Yibo Ding:** School of Astronautics, Northwestern Polytechnical University, Xi'an 710072, China; Tel: +86 18603615652; E-mails: dingyibo@nwpu.edu.cn and liwenbo_502@163.com

Chuang Liu, Honghua Dai, Xiaokui Yue & Yiqing Ma (Eds.)

time convergence and robustness. Considering FAHV is statically unstable, system states should be rapidly stabilized before they escape to infinity under abrupt actuator faults. Therefore, in this paper, an anti-disturbance continuous fixed-time control algorithm (ACFTC) is proposed to achieve high-precision fast control for FAHV under actuator faults with three components: novel fast fixed-time integral sliding surface (FFIS), continuous fixed-time super-twisting-like reaching law (CFSTL) and uniformly convergent observer.Many sliding surfaces have been proposed such as , linear sliding surfaces, terminal sliding surfaces, integral sliding surfaces *etc.* The linear sliding surface could only drive states to converge exponentially. The terminal sliding surface is able to realize finite-time convergence of states, but the singularity phenomenon is inevitable, which restricts its application. In view of it, a non-singular sliding surface is proposed by Feng [8], but its superiority is lost when applied to a system with an order higher than two. An integral sliding surface is a kind of sliding surface that can avoid singularity by means of using a finite-time regulator [9, 10]. In [11], Basin proposes a fixed-time high-order regulator to improve the conventional finite-time regulator, which can achieve fixed-time convergence. In this paper, we present a fast fixed-time high-order regulator (FFTR) based on the method in [11]. FFTR can accelerate the system's response speed by simply adjusting the values of two gains. Then, FFIS is established to make tracking errors of FAHV to achieve fixed-time convergence based on FFTR.

Traditional reaching laws mainly include: constant rate reaching law, exponential reaching law, power rate reaching law and so on. The super twisting algorithm can be used as a reaching law to realize finite-time convergence. Basin in [12] designs a continuous fixed-time super-twisting-like reaching law (CFSTL). It is more simplified in form than the conventional method in [13]. By combing FFIS, CFSTL and uniformly convergent observer, an ACFTC is designed to enhance the fault-tolerant ability, in which uniformly convergent observer can estimate lumped disturbances accurately in fixed time.

The paper's organization is as follows: Section 2 provides the model of FAHV. In Section 3, the input/output feedback linearization technique is designed. Section 4 presents ACFTC. In Section 5, simulations are shown to verify the effectiveness of ACFTC.

FLEXIBLE HYPERSONIC VEHICLE MODEL

The longitudinal dynamics of FAHV are given below

$$\begin{aligned}
&\dot{V} = (T\cos\alpha - D)/m - g\sin\gamma \\
&\dot{h} = V\sin\gamma \\
&\dot{\gamma} = (L + T\sin\alpha)/(mV) - g/V\cos\gamma \\
&\dot{\alpha} = Q - \dot{\gamma} \\
&\dot{Q} = (M + \tilde{\psi}_1\ddot{\eta}_1 + \tilde{\psi}_2\ddot{\eta}_2)/I_{yy} \\
&k_1\ddot{\eta}_1 = -2\zeta_1\omega_1\dot{\eta}_1 - \omega_1^2\eta_1 + N_1 - \tilde{\psi}_1 M/I_{yy} - \tilde{\psi}_1\tilde{\psi}_2\ddot{\eta}_2/I_{yy} \\
&k_2\ddot{\eta}_2 = -2\zeta_2\omega_2\dot{\eta}_2 - \omega_2^2\eta_2 + N_2 - \tilde{\psi}_2 M/I_{yy} - \tilde{\psi}_2\tilde{\psi}_1\ddot{\eta}_1/I_{yy}
\end{aligned} \tag{1}$$

where $V,h,\gamma,\alpha,Q,\eta_1,\eta_2$ are vehicle velocity, altitude, flight path angle, angle of attack, pitch rate, and generalized modal coordinates respectively. m,g,I_{yy} are mass, gravitational acceleration and moment of inertia. $\zeta_i,\omega_i,\tilde{\psi}_i$ are damping ratio, natural frequency and inertial coupling parameter. k_i satisfy with $k_i = 1 + \tilde{\psi}_i^2/I_{yy}$. T,D,L,M,N_i are thrust, drag, lift, pitching moment and generalized forces which are expressed using curve-fitted approximations as follows

$$\begin{aligned}
&T = C_T^{\alpha^3}\alpha^3 + C_T^{\alpha^2}\alpha^2 + C_T^{\alpha}\alpha + C_T^0 \\
&D = \bar{q}S\left(C_D^{\alpha^2}\alpha^2 + C_D^{\alpha}\alpha + C_D^{\delta_e^2}\delta_e^2 + C_D^{\delta_e}\delta_e + C_D^0\right) \\
&L = \bar{q}S\left(C_L^{\alpha}\alpha + C_L^{\delta_e}\delta_e + C_L^0\right) \\
&M = z_T T + \bar{q}S\bar{c}\left(C_{M,\alpha}^{\alpha^2}\alpha^2 + C_{M,\alpha}^{\alpha}\alpha + C_{M,\alpha}^0 + c_e\delta_e\right) \\
&N_1 = N_1^{\alpha^2}\alpha^2 + N_1^{\alpha}\alpha + N_1^0 \\
&N_2 = N_2^{\alpha^2}\alpha^2 + N_2^{\alpha}\alpha + N_2^{\delta_e}\delta_e + N_2^0 \\
&\bar{q} = \rho V^2/2 \quad \rho = \rho_0\exp\left[-(h-h_0)/h_s\right] \\
&C_T^{\alpha^3} = \beta_1(h,\bar{q})\Phi + \beta_2(h,\bar{q}) \quad C_T^{\alpha^2} = \beta_3(h,\bar{q})\Phi + \beta_4(h,\bar{q}) \\
&C_T^{\alpha} = \beta_5(h,\bar{q})\Phi + \beta_6(h,\bar{q}) \quad C_T^0 = \beta_7(h,\bar{q})\Phi + \beta_8(h,\bar{q})
\end{aligned} \tag{2}$$

where δ_e,Φ are elevator angular deflection and fuel-to-air ratio. In order to compensate for non-minimum phase behavior, the canard δ_c is responded to ensure $C_L^{\delta_e}\delta_e + C_L^{\delta_c}\delta_c = 0$. Therefore, the lift L is expressed as

$$L = \bar{q}S\left(C_L^{\alpha}\alpha + C_L^0\right) \tag{3}$$

INPUT/OUTPUT FEEDBACK LINEARIZATION

By neglecting elevator couplings and treating rigid-flexible couplings as disturbances, the Equation (1) is changed as:

$$\begin{aligned}
\dot{V} &= \left(T\cos\alpha - \bar{D}\right)/m - g\sin\gamma \\
\dot{h} &= V\sin\gamma \\
\dot{\gamma} &= \left(L + T\sin\alpha\right)/\left(mV\right) - g/V\cos\gamma \\
\dot{\alpha} &= Q - \dot{\gamma} \\
\dot{Q} &= f_Q + g_Q\delta_e + d_c
\end{aligned} \tag{4}$$

in which

$$\begin{aligned}
\bar{D} &= \bar{q}S\left(C_D^{\alpha^2}\alpha^2 + C_D^{\alpha}\alpha + C_D^0\right) \\
f_Q &= z_T T + \bar{q}S\bar{c}\left(C_{M,\alpha}^{\alpha^2}\alpha^2 + C_{M,\alpha}^{\alpha}\alpha + C_{M,\alpha}^0\right) \\
g_Q &= \bar{q}S\bar{c}c_e \\
d_c &= \left(\tilde{\psi}_1\ddot{\eta}_1 + \tilde{\psi}_2\ddot{\eta}_2\right)/I_{yy}
\end{aligned} \tag{5}$$

The dynamics of Φ and Φ_c are given as follows

$$\ddot{\Phi} = -2\zeta\omega\dot{\Phi} - \omega^2\Phi + \omega^2\Phi_c \tag{6}$$

where $\zeta = 0.7, \omega = 20$. In order to modulate the desired commands V_c and h_c into continuous smooth reference commands V_d and h_d, we design two second-order pre-filters as follows. The changes of damping ratio and angular frequency can change the speed of reference command response. Therefore, they can be designed according to the needs.

$$V_d = V_c \times 0.03^2/\left(s^2 + 2\times 0.95\times 0.03s + 0.03^2\right) \tag{7}$$

$$h_d = h_c \times 0.03^2/\left(s^2 + 2\times 0.95\times 0.03s + 0.03^2\right) \tag{8}$$

The desired value of γ is [14]

$$\gamma_d = \arcsin\left[\left(\dot{h}_d - k_P e_h\right)/V\right] \tag{9}$$

where $k_P > 0$, $e_h = h - h_d$. If γ is controlled to follow γ_d, then e_h can satisfy

$$\dot{e}_h + k_P e_h = 0 \tag{10}$$

Select $\boldsymbol{u} = [\delta_e \quad \Phi_c]^T$ and $\boldsymbol{y} = [V \quad \gamma]^T$. Control signal under faults denoted by $\boldsymbol{u}_f$ is:

$$\boldsymbol{u}_f = \boldsymbol{F}_g \boldsymbol{u} + \boldsymbol{F}_d \tag{11}$$

in which $\boldsymbol{F}_g$ and $\boldsymbol{F}_d$ are:

$$\begin{aligned} \boldsymbol{F}_g &= diag\left(F_{g1} \quad F_{g2}\right) \\ \boldsymbol{F}_d &= \left[F_{d1} \quad F_{d2}\right]^T \end{aligned} \tag{12}$$

where $0 < F_{g1}, F_{g2} \le 1$. If $\boldsymbol{F}_g = \boldsymbol{E}_2$ and $\boldsymbol{F}_d = [0 \quad 0]^T$, the actuators are working in normal mode.

The output vector is defined as $\boldsymbol{e} = \boldsymbol{y} - \boldsymbol{y}_d = [e_1 \quad e_2]^T = [V - V_d \quad \gamma - \gamma_d]^T$. Therefore, an affine nonlinear form of error dynamics under actuator faults is derived from COM *via* input/output feedback linearization [15].

$$\begin{aligned} \dddot{\boldsymbol{e}} &= \dddot{\boldsymbol{y}} - \dddot{\boldsymbol{y}}_d \\ &= \boldsymbol{F} + \boldsymbol{G}\boldsymbol{u}_f + \boldsymbol{G}_d d_c - \dddot{\boldsymbol{y}}_d \\ &= \boldsymbol{F} + \boldsymbol{G}\left(\boldsymbol{F}_g \boldsymbol{u} + \boldsymbol{F}_d\right) + \boldsymbol{G}_d d_c - \dddot{\boldsymbol{y}}_d \\ &= \boldsymbol{F} + \boldsymbol{G}\boldsymbol{u} - \dddot{\boldsymbol{y}}_d + \boldsymbol{G}\left[\left(\boldsymbol{F}_g - \boldsymbol{E}_2\right)\boldsymbol{u} + \boldsymbol{F}_d\right] + \boldsymbol{G}_d d_c \end{aligned} \tag{13}$$

where

$$\begin{aligned} \boldsymbol{F} &= \left[f_V \quad f_\gamma\right]^T \\ \boldsymbol{G} &= \begin{bmatrix} g_{V_1} & g_{V_2} \\ g_{\gamma_1} & g_{\gamma_2} \end{bmatrix} \\ \boldsymbol{G}_d &= \left[g_{V_3} \quad g_{\gamma_3}\right]^T \\ f_V &= \boldsymbol{\omega}_1 \ddot{\boldsymbol{w}} + \dot{\boldsymbol{w}}^T \boldsymbol{\Omega}_2 \dot{\boldsymbol{w}} \\ g_{V_1} &= g_Q\left(\partial T/\partial\alpha \cos\alpha - T\sin\alpha - \partial\bar{D}/\partial\alpha\right)/m \\ g_{V_2} &= \omega^2\left(\partial T/\partial\Phi \cos\alpha\right)/m \\ g_{V_3} &= \left(\partial T/\partial\alpha \cos\alpha - T\sin\alpha - \partial\bar{D}/\partial\alpha\right)/m \end{aligned}$$

$$\begin{aligned}
f_\gamma &= \boldsymbol{\pi}_1 \dddot{\boldsymbol{w}} + \dot{\boldsymbol{w}}^T \boldsymbol{\Pi}_2 \dot{\boldsymbol{w}} \\
g_{\gamma_1} &= g_Q \left(\partial T/\partial\alpha \sin\alpha + T\cos\alpha + \partial L/\partial\alpha\right)/(mV) \\
g_{\gamma_2} &= \omega^2 \left(\partial T/\partial\Phi \sin\alpha\right)/(mV) \\
g_{\gamma_3} &= \left(\partial T/\partial\alpha \sin\alpha + T\cos\alpha + \partial L/\partial\alpha\right)/(mV)
\end{aligned} \tag{14}$$

in which

$$\begin{aligned}
\dot{\boldsymbol{w}} &= \begin{bmatrix} \dot{V} & \dot{\alpha} & \dot{\gamma} & \dot{\Phi} & \dot{h} \end{bmatrix}^T \\
\ddot{\boldsymbol{w}} &= \begin{bmatrix} \boldsymbol{\omega}_1 \dot{\boldsymbol{w}} \\ -\boldsymbol{\pi}_1 \dot{\boldsymbol{x}} + f_Q \\ \boldsymbol{\pi}_1 \dot{\boldsymbol{w}} \\ -2\zeta\omega\dot{\Phi} - \omega^2\Phi \\ \dot{V}\sin\gamma + V\dot{\gamma}\cos\gamma \end{bmatrix}
\end{aligned} \tag{15}$$

The expressions of $\boldsymbol{\omega}_1, \boldsymbol{\Omega}_2, \boldsymbol{\pi}_1, \boldsymbol{\Pi}_2$ are shown in Appendix A. The lumped disturbances $\boldsymbol{D}_f$ are denoted as

$$\boldsymbol{D}_f = \begin{bmatrix} D_{f1} & D_{f2} \end{bmatrix}^T = \boldsymbol{G}\left[\left(\boldsymbol{F}_g - \boldsymbol{E}_2\right)\boldsymbol{u} + \boldsymbol{F}_d\right] + \boldsymbol{G}_d d_c \tag{16}$$

The derivatives of $\boldsymbol{y}$ are calculated as

$$\begin{aligned}
\dot{\boldsymbol{y}} &= \begin{bmatrix} \dot{V} & \dot{\gamma} \end{bmatrix}^T \\
\ddot{\boldsymbol{y}} &= \begin{bmatrix} \boldsymbol{\omega}_1 \dot{\boldsymbol{w}} & \boldsymbol{\pi}_1 \dot{\boldsymbol{w}} \end{bmatrix}^T
\end{aligned} \tag{17}$$

ANTI-DISTURBANCE CONTINUOUS FIXED-TIME CONTROLLER DESIGN

Novel Fast Fixed-Time Integral Sliding Surface Design

Theorem 1: *Consider a general 3-order chain of integrators*

$$\begin{aligned}
\dot{x}_1 &= x_2 \\
\dot{x}_2 &= x_3 \\
\dot{x}_3 &= u
\end{aligned} \tag{18}$$

with a fast fixed-time high-order regulator (FFTR) given as

$$\begin{aligned}u = &-k_{L1}L_L^{1-\alpha_1}sig^{\alpha_1}(x_1)-k_{L2}L_L^{1-\alpha_2}sig^{\alpha_2}(x_2)-k_{L3}L_L^{1-\alpha_3}sig^{\alpha_3}(x_3)\\ &-L_H sig^{\beta_1}(x_1)-2L_H sig^{\beta_2}(x_2)-L_H sig^{\beta_3}(x_3)\end{aligned} \tag{19}$$

in which $L_L \geq 1$, $L_H \geq 1$ and polynomial $s^3+k_{L3}s^2+k_{L2}s+k_{L1}$ is Hurwitz. Origin of (18) is fixed-time stable with α_i and β_i satisfy

$$\begin{aligned}\alpha_{i-1} &= \frac{\alpha_i \alpha_{i+1}}{2\alpha_{i+1}-\alpha_i}\\ \beta_{i-1} &= \frac{\beta_i \beta_{i+1}}{2\beta_{i+1}-\beta_i}\\ i &= 1,2,3\end{aligned} \tag{20}$$

where $\alpha_4=\beta_4=1$, $\alpha_3=\alpha_\varepsilon \in (1-\varepsilon,\ 1)$ and $\beta_3=\beta_\varepsilon \in (1,\ 1+\varepsilon_1)$ for sufficiently small positive numbers ε and ε_1. The upper bound of convergence time for (18) will decrease along with the increasing of L_L and L_H.

Proof: Substitute (19) into (18) and separate it into two subsystems

$$\begin{aligned}\dot{x}_1 &= x_2\\ \dot{x}_2 &= x_3\\ \dot{x}_3 &= -k_{L1}L_L^{1-\alpha_1}sig^{\alpha_1}(x_1)-k_{L2}L_L^{1-\alpha_2}sig^{\alpha_2}(x_2)-k_{L3}L_L^{1-\alpha_3}sig^{\alpha_3}(x_3)\end{aligned} \tag{21}$$

$$\begin{aligned}\dot{x}_1 &= x_2\\ \dot{x}_2 &= x_3\\ \dot{x}_3 &= -L_H sig^{\beta_1}(x_1)-2L_H sig^{\beta_2}(x_2)-L_H sig^{\beta_3}(x_3)\end{aligned} \tag{22}$$

Step 1.

Define matrix $\boldsymbol{A}_L$ as

$$\boldsymbol{A}_L = \begin{bmatrix} 0 & 1 & 0\\ 0 & 0 & 1\\ -k_{L1}L_L^{1-\alpha_1} & -k_{L2}L_L^{1-\alpha_2} & -k_{L3}L_L^{1-\alpha_3}\end{bmatrix} \tag{23}$$

The characteristic polynomial of $\boldsymbol{A}_L$ is $s^3+k_{L3}L_L^{1-\alpha_3}s^2+k_{L2}L_L^{1-\alpha_2}s+k_{L1}L_L^{1-\alpha_1}$. $\boldsymbol{A}_L$ is Hurwitz if the following inequations are satisfied

$$\begin{aligned} &k_{L3}L_L^{1-\alpha_3} > 0 \\ &k_{L1}L_L^{1-\alpha_1} > 0 \\ &\frac{k_{L3}L_L^{1-\alpha_3} \cdot k_{L2}L_L^{1-\alpha_2} - k_{L1}L_L^{1-\alpha_1}}{k_{L3}L_L^{1-\alpha_3}} > 0 \end{aligned} \tag{24}$$

in which

$$\frac{k_{L3}L_L^{1-\alpha_3} \cdot k_{L2}L_L^{1-\alpha_2} - k_{L1}L_L^{1-\alpha_1}}{k_{L3}L_L^{1-\alpha_3}} = k_{L2}L_L^{1-\alpha_2} - \frac{k_{L1}}{k_{L3}}L_L^{\alpha_3-\alpha_1} = k_{L2}L_L^{\frac{2(1-\alpha_\varepsilon)}{2-\alpha_\varepsilon}} - \frac{k_{L1}}{k_{L3}}L_L^{\frac{2\alpha_\varepsilon(1-\alpha_\varepsilon)}{3-2\alpha_\varepsilon}} \tag{25}$$

When $\alpha_\varepsilon \in (1-\varepsilon,\ 1)$, variables $\alpha_3 = \alpha_\varepsilon$, $\alpha_2 = \alpha_\varepsilon/(2-\alpha_\varepsilon)$ and $\alpha_1 = \alpha_\varepsilon/(3-2\alpha_\varepsilon)$ belong to $(0,\ 1)$. The following inequality holds

$$\frac{2(1-\alpha_\varepsilon)}{2-\alpha_\varepsilon} - \frac{2\alpha_\varepsilon(1-\alpha_\varepsilon)}{3-2\alpha_\varepsilon} = (1-\alpha_\varepsilon)\frac{2\alpha_\varepsilon^2 - 8\alpha_\varepsilon + 6}{(2-\alpha_\varepsilon)(3-2\alpha_\varepsilon)} > 0 \tag{26}$$

Considering $s^3 + k_{L3}s^2 + k_{L2}s + k_{L1}$ is Hurwitz, it is known that $k_{L2} > k_{L1}/k_{L3}$. Therefore, when $L_L \geq 1$, the conditions in (24) are satisfied. There exists a symmetric positive definite matrix $\boldsymbol{P}_L$ satisfied with

$$\boldsymbol{P}_L\boldsymbol{A}_L + \boldsymbol{A}_L^T\boldsymbol{P}_L = -\boldsymbol{Q}_L \tag{27}$$

where $\boldsymbol{Q}_L$ is an arbitrary positive definite and symmetric matrix. A Lyapunov function for (21) is defined as

$$V_L(\varphi) = \varphi^T(\chi)\boldsymbol{P}_L\varphi(\chi) \tag{28}$$

in which $\varphi = \begin{bmatrix} x_1^{1/r_1} & x_2^{1/r_2} & x_3^{1/r_3} \end{bmatrix}^T$, $\chi = \begin{bmatrix} x_1 & x_2 & x_3 \end{bmatrix}^T$ $r_1 = 3-2\alpha_\varepsilon$, $r_2 = 2-\alpha_\varepsilon$, and $r_3 = 1$. Considering $\dot{\chi} = \boldsymbol{A}_L\chi$ is asymptotically stable. Selecting $V_L(\chi) = \chi^T\boldsymbol{P}_L\chi$ as a Lyapunov function for $\dot{\chi} = \boldsymbol{A}_L\chi$, $V_L(\chi)$ is satisfied with

$$\dot{V}_L(\chi) = \dot{\chi}^T\boldsymbol{P}_L\chi + \chi^T\boldsymbol{P}_L\dot{\chi} = \chi^T\left(\boldsymbol{A}_L^T\boldsymbol{P}_L + \boldsymbol{P}_L\boldsymbol{A}_L\right)\chi = -\chi^T\boldsymbol{Q}_L\chi < 0 \tag{29}$$

The derivative of Lyapunov function $V_L(\varphi)$ is also negative definite with $\alpha_\varepsilon \in (1-\varepsilon,\ 1)$ for a sufficiently small $\varepsilon > 0$ [16]. Thus, (21) is asymptotically stable. Let the homogeneity degree of x_1, x_2, x_3 be r_1, r_2 and r_3 respectively. Then the system (21) is homogeneous of degree $\alpha_\varepsilon - 1$ with respect to the vector χ according to the definition of homogeneous vector fields in [17]. The system (21) could converge to the origin in finite time. A dilation rescaling of states in the system (21) is taken as

$$z_1 = x_1/L_L \quad z_2 = x_2/L_L \quad z_3 = x_3/L_L \tag{30}$$

The system (21) can be expressed as:

$$\begin{aligned} \dot{z}_1 &= z_2 \\ \dot{z}_2 &= z_3 \\ \dot{z}_3 &= -k_{L1}sig^{\alpha_1}(z_1) - k_{L2}sig^{\alpha_2}(z_2) - k_{L3}sig^{\alpha_3}(z_3) \end{aligned} \tag{31}$$

System (21) and system (31) have the same convergence time. There exists a symmetric positive definite matrix $\boldsymbol{P}_{LI}$ satisfied with:

$$\boldsymbol{P}_{LI}\boldsymbol{A}_{LI} + \boldsymbol{A}_{LI}^T\boldsymbol{P}_{LI} = -\boldsymbol{Q}_{LI} \tag{32}$$

in which $\boldsymbol{Q}_{LI}$ is symmetric positive definite and $\boldsymbol{A}_{LI}$ is defined as

$$\boldsymbol{A}_{LI} = \begin{bmatrix} 0 & 1 & 0 \\ 0 & 0 & 1 \\ -k_{L1} & -k_{L2} & -k_{L3} \end{bmatrix} \tag{33}$$

Define a Lyapunov function as $V_{L1}(\varphi_I) = \varphi_I^T(\zeta)\boldsymbol{P}_{LI}\varphi_I(\zeta)$, where $\varphi_I = \begin{bmatrix} z_1^{1/r_1} & z_2^{1/r_2} & z_3^{1/r_3} \end{bmatrix}^T$ and $\zeta = \begin{bmatrix} z_1 & z_2 & z_3 \end{bmatrix}^T$. The proof process of the system (31) is similar to the system(21). Function $V_{L1}(\varphi_I)$ and $\dot{V}_{L1}(\varphi_I)$ are homogeneous in vector ζ of degree 2 and $1+\alpha_\varepsilon$ respectively with respect to the same weights r_i. Thus, the following inequality holds [18]:

$$\dot{V}_{L1}(\varphi_I) \le -c_1 V_{L1}^{\frac{1+\alpha_\varepsilon}{2}}(\varphi_I) \tag{34}$$

where c_1 is a positive constant. Considering $\dot{\zeta} = \boldsymbol{A}_{LI}\zeta$ is stable, it can be known that $V_{L1}(\zeta) \le \lambda_{\max}(\boldsymbol{P}_{LI})\|\zeta\|^2$ and $\dot{V}_{L1}(\zeta) \le -\lambda_{\min}(\boldsymbol{Q}_{LI})\|\zeta\|^2$. Therefore,

$$\dot{V}_{L1}(\zeta) \le -\frac{\lambda_{\min}(\boldsymbol{Q}_{LI})}{\lambda_{\max}(\boldsymbol{P}_{LI})} V_{L1}(\zeta) < -\frac{\lfloor \lambda_{\min}(\boldsymbol{Q}_{LI}) - \delta \rfloor}{\lambda_{\max}(\boldsymbol{P}_{LI})} V_{L1}(\zeta) \tag{35}$$

in which δ is an arbitrarily small positive number. Since (31) is continuous with respect to α_ε, the following relationship can be obtained [19]:

$$\dot{V}_{L1}(\varphi_I) \le -\frac{\lambda_{\min}(\boldsymbol{Q}_{LI})}{\lambda_{\max}(\boldsymbol{P}_{LI})} V_{L1}^{\frac{1+\alpha_\varepsilon}{2}}(\varphi_I) \tag{36}$$

Thus, the finite convergence time of (31) satisfies [20]:

$$T_L \le \frac{\lambda_{\max}(\boldsymbol{P}_{LI})(1-\alpha_\varepsilon)}{2\lambda_{\min}(\boldsymbol{Q}_{LI})} V_{L1}^{\frac{1-\alpha_\varepsilon}{2}}(\varphi_I(t_0)) \le \frac{\lambda_{\max}(\boldsymbol{P}_{LI})(1-\alpha_\varepsilon)}{2\lambda_{\min}(\boldsymbol{Q}_{LI})} \lambda_{\max}^{\frac{1-\alpha_\varepsilon}{2}}(\boldsymbol{P}_{LI}) \|\varphi_I(t_0)\|^{1-\alpha_\varepsilon} \tag{37}$$

in which $\varphi_I(t_0)$ is the initial condition of (31). Since $L_L \ge 1$, $0 < \alpha_\varepsilon < 1$ and $1 = r_3 < r_2 < r_1$, the following inequality holds:

$$T_L \le \frac{(1-\alpha_\varepsilon)}{2\lambda_{\min}(\boldsymbol{Q}_{LI})} \lambda_{\max}^{\frac{3-\alpha_\varepsilon}{2}}(\boldsymbol{P}_{LI}) \left(\frac{\|\varphi(t_0)\|}{L_L^{1/r_1}}\right)^{1-\alpha_\varepsilon} \le \frac{(1-\alpha_\varepsilon)}{2\lambda_{\min}(\boldsymbol{Q}_{LI})} \lambda_{\max}^{\frac{3-\alpha_\varepsilon}{2}}(\boldsymbol{P}_{LI}) \left(\frac{\|\varphi(t_0)\|}{L_L^{1/(3-2\alpha_\varepsilon)}}\right)^{1-\alpha_\varepsilon} \tag{38}$$

where $\varphi(t_0)$ is the initial condition of (21). It can be seen T_L will decrease with the increase of L_L.

Step 2.

Define $\boldsymbol{A}_H$ as

$$\boldsymbol{A}_H = \begin{bmatrix} 0 & 1 & 0 \\ 0 & 0 & 1 \\ -L_H & -2L_H & -L_H \end{bmatrix} \tag{39}$$

The characteristic polynomial of $\boldsymbol{A}_H$ is $s^3 + L_H s^2 + 2L_H s + L_H$. There exists a symmetric positive definite matrix $\boldsymbol{P}_H$ satisfied with

$$\boldsymbol{P}_H \boldsymbol{A}_H + \boldsymbol{A}_H^T \boldsymbol{P}_H = -\boldsymbol{Q}_H \tag{40}$$

in which Q_H is an arbitrary positive definite and symmetric matrix. A Lyapunov function for (22) is given as:

$$V_H(\xi)=\xi^T(\chi)P_H\xi(\chi) \quad \textbf{(41)}$$

where $\xi=\left[x_1^{1/\gamma_1} \quad x_2^{1/\gamma_2} \quad x_3^{1/\gamma_3}\right]^T$, $\chi=\left[x_1 \quad x_2 \quad x_3\right]^T$ $\gamma_1=3-2\beta_\varepsilon$, $\gamma_2=2-\beta_\varepsilon$, and $\gamma_3=1$. Since A_H is Hurwitz, $\dot{\chi}=A_H\chi$ is asymptotically stable. Selecting $V_H(\chi)=\chi^T P_H\chi$ as a Lyapunov function for $\dot{\chi}=A_H\chi$, the derivative of $V_H(\chi)$ satisfies:

$$\dot{V}_H(\chi)=\dot{\chi}^T P_H\chi+\chi^T P_H\dot{\chi}=\chi^T\left(A_H^T P_H+P_H A_H\right)\chi=-\chi^T Q_H\chi<0 \quad \textbf{(42)}$$

In view of β_ε is selected in the interval $(1,\ 1+\varepsilon)$ with a sufficiently small positive number ε_1, the derivative of $V_H(\xi)$ is also negative [16]. Therefore, system (22) is asymptotically stable. Selecting homogeneity degree of x_1, x_2, x_3 be γ_1, γ_2 and γ_3, $V_H(\xi)$ and $\dot{V}_H(\xi)$ are homogeneous in vector ξ of degree 2 and $1+\beta_\varepsilon$ respectively. Then the inequality holds:

$$\dot{V}_H(\xi)\le -c_2 V_H^{\frac{1+\beta_\varepsilon}{2}}(\xi) \quad \textbf{(43)}$$

in which c_2 is a positive constant. The following inequality can be obtained:

$$\dot{V}_H(\xi)\le -\frac{\lambda_{\min}(Q_H)}{\lambda_{\max}(P_H)}V_H^{\frac{1+\beta_\varepsilon}{2}}(\xi) \quad \textbf{(44)}$$

Defining $\dot{V}_H$ as the derivative of $V_H(\xi)$ along original entire system (18), it can be obtained [19]:

$$\dot{V}_H \le \dot{V}_H(\xi)\le -\frac{\lambda_{\min}(Q_H)}{\lambda_{\max}(P_H)}V_H^{\frac{1+\beta_\varepsilon}{2}}(\xi) \quad \textbf{(45)}$$

Therefore, it can be obtained (18) is stable and $V_H(\xi)$ is a Lyapunov function of it. Consider a positive number Ψ that satisfies with $\Psi<V_H(\xi(t_0))$ and $\Psi\le\lambda_{\min}(P_H)$, in which $\xi(t_0)$ is initial condition of (18). $V_H(\xi)$ will decrease along (18) until less than Ψ within a time T_H:

$$T_H = \frac{2\lambda_{\max}(\boldsymbol{P}_H)}{(\beta_\varepsilon - 1)\lambda_{\min}(\boldsymbol{Q}_H)} \Psi^{\frac{1-\beta_\varepsilon}{2}} \tag{46}$$

Selecting $\boldsymbol{Q}_H$ as $\boldsymbol{E}_3$, $\boldsymbol{P}_H$ can be solved:

$$\boldsymbol{P}_H = \frac{1}{L_H(4L_H - 2)} \begin{bmatrix} 7L_H^2 - 2L_H & 4L_H^2 + L_H & 2L_H - 1 \\ 4L_H^2 + L_H & 7L_H^2 + 2L_H + 1 & 3L_H \\ 2L_H - 1 & 3L_H & 2L_H + 2 \end{bmatrix} \tag{47}$$

Characteristic equation of $\boldsymbol{P}_H$ is

$$s^3 - \frac{14L_H^2 + 2L_H + 3}{4L_H^2 - 2L_H} s^2 + \frac{33L_H^4 + 20L_H^3 + 17L_H^2 + 4L_H + 1}{16L_H^4 - 16L_H^3 + 4L_H^2} \left(s - \frac{1}{2L_H} \right) = 0 \tag{48}$$

The expressions of roots are so long that they are omitted here. The following Fig. (**1**) shows the change of maximum eigenvalue for $\boldsymbol{P}_H$ with respect to L_H.

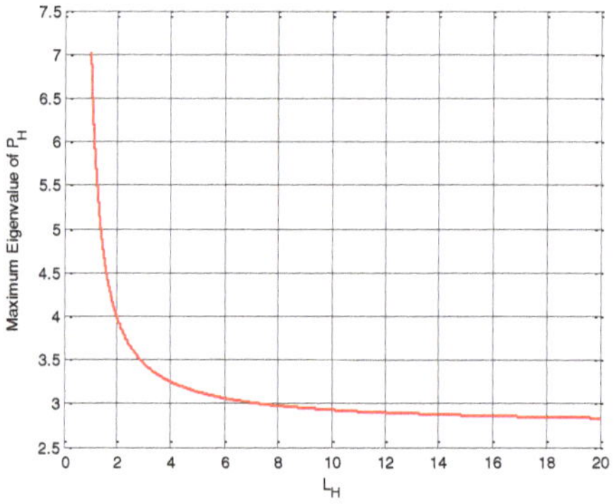

Fig. (1). The curve of maximum eigenvalue for $\boldsymbol{P}_H$ with respect to L_H.

It is observed from Fig. (**1**) that with the increase of gain L_H, the maximum eigenvalue of $\boldsymbol{P}_H$ will decrease, causing the decrease of T_H. Defining $\dot{V}_L$ as the derivative of $V_L(\chi)$ along original entire system (18), it can be obtained that:

$$\dot{V}_L \le \dot{V}_L(\varphi) \le -\frac{\lambda_{\min}(\boldsymbol{Q}_L)}{\lambda_{\max}(\boldsymbol{P}_L)} V_L^{\frac{1+\alpha_\varepsilon}{2}}(\varphi) \tag{49}$$

which means the system (18) could converge to the origin in finite time, and the convergence time is not greater than the convergence time T_L of the system (18).

When $V_H(\xi)$ converges to be less than Ψ, it follows:

$$\|\xi\|^2 \le \frac{V_H(\xi)}{\lambda_{\min}(\boldsymbol{P}_H)} \le \frac{\Psi}{\lambda_{\min}(\boldsymbol{P}_H)} \le 1 \tag{50}$$

Since $1/\gamma_i < 1/r_i$, we can get $\|\varphi\| \le \|\xi\| \le 1$. According to (37) and (38), convergence time T_c of (18) satisfies

$$T_c \le \frac{2\lambda_{\max}(\boldsymbol{P}_H)}{(\beta_\varepsilon - 1)\lambda_{\min}(\boldsymbol{Q}_H)}\Psi^{\frac{1-\beta_\varepsilon}{2}} + \frac{\lambda_{\max}^{\frac{3-\alpha_\varepsilon}{2}}(\boldsymbol{P}_{LI})(1-\alpha_\varepsilon)}{2\lambda_{\min}(\boldsymbol{Q}_{LI})L_L^{(1-\alpha_\varepsilon)/(3-2\alpha_\varepsilon)}} \tag{51}$$

It can be seen from (51) that states of (18) can converge to their origin in a fixed time T_c. In addition, the upper bound of convergence time T_c will decrease along with the increasing of gains L_L and L_H.

Define a novel FFIS for (13) as follows

$$\boldsymbol{S} = \ddot{\boldsymbol{e}} + \int_0^t \boldsymbol{v}_{FT} + \boldsymbol{w}_{FT} d\tau \tag{52}$$

where $\boldsymbol{v}_{FT} = [v_{FT1} \quad v_{FT2}]^T$ and $\boldsymbol{w}_{FT} = [w_{FT1} \quad w_{FT2}]^T$ are expressed as

$$\begin{aligned} v_{FTi} &= k_{L1i}L_{Li}^{1-\alpha_{1i}} sig^{\alpha_{1i}}(e_i) + k_{L2i}L_{Li}^{1-\alpha_{2i}} sig^{\alpha_{2i}}(\dot{e}_i) + k_{L3i}L_{Li}^{1-\alpha_{3i}} sig^{\alpha_{3i}}(\ddot{e}_i) \\ w_{FTi} &= L_{Hi} sig^{\beta_{1i}}(e_i) + 2L_{Hi} sig^{\beta_{2i}}(\dot{e}_i) + L_{Hi} sig^{\beta_{3i}}(\ddot{e}_i) \end{aligned} \tag{53}$$

in which $L_{Hi}, L_{Li} \in (1, \ \infty)$, $s^3 + k_{L3i}s^2 + k_{L2i}s + k_{L1i}$ is Hurwitz for $i = 1,2$. Parameters $\alpha_{1i}, \alpha_{2i}, \alpha_{3i}, \beta_{1i}, \beta_{2i}, \beta_{3i}, (i = 1,2)$ are defined as

$$\begin{aligned} &\alpha_{3i} = \alpha_{\varepsilon i}, \quad \alpha_{2i} = \alpha_{\varepsilon i}/(2-\alpha_{\varepsilon i}), \quad \alpha_{1i} = \alpha_{\varepsilon i}/(3-2\alpha_{\varepsilon i}) \\ &\beta_{3i} = \beta_{\varepsilon i}, \quad \beta_{2i} = \beta_{\varepsilon i}/(2-\beta_{\varepsilon i}), \quad \beta_{1i} = \beta_{\varepsilon i}/(3-2\beta_{\varepsilon i}) \end{aligned} \tag{54}$$

with $\alpha_{\varepsilon i} \in (1-\varepsilon_i, \ 1)$ and $\beta_{\varepsilon i} \in (1, \ 1+\varepsilon_{1i})$ for sufficiently small positive numbers ε_i and ε_{1i}.

When the derivative of S is equal to zero, it can be obtained:

$$\ddot{\boldsymbol{e}} = -\boldsymbol{v}_{FT} - \boldsymbol{w}_{FT} \tag{55}$$

Based on the conclusion of **Theorem 1**, $\boldsymbol{e}$ will converge to the origin in a fixed time.

Continuous Fixed-Time Super-Twisting-Like Reaching Law Scheme

In this section, continuous fixed-time super-twisting-like reaching law (CFSTL) is designed to combine with FFIS to establish a continuous fixed-time controller (FSMC):

$$\boldsymbol{u} = \boldsymbol{G}^{-1}\left[-\boldsymbol{F} + \ddot{\boldsymbol{y}}_d - \boldsymbol{v}_{FT} - \boldsymbol{w}_{FT} - \lambda_1 sig^{1/2}(\boldsymbol{S}) - \lambda_2 sig^{p}(\boldsymbol{S}) - \lambda_3 \int_0^t \operatorname{sgn}(\boldsymbol{S})\, d\tau\right] \tag{56}$$

in which $\lambda_1, \lambda_2, \lambda_3 \in \mathbb{R}^{2\times 2}$ are positive definite diagonal matrices and $p > 1$ is satisfied.

Combing (13), (52) and (56), it follows that

$$\begin{aligned}\dot{\boldsymbol{S}} &= \ddot{\boldsymbol{e}} + \boldsymbol{v}_{FT} + \boldsymbol{w}_{FT} = \boldsymbol{F} + \boldsymbol{G}\boldsymbol{u} + \boldsymbol{D}_f - \ddot{\boldsymbol{y}}_d + \boldsymbol{v}_{FT} + \boldsymbol{w}_{FT} \\ &= -\lambda_1 sig^{1/2}(\boldsymbol{S}) - \lambda_2 sig^{p}(\boldsymbol{S}) - \lambda_3 \int_{t_0}^t \operatorname{sgn}(\boldsymbol{S})\, d\tau + \boldsymbol{D}_f\end{aligned} \tag{57}$$

The derivatives of lumped disturbances D_{f1} and D_{f2} are norm-bounded before and after actuator faults occur. Therefore, the lumped disturbances have Lipschitz constants L_{d1} and L_{d2} which are satisfied with: $0 < |\dot{D}_{f1}| \le L_{d1}$, $0 < |\dot{D}_{f1}| \le L_{d2}$. Parameters $\lambda_3 = diag(\lambda_{31} \quad \lambda_{32})$ and $\lambda_1 = diag(\lambda_{11} \quad \lambda_{12})$ are selected to be satisfied with:

$$\begin{aligned}\lambda_{31} > 4L_{d1} \quad \lambda_{11} > \sqrt{2\lambda_{31}} \\ \lambda_{32} > 4L_{d2} \quad \lambda_{12} > \sqrt{2\lambda_{32}}\end{aligned} \tag{58}$$

Therefore, S and its derivative can converge to the origin in fixed time.

Uniformly Convergent Observer Design

In order to enhance the robustness of the controller further, a uniformly convergent observer is introduced to realize accurate estimation of lumped disturbances $\boldsymbol{D}_f$.

$$
\begin{aligned}
\dot{\hat{e}}_1 &= -\kappa_{A1} L_A^{1/5} \theta sig^{4/5}(e_1 - e) - k_{A1}(E_2 - \theta) sig^{(5+\alpha_A)/5}(e_1 - e) + \hat{e}_2 \\
\dot{\hat{e}}_2 &= -\kappa_{A2} L_A^{2/5} \theta sig^{3/5}(e_1 - e) - k_{A2}(E_2 - \theta) sig^{(5+2\alpha_A)/5}(e_1 - e) + \hat{e}_3 \\
\dot{\hat{e}}_3 &= -\kappa_{A3} L_A^{3/5} \theta sig^{2/5}(e_1 - e) - k_{A3}(E_2 - \theta) sig^{(5+3\alpha_A)/5}(e_1 - e) + \hat{e}_4 + (F + Gu - \ddot{y}_d) \\
\dot{\hat{e}}_4 &= -\kappa_{A4} L_A^{4/5} \theta sig^{1/5}(e_1 - e) - k_{A4}(E_2 - \theta) sig^{(5+4\alpha_A)/5}(e_1 - e) + \hat{e}_5 \\
\dot{\hat{e}}_5 &= -\kappa_{A5} L_A \theta \operatorname{sgn}(e_1 - e) - k_{A5}(E_2 - \theta) sig^{1+\alpha_A}(e_1 - e)
\end{aligned}
\qquad \textbf{(59)}
$$

where $L_A = diag(L_{A1} \quad L_{A2}) > 0$ and $k_{Ai} = diag(k_{Ai1} \quad k_{Ai2})$ for $i = 1,\ldots,5$. κ_{Ai} are selected as

$$
\kappa_{A1} = 5E_2 \quad \kappa_{A2} = 10.03E_2 \quad \kappa_{A3} = 9.3E_2 \quad \kappa_{A4} = 4.57E_2 \quad \kappa_{A5} = 1.1E_2 \qquad \textbf{(60)}
$$

The coefficient α_A is a sufficiently small positive number and k_{Ai} are selected to make $s^5 + k_{A1j}s^4 + k_{A2j}s^3 + k_{A3j}s^2 + k_{A4j}s + k_{A5j}$ $(j = 1,2)$ be Hurwitz. $\theta = diag(\theta_1 \quad \theta_2)$ is:

$$
\theta_j = \begin{cases} 0 & if\ t \le T_{Aj} \\ 1 & otherwise \end{cases} \quad j = 1,2 \qquad \textbf{(61)}
$$

where T_{Aj} is the positive switching time parameter.

The uniformly convergent observer switches from a uniform convergent observer to a conventional homogenous higher order sliding mode observer. Therefore, outputs $\hat{e}_1$ and $\hat{e}_4$ will exactly be equal to e and D_f in fixed time. Thus, $\hat{e}_4$ can be compensated into FSMC (56) to design an anti-disturbance continuous fixed-time controller (ACFTC) as follow:

$$
u = G^{-1}\left[-F + \ddot{y}_d - \hat{e}_4 - v_{FT} - w_{FT} - \lambda_1 sig^{1/2}(S) - \lambda_2 sig^{p}(S) - \lambda_3 \int_0^t \operatorname{sgn}(S) d\tau\right] \qquad \textbf{(62)}
$$

SIMULATION RESULTS

In this section, three sets of simulations are performed. The first one applies the FFTR presented to show its advantages. The second one applies the FSMC in (56) without observer for a faulted FHAV as contrast simulation. Two cases are included to show the results using different design parameters of reaching law. The third one simulates the ACFTC proposed in this paper for a faulted FHAV to demonstrate its superiority of it compared with the second set of simulations.

Simulations for a Fast Fixed-Time High-Order Regulator

For a third-order system (18), parameters of FFTR are selected as: $k_{L1}=1.1$, $k_{L2}=2.12$, $k_{L3}=2$, $\alpha_3=0.9$, $\alpha_2=9/11$, $\alpha_1=0.75$, $\beta_3=1.1$, $\beta_2=11/9$, $\beta_1=1.375$.

Case A: Initial conditions are set as: $x_1(0)=-5$, $x_2(0)=10$ and $x_3(0)=5$. Simulations are performed in four cases with different gains L_L and L_H:

$$\begin{array}{lll} \text{Case A.1}: & L_L=1 & L_H=1 \\ \text{Case A.2}: & L_L=20 & L_H=1 \\ \text{Case A.3}: & L_L=1 & L_H=1.5 \\ \text{Case A.4}: & L_L=20 & L_H=1.5 \end{array} \tag{63}$$

Case B: Set $L_L=20$ and $L_H=2$, different initial values are applied:

$$\begin{array}{llll} \text{Case B.1}: & x_1(0)=-5 & x_2(0)=10 & x_3(0)=5 \\ \text{Case B.2}: & x_1(0)=-50 & x_2(0)=100 & x_3(0)=50 \end{array} \tag{64}$$

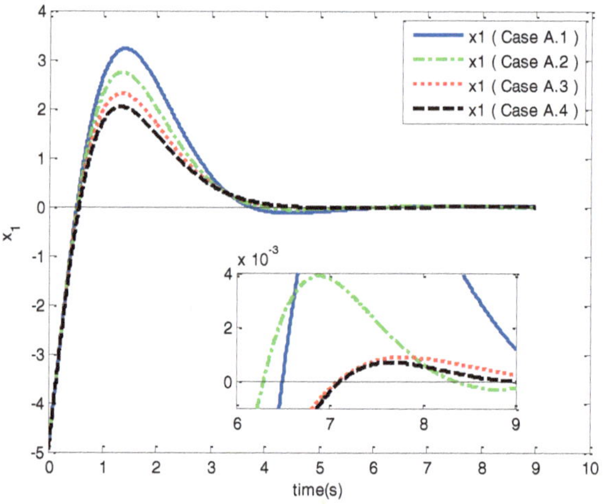

Fig. (2). System response to different control gains in Case A.

It can be seen from Fig. (**2**) that the convergence speed of FFTR can be accelerated *via* increasing values of L_L and L_H. Fig. (**3**) shows that convergence time will not grow with larger initial values, which demonstrates the property of fixed-time convergence.

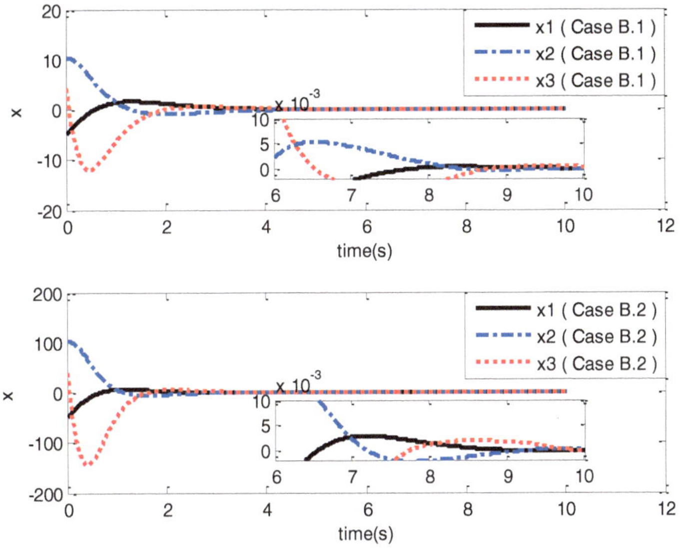

Fig. (3). System response to different initial values in Case B.

Simulations of FAHV using FSMC

In this subsection, FSMC in (56) without observer is applied to a high-fidelity nonlinear CFM stated in Equation (1).

Actuator faults are assumed to occur after simulation begins 30 seconds: $\boldsymbol{F}_g = diag(0.6 \quad 0.8)$, $\boldsymbol{F}_d = [0.1 \quad 0.02]^T$. The parameters are: $m = 4378\,\mathrm{kg}$, $I_{yy} = 6.8\times10^5\,\mathrm{kg\cdot m^2}$, $\tilde{\psi}_1 = -423$, $\tilde{\psi}_2 = 383$, $\omega_1 = 16$, $\omega_2 = 20$, $\zeta_1 = \zeta_2 = 0.02$, $V_c = 2408\,\mathrm{m/s}$, h_c=26212 m, $k_P = 0.3$, $k_{L11} = k_{L12} = 1.1$, $k_{L21} = k_{L22} = 2.12$, $k_{L31} = k_{L32} = 2$, $\alpha_{31} = \alpha_{32} = 0.9$, $\alpha_{21} = \alpha_{22} = 9/11$, $\alpha_{11} = \alpha_{12} = 0.75$, $\beta_{31} = \beta_{32} = 1.1$, $\beta_{21} = \beta_{22} = 11/9$, $\beta_{11} = \beta_{12} = 1.375$, $L_{L1} = 2000$, $L_{L2} = 2$, $L_{H1} = L_{H2} = 2$. The initial conditions are shown in Table **1**:

Table 1. Initial conditions.

Parameter	Value	Parameter	Value	Parameter	Value
h	25908 m	Q	0 deg/s	η_1	1.8
V	2347.6 m/s	Φ	0.25	$\dot{\eta}_1$	0
α	1.5 deg	$\dot{\Phi}$	0	η_2	0.4
γ	0 deg	δ_e	11.5 deg	$\dot{\eta}_2$	0

Case A: The parameters of CFSTL are selected as: $\lambda_1 = \lambda_2 = diag(4.75 \quad 10.61) \times 10^{-5}$, $\lambda_3 = diag(1.1 \quad 5.5) \times 10^{-9}$, $p = 3/2$.

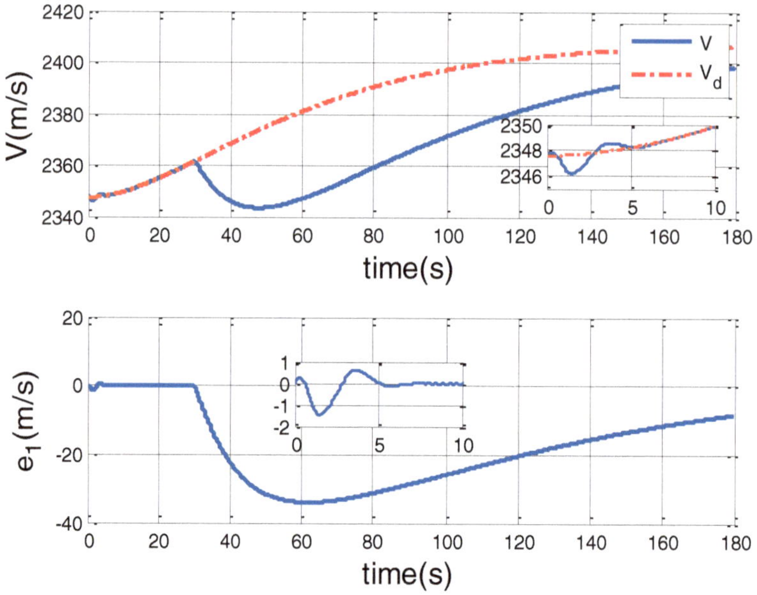

Fig. (4). Velocity tracking performance.

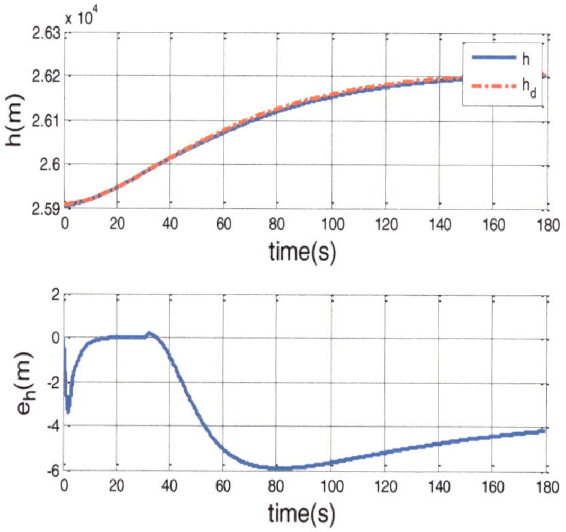

Fig. (5). Height tracking performance.

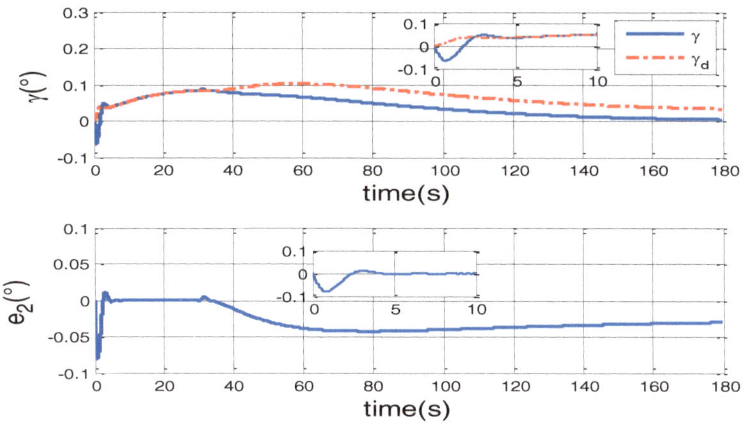

Fig. (6). Flight path angle tracking performance.

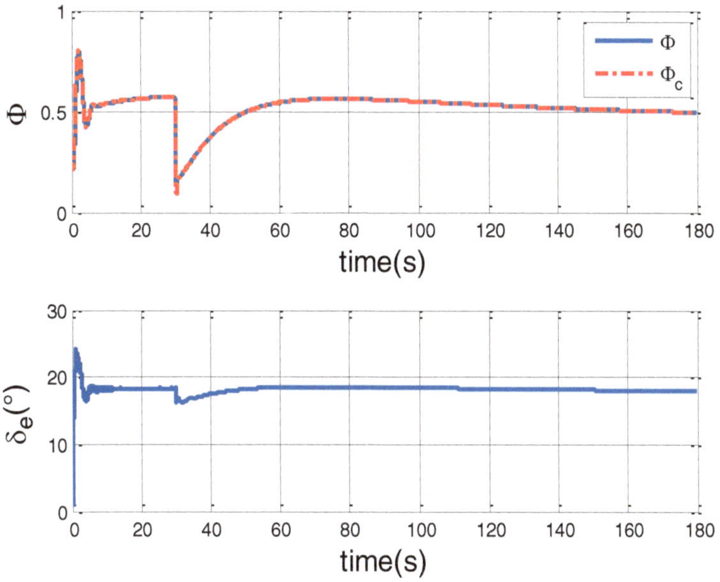

Fig. (7). Fuel-to-air ratio and elevator angular deflection.

It can be seen from Figs. (**4-7**) that the tracking errors of velocity and flight path angle can converge to zero quickly at the beginning of the simulation. However, the control performance deteriorates after actuator faults occur, because the robustness is not strong enough. Velocity, height and flight path angle is not able to follow the desired signals

Case B: For the sake of overcoming the effects of actuator faults, the parameters of CFSTL are selected larger than that in **Case A** as: $\lambda_1 = \lambda_2 = diag(47.5 \quad 3.35) \times 10^{-2}$, $\lambda_3 = diag(0.11 \quad 5.5 \times 10^{-4})$, $p = 3/2$

The results in Figs (**8-13**) demonstrate the effectiveness of FSMC. By means of selecting larger parameters $\lambda_1, \lambda_2, \lambda_3$ to ensure robustness, velocity and flight path angle are able to follow the desired signals in fixed time when actuator faults occur. However, chattering is inevitably obvious owing to relatively larger values of gains selected in CFSTL to suppress the effects of actuator faults.

Simulations of FAHV using ACFTC

The ACFTC proposed is utilized on FAHV to illustrate its performance and superiority of it.

Parameters in uniformly convergent observer are given as: $\alpha_A = 0.06$, $T_{A1} = T_{A2} = 1$, $\boldsymbol{L}_A = diag(700 \quad 0.02)$, $\boldsymbol{k}_{A1} = 5\boldsymbol{E}_2$, $\boldsymbol{k}_{A2} = 10.03\boldsymbol{E}_2$, $\boldsymbol{k}_{A3} = 9.3\boldsymbol{E}_2$, $\boldsymbol{k}_{A4} = 4.57\boldsymbol{E}_2$, $\boldsymbol{k}_{A5} = 1.1\boldsymbol{E}_2$. Due to the introduction of a uniformly convergent observer, parameters of CFSTL are permitted to be chosen smaller as: $\lambda_1 = \lambda_2 = diag(1.5 \quad 3.35) \times 10^{-6}$, $\lambda_3 = diag(1.1 \quad 5.5) \times 10^{-12}$. The remaining parameters of FFIS and CFSTL are set as the same as the selection in Subsection 5.2.

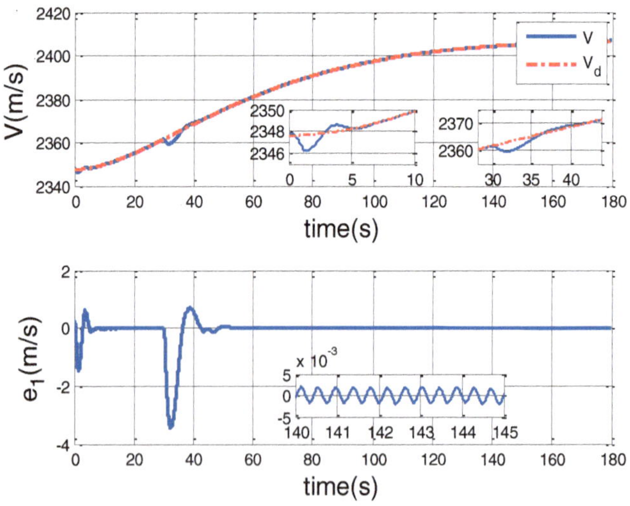

Fig. (8). Velocity tracking performance.

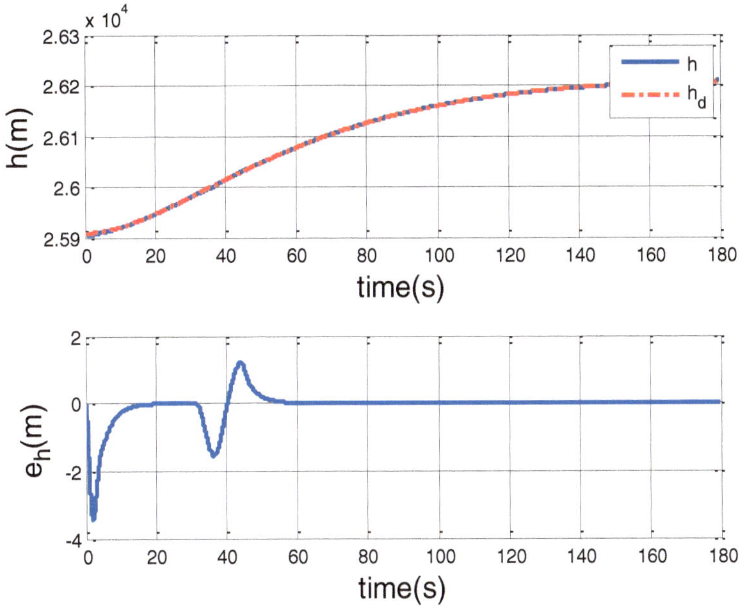

Fig. (9). Height tracking performance.

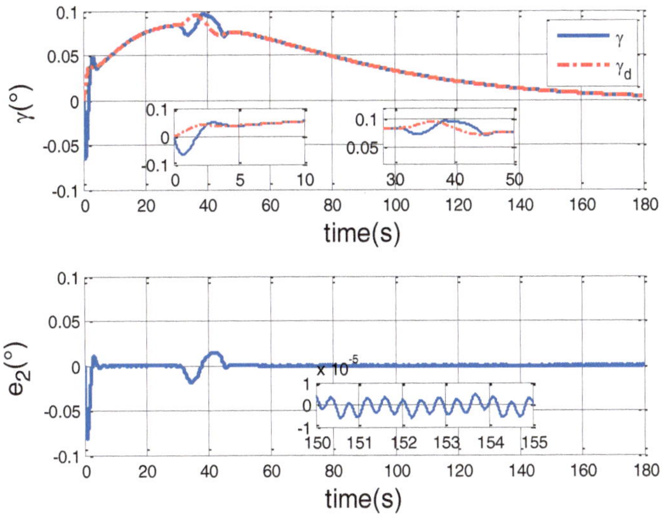

Fig. (10). Flight path angle tracking performance.

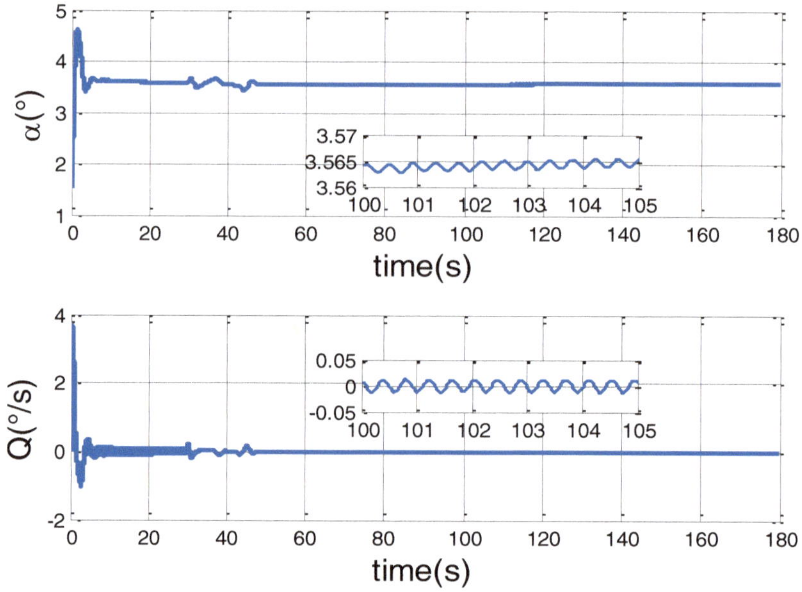

Fig. (11). Fuel-to-air ratio and elevator angular deflection.

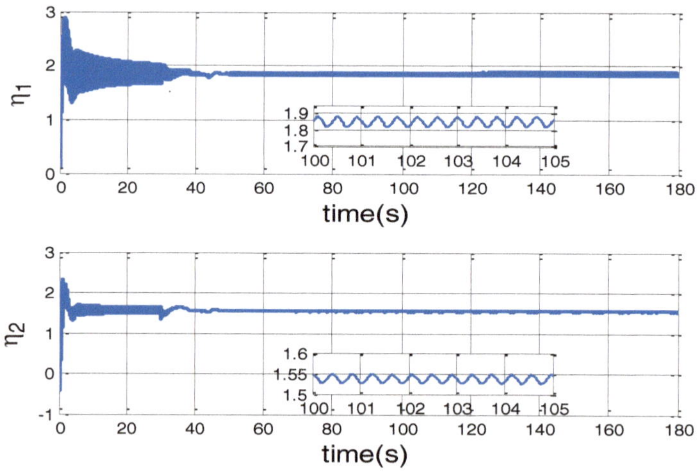

Fig. (12). Generalized modal coordinates.

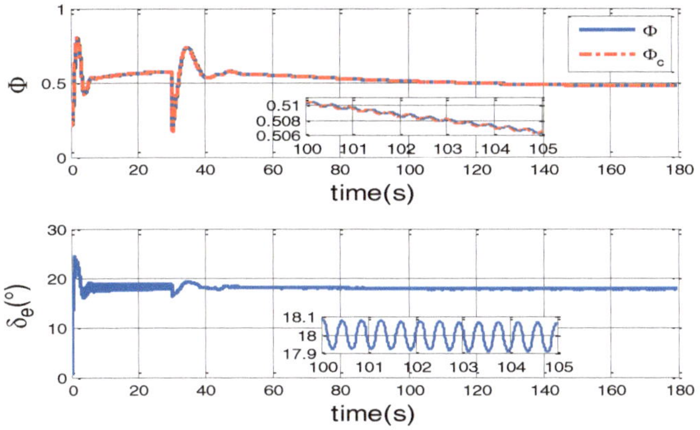

Fig. (13). Fuel-to-air ratio and elevator angular deflection.

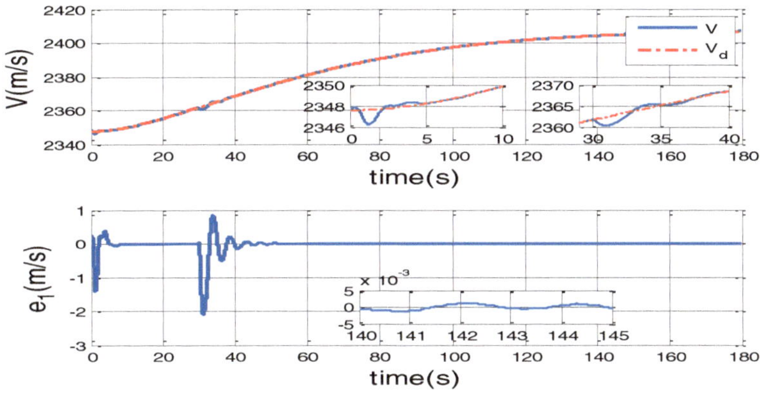

Fig. (14). Velocity tracking performance.

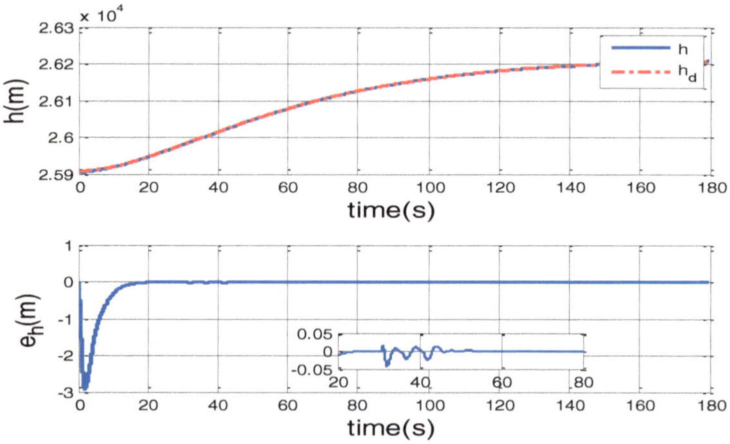

Fig. (15). Height tracking performance.

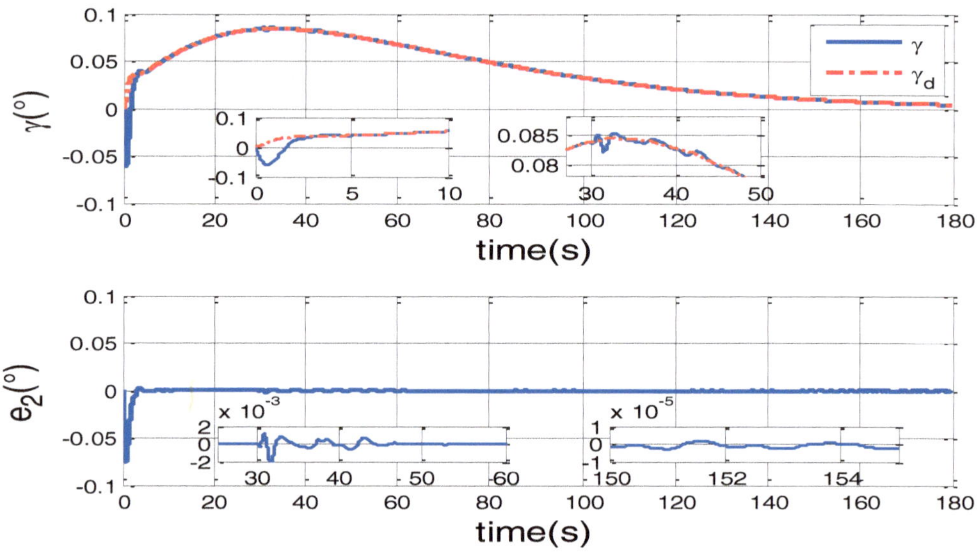

Fig. (16). Flight path angle tracking performance.

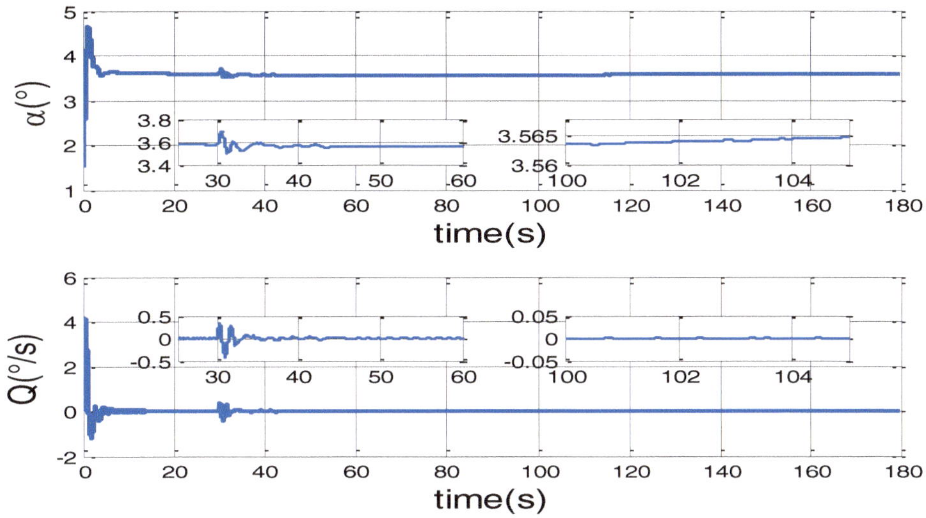

Fig. (17). Angle of attack and pitch rate.

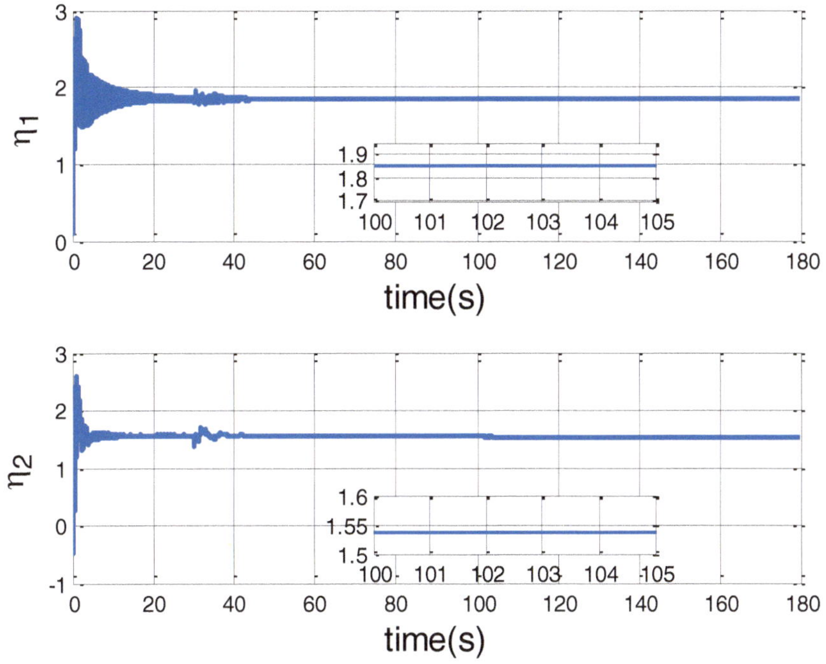

Fig. (18). Generalized modal coordinates.

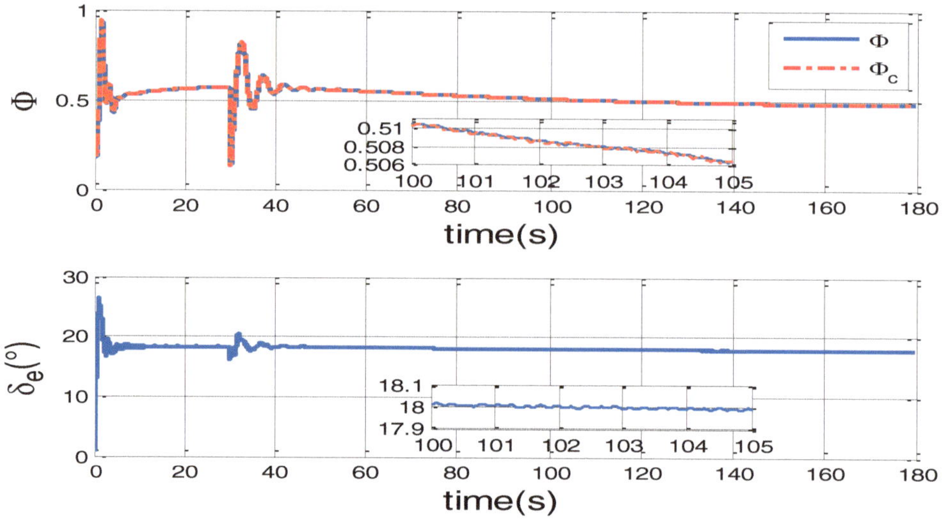

Fig. (19). Fuel-to-air ratio and elevator angular deflection.

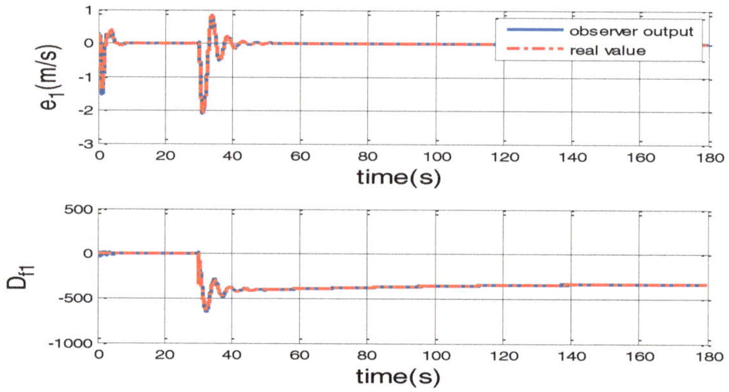

Fig. (20). Observer estimation performance for velocity channel.

Figs. (**14-16**) show the excellent tracking performance of FAHV using ACFTC. The velocity and flight path angle tracking error can realize higher-precision fixed-time convergence under the influences of actuator faults. The angle of attack and pitch rate are displayed in Fig. (**17**). Generalized modal coordinates in Fig. (**18**) are significantly small. In Fig. (**19**), the Fuel-to-air ratio and elevator angular deflection are shown to be within their reasonable bounds. Fig. (**20** & **21**) show the performance of the observer, in which $\hat{e}_1$ and $\hat{e}_4$ can achieve accurate tracking for $\boldsymbol{e}$ and $\boldsymbol{D}_f$.

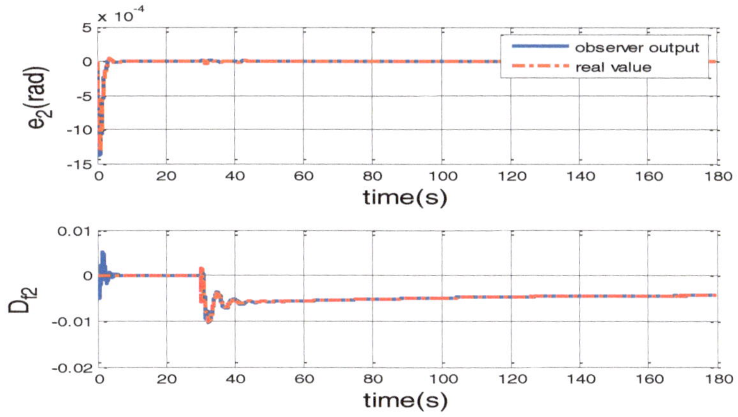

Fig (21). Observer estimation performance for flight path angle channel.

CONCLUSION

An ACFTC is presented for FAHV with actuator faults, which is composed of FFIS, CFSTL and uniformly convergent observer. The novel FFTR is proposed to realize fixed-time convergence without complicated parameter adjustment. CFSTL can guarantee fixed-time convergence of the sliding mode vector and its derivative with a more simplified form. The lumped disturbances can be estimated accurately by a uniformly convergent observer in a fixed time. By means of combing three components, the ACFTC could make tracking errors of FAHV realize fixed-time convergence under actuator faults with less chattering.

CONSENT FOR PUBLICATION

Not applicable.

CONFLICT OF INTEREST

The authors declare no conflict of interest, financial or otherwise.

ACKNOWLEDGEMENTS

This study was co-supported by the National Key R&D Program of China under Grant 2021YFB1715000, the National Natural Science Foundation of China (No. 12102343), the Key Program of the National Natural Science Foundation of China (No. U2013206), Shanghai Space Science and Technology Innovation Fund (No.

SAST2020-072), Science and Industry Bureau Stable Support Project (HTKJ2021KL502013) and the Fundamental Research Funds for the Central Universities (No. D5000210833).

APPENDIX A: DEFINITIONS OF

$$
\begin{aligned}
&\omega_1 = \frac{1}{m}\begin{bmatrix} -\partial\bar{D}/\partial V \\ \partial T/\partial\alpha\cos\alpha - T\sin\alpha - \partial\bar{D}/\partial\alpha \\ -mg\cos\gamma \\ \partial T/\partial\Phi\cos\alpha \\ -\partial\bar{D}/\partial h - \partial g/\partial h\, m\sin\gamma \end{bmatrix}^T \\
&\boldsymbol{\Omega}_2 = [\boldsymbol{\omega}_{21} \quad \boldsymbol{\omega}_{22} \quad \boldsymbol{\omega}_{23} \quad \boldsymbol{\omega}_{24} \quad \boldsymbol{\omega}_{25}]/m \\
&\boldsymbol{\omega}_{21} = \left[-\partial^2\bar{D}/\partial V^2 \quad -\partial^2\bar{D}/(\partial\alpha\partial V) \quad 0 \quad 0 \quad -\partial^2\bar{D}/(\partial V\partial h)\right]^T \\
&\boldsymbol{\omega}_{22} = \begin{bmatrix} -\partial^2\bar{D}/(\partial\alpha\partial V) \\ \left(\partial^2 T/\partial\alpha^2 - T\right)\cos\alpha - 2\partial T/\partial\alpha\sin\alpha - \partial^2\bar{D}/\partial\alpha^2 \\ 0 \\ \partial^2 T/(\partial\alpha\partial\Phi)\cos\alpha - \partial T/\partial\Phi\sin\alpha \\ -\partial^2\bar{D}/(\partial\alpha\partial V) \end{bmatrix} \\
&\boldsymbol{\omega}_{23} = [0 \quad 0 \quad mg\sin\gamma \quad 0 \quad -m\cos\gamma\,\partial g/\partial h]^T \\
&\boldsymbol{\omega}_{24} = \left[0 \quad \partial^2 T/(\partial\alpha\partial\Phi)\cos\alpha - \partial T/\partial\Phi\sin\alpha \quad 0 \quad 0 \quad 0\right]^T \\
&\boldsymbol{\omega}_{25} = \left[-\partial^2\bar{D}/(\partial V\partial h) \quad -\partial^2\bar{D}/(\partial\alpha\partial V) \quad -m\cos\gamma\,\partial g/\partial h \quad 0 \quad 0\right]^T
\end{aligned}
\tag{65}
$$

$$
\begin{aligned}
&\boldsymbol{\pi}_1 = \begin{bmatrix} (\partial L/\partial V + \partial T/\partial V\sin\alpha)/(mV) - (L + T\sin\alpha)/(mV^2) + g\cos\gamma/V^2 \\ (\partial L/\partial\alpha + \partial T/\partial\alpha\sin\alpha + T\cos\alpha)/(mV) \\ g\sin\gamma/V \\ \partial T/\partial\Phi\sin\alpha/(mV) \\ (\partial L/\partial h - m\cos\gamma\,\partial g/\partial h)/(mV) \end{bmatrix}^T \\
&\boldsymbol{\Pi}_2 = [\boldsymbol{\pi}_{21} \quad \boldsymbol{\pi}_{22} \quad \boldsymbol{\pi}_{23} \quad \boldsymbol{\pi}_{24} \quad \boldsymbol{\pi}_{25}] \\
&\boldsymbol{\pi}_{21} = \begin{bmatrix} \partial^2 L/\partial V^2/(mV) - 2\partial L/\partial V/(mV^2) + 2(L + T\sin\alpha)/(mV^3) - 2g\cos\gamma/V^3 \\ \partial^2 L/\partial V\partial\alpha/(mV) - (\partial L/\partial\alpha + \partial T/\partial\alpha\sin\alpha + T\cos\alpha)/(mV^2) \\ -g\sin\gamma/V^2 \\ -\partial T/\partial\Phi\sin\alpha/(mV^2) \\ \partial^2 L/(\partial V\partial h)/(mV) - \partial L/\partial h/(mV^2) + \partial g/\partial h\cos\gamma/V^2 \end{bmatrix}
\end{aligned}
$$

$$\boldsymbol{\pi}_{22}=\begin{bmatrix}\partial^2 L/(\partial V\partial\alpha)/(mV)-(\partial L/\partial\alpha+\partial T/\partial\alpha\sin\alpha+T\cos\alpha)/(mV^2)\\ \left[\partial^2 L/\partial\alpha^2+\left(\partial^2 T/\partial\alpha^2-T\right)\sin\alpha+2\partial T/\partial\alpha\cos\alpha\right]/(mV)\\ 0\\ \left[\partial^2 T/(\partial\alpha\partial\Phi)\sin\alpha+\partial T/\partial\Phi\cos\alpha\right]/(mV)\\ \partial^2 L/(\partial\alpha\partial h)/(mV)\end{bmatrix}$$

$$\boldsymbol{\pi}_{23}=\left[-g\sin\gamma/V^2\quad 0\quad g\cos\gamma/V\quad 0\quad \partial g/\partial h\sin\gamma/V\right]^T \tag{66}$$

$$\boldsymbol{\pi}_{24}=\left[-\partial T/\partial\Phi\sin\alpha/(mV^2)\quad \left[\partial^2 T/(\partial\alpha\partial\Phi)\sin\alpha+\partial T/\partial\Phi\cos\alpha\right]/(mV)\quad 0\quad 0\quad 0\right]^T$$

$$\boldsymbol{\pi}_{25}=\begin{bmatrix}\partial^2 L/(\partial V\partial h)/(mV)-\partial L/\partial h/(mV^2)+\partial g/\partial h\cos\gamma/V^2\\ \partial^2 L/(\partial\alpha\partial h)/(mV)\\ \partial g/\partial h\sin\gamma/V\\ 0\\ \partial^2 L/\partial h^2/(mV)-\partial^2 g/\partial h^2\cos\gamma/V\end{bmatrix}$$

REFERENCES

[1] Bolender MA, Doman DB. Nonlinear Longitudinal Dynamical Model of an Air-Breathing Hypersonic Vehicle. Journal of Spacecraft and Rockets. 2007, 44, 374-87.

[2] Pu Z, Yuan R, Tan X, Yi J. Active robust control of uncertainty and flexibility suppression for air-breathing hypersonic vehicles. Aerospace Science and Technology. 2015, 42, 429-41.

[3] Xu B, Guo Y, Yuan Y, Fan Y, Wang D. Fault-tolerant control using command-filtered adaptive back-stepping technique: Application to hypersonic longitudinal flight dynamics. International Journal of Adaptive Control and Signal Processing. 2016, 30, 553-77.

[4] Xu B, Huang X, Wang D, Sun F. Dynamic Surface Control of Constrained Hypersonic Flight Models with Parameter Estimation and Actuator Compensation. Asian Journal of Control. 2014, 16, 162-74.

[5] B.O. Mushage, J.C. Chedjou, K. Kyamakya, Observer-based fuzzy adaptive fault-tolerant nonlinear control for uncertain strict-feedback nonlinear systems with unknown control direction and its applications, Nonlinear Dynamics, 88 (2017) 2553-2575.

[6] Sun J-G, Song S-M, Wu G-Q. Fault-Tolerant Track Control of Hypersonic Vehicle Based on Fast Terminal Sliding Mode. Journal of Spacecraft and Rockets. 2017, 54, 1304-16.

[7] Wang J, Zong Q, He X, Karimi HR. Adaptive Finite-Time Control for a Flexible Hypersonic Vehicle with Actuator Fault. Mathematical Problems in Engineering. 2013, 2013, 1-10.

[8] Feng Y, Yu X, Man Z. Non-singular terminal sliding mode control of rigid manipulators. Automatica. 2002, 38, 2159-67.

[9] Yu P, Shtessel Y, Edwards C. Continuous higher order sliding mode control with adaptation of air breathing hypersonic missile. International Journal of Adaptive Control and Signal Processing. 2016, 30, 1099-117.

[10] Li P, Yu X, Zhang Y, Peng X. Adaptive Multivariable Integral TSMC of a Hypersonic Gliding Vehicle With Actuator Faults and Model Uncertainties. IEEE/ASME Transactions on Mechatronics. 2017, 22, 2723-35.

[11] Basin M, Shtessel Y, Aldukali F. Continuous finite- and fixed-time high-order regulators. Journal of the Franklin Institute. 2016, 353, 5001-12.

[12] Basin M, Bharath Panathula C, Shtessel Y. Multivariable continuous fixed-time second-order sliding mode control: design and convergence time estimation. IET Control Theory & Applications. 2017, 11, 1104-11.

[13] Cruz-Zavala E, Moreno JA, Fridman LM. Uniform robust exact differentiator. IEEE Transactions on Automatic Control. 2011, 56, 2727-33.

[14] Bu X, Wu X, He G, Huang J. Novel adaptive neural control design for a constrained flexible air-breathing hypersonic vehicle based on actuator compensation. Acta Astronautica. 2016, 120, 75-86.

[15] Wang N, Wu H-N, Guo L. Coupling-observer-based nonlinear control for flexible air-breathing hypersonic vehicles. Nonlinear Dynamics. 2014, 78, 2141-59.

[16] Perruquetti W, Floquet T, Moulay E. Finite-time observers: application to secure communication. IEEE Transactions on Automatic Control. 2008, 53, 356-60.

[17] Ríos H, Efimov D, Fridman LM, Moreno JA, Perruquetti W. Homogeneity based uniform stability analysis for time-varying systems. IEEE Transactions on Automatic Control. 2016, 61, 725-34.

[18] Moreno JA, Negrete DY, Torres-González V, Fridman L. Adaptive continuous twisting algorithm. International Journal of Control. 2015, 89, 1798-806.

[19] Basin M, Yu P, Shtessel Y. Finite- and fixed-time differentiators utilising HOSM techniques. IET Control Theory & Applications. 2017, 11, 1144-52.

[20] Bhat SP, Bernstein DS. Continuous finite-time stabilization of the translational and rotational double integrators. IEEE Transactions on automatic control. 1998, 43, 678-82.

CHAPTER 2

Fast and Parallel Algorithms for Orbit and Attitude Computation

Xuechuan Wang[1*], Haoyang Feng[1] and **Wei He[1]**

[1] *Northwestern Polytechnical University, Xi'an 710072, China*

Abstract: This chapter provides a simple Adaptive Local Variational Iteration Method (ALVIM) that can efficiently solve nonlinear differential equations and orbital problems of spacecrafts. Based on a general first-order form of nonlinear differential equations, the iteration formula is analytically derived and then discretized using Chebyshev polynomials as basis functions in the time domain. It leads to an iterative numerical algorithm that only involves the addition and multiplication of sparse matrices. Moreover, the Jacobian matrix is free from inversing. Apart from that, a straightforward adaptive scheme is proposed to refine the configuration of the algorithm, involving the length of time steps and the number of collocation nodes in a time step. With the adaptive scheme, the prescribed accuracy can be guaranteed without manually tuning the configuration of the algorithm. Since the refinement is adjusted automatically, our algorithm reduces over-calculation for smooth and slowly changing problems. Examples such as large amplitude pendulum and perturbed two-body problem are used to verify this easy-to-use adaptive method's high accuracy and efficiency.

Keywords: Orbital propagation, orbit-attitude coupling, Local Variational Iteration Method.

INTRODUCTION

In orbital mechanics, the accuracy and efficiency of numerical methods are mostly concerned. Highly accurate results require a very small step size for classical integration methods such as finite-difference methods. In this aspect, the collocation method shows superiority because the solution of collocation can be semi-analytical and the approximation can be made in large intervals. No need to say that the collocation method only results in a few coefficients that are related to

[*] **Corresponding author Xuechuan Wang:** Northwestern Polytechnical University, Xi'an 710072, China; Tel: +86 02988492338; E-mails: xcwang@nwpu.edu.cn; fhaoyang@126.com; 1778784694@qq.com

Chuang Liu, Honghua Dai, Xiaokui Yue & Yiqing Ma (Eds.)

basic functions, while the classical integration methods need to save a large amount of discrete data [1]. The governing system and the related boundary conditions can be reduced as algebraic equations of the states at collocation nodes for either the initial value problem or the two-point boundary value problem. Applications of the collocation method have been very successful in orbital problems of spacecraft. However, there is a need to construct nonlinear algebraic equations, which are usually solved by Newton-like algorithm that involves the inversion of Jacobian matrices. This process is made much simpler by introducing the Picard iteration with orthogonal basis functions to approximate the solution. In the Picard-based algorithms, there is no need to invert a Jacobian matrix. Such a method was named the modified Chebyshev-Picard iteration (MCPI) method in literature [2-5].

In this chapter, we show that the Local Variational Iteration Method (LVIM) method can be a highly efficient and accurate alternative to traditional numerical methods. Numerical examples such as the orbit propagation problem and Lambert's problem are used to verify the convergence, accuracy and efficiency of LVIM.

In addition, for solving two-point boundary value problems (TPBVPs), a novel quasi-linearization method is introduced to transform TPBVPs into initial value problems (IVPs). Particularly, in conservative problems, a so-called Fish-Scales-Growing Method (FSGM) is further proposed. Compared with the multiple shooting method and the modified simple shooting method, the proposed Fish-Scales-Growing Method is based on the principle of least action for a Hamiltonian system, instead of using Newton's iterative algorithm or a an asymptotic homotopy strategy. In a conservative system, the solution of a two-point boundary value problem is determined by the principle of least action. A piecewise solution obtained by solving multiple two-point boundary value problems is used as an approximation of the true solution, using the Fish-Scales-Growing Method. Then by changing the boundary conditions of the multiple two-point boundary value problems, the piecewise solution will iteratively evolve in a pattern that looks like the growing of fish scales. In this pattern, the action along the piecewise trajectory is guaranteed to decrease. Theoretically, the piecewise solution will eventually approach the true solution in a conservative system. In the Fish-Scales-Growing Method, the two-point boundary value problem to be solved is converted into multiple two-point boundary value problems that can be solved much more easily. Since the principle of least action theoretically guarantees the convergence of the piecewise solutions obtained by Fish-Scales-Growing Method, the initial guess can be selected rather arbitrarily, although different initial guesses may affect the convergence speed. Moreover, in each iteration step of the Fish-Scales-Growing Method, the multiple two-point boundary value problems are independent to each other. Thus, it can be conveniently coded for parallel computation to enhance computational efficiency. At last, for each solution of the multiple two-point boundary value problems, only one time instant and its corresponding positions and velocities need to be recorded for the next iteration, so the Fish-Scales-Growing method is very efficient and memory-saving.

In addition, this chapter also discusses the relative motion modeling of spacecraft. The relative motion of spacecraft in space is a six degree of freedom system, including relative attitude and orbit motions. When describing the motion of a spacecraft in space, generally speaking, the rotation relative to the center of mass and translation of the center of mass can be used to describe the whole motion of a spacecraft in three-dimensional space. The relative motion of two spacecraft can be described by relative attitude and centroid position, or by the relative position of any two points on the spacecraft to establish the corresponding motion model according to the corresponding description method. In order to accurately predict and control the motion of a spacecraft in space, an accurate and practical model is essential. A good motion model can be more convenient for the study and reveal the essence of the problem.

At present, considerable progress has been made in establishing the motion model of spacecraft. In traditional methods, attitude and orbit are modeled separately. Attitude can be described by Euler angle, directional cosine matrix and quaternion; the orbit can be described by a rectangular coordinate system and orbital elements. The traditional method is simple, but it is often difficult to accurately describe the actual motion of spacecraft, and it is not convenient for the controller design. Therefore, we need a more accurate model. In terms of attitude orbit coupling modeling, there are two main methods. One is to model the attitude and orbit separately, and then introduce the coupling term to modify the attitude orbit model to achieve the description and control of attitude orbit coupling; The other is the integrated modeling of attitude and orbit, and the attitude and orbit are described by the same tool. This method can not only reflect the actual motion of spacecraft, but also facilitate the design of the controller. This chapter mainly discusses the attitude orbit integration modeling method. Therefore, we introduce two mathematical tools: dual quaternion and spinor. Using the advantages of these two tools, we describe all the state variables of spacecraft in space, and deduce the motion model and dynamic model

.

LVIM AND QUASI-LINEARIZATION METHOD

Local Variational Iteration Method

A first order form of ordinary differential equations is expressed as follows

$$\dot{\mathbf{x}} = \mathbf{f}(\mathbf{x},\tau),\ \tau \in [t_0, t] \tag{1}$$

The variational iteration method (VIM) can approximate the solution of the equation by using an initial approximation $\mathbf{x}_0(\tau)$ and then successively correct it. The correctional formula is expressed as

$$\mathbf{x}_{n+1}(t) = \mathbf{x}_n(\tau)\big|_{\tau=t} + \int_{t_0}^{t} \boldsymbol{\lambda}(\tau)\{\dot{\mathbf{x}}_n(\tau) - \mathbf{f}[\mathbf{x}_n(\tau),\tau]\}d\tau\ ,\ t_0 \le \tau \le t \tag{2}$$

It indicates that the $(n+1)$ th correction is the addition of $\mathbf{x}_n$ and integration of weighted error of the previous solution $\mathbf{x}_n$ up to the current time t. Note that $\boldsymbol{\lambda}(\tau)$ is a matrix of Lagrange multipliers that can be optimally determined by making the right-hand side of Eq.(2) stationary about $\delta\mathbf{x}_n(\tau)$, *i.e.* the variation of $\mathbf{x}_n(\tau)$. From that, we have the following stationary condition.

$$\delta\mathbf{x}_n(\tau)\big|_{\tau=t} + \boldsymbol{\lambda}(\tau)\delta\mathbf{x}_n(\tau)\Big|_{\tau=t_0}^{\tau=t} - \int_{t_0}^{t}[\dot{\boldsymbol{\lambda}}(\tau) + \boldsymbol{\lambda}(\tau)\frac{\partial\mathbf{f}(\mathbf{x}_n,\tau)}{\partial\mathbf{x}_n}]\delta\mathbf{x}_n(\tau)d\tau = 0 \tag{3}$$

Then we collect the terms including $\delta\mathbf{x}_n(\tau)\big|_{\tau=t}$ and $\delta\mathbf{x}_n(\tau)$. Since the boundary value of $\mathbf{x}(\tau)$ at $\tau = t_0$ is unchanged, we have $\delta\mathbf{x}(\tau)\big|_{\tau=t_0} = \mathbf{0}$. Eq. (3) leads to the constraints of $\boldsymbol{\lambda}(\tau)$.

$$\begin{cases} \delta\mathbf{x}_n(\tau)\big|_{\tau=t} : \mathbf{I} + \boldsymbol{\lambda}(\tau)\big|_{\tau=t} = \mathbf{0} \\ \delta\mathbf{x}_n(\tau) : \dot{\boldsymbol{\lambda}}(\tau) + \boldsymbol{\lambda}(\tau)\dfrac{\partial\mathbf{f}(\mathbf{x}_n,\tau)}{\partial\mathbf{x}_n} = \mathbf{0} \end{cases} \tag{4}$$

From the variational iteration method we obtained two modifications. By differentiating Eq. (2) with the utilization of the constraints in Eq. (4), the first modified iteration formula is obtained.

$$\begin{aligned}\frac{d\mathbf{x}_{n+1}}{dt} &= \frac{d\mathbf{x}_n}{dt} + \boldsymbol{\lambda}(\tau)\big|_{\tau=t}[\frac{d\mathbf{x}_n}{dt} - \mathbf{f}(\mathbf{x}_n,t)] + \int_{t_i}^{t} \frac{\partial \boldsymbol{\lambda}}{\partial t}[\dot{\mathbf{x}}_n - \mathbf{f}(\mathbf{x}_n,\tau)]d\tau \\ &= \mathbf{f}(\mathbf{x}_n,t) + \int_{t_i}^{t} \frac{\partial \boldsymbol{\lambda}}{\partial t}[\dot{\mathbf{x}}_n - \mathbf{f}(\mathbf{x}_n,\tau)]d\tau\end{aligned} \tag{5}$$

Note that the generalized Lagrange multipliers $\boldsymbol{\lambda}(\tau)$ are exactly the solution $\bar{\boldsymbol{\lambda}}(t)$ of the following equations [6]:

$$\begin{cases}\mathbf{I} + \bar{\boldsymbol{\lambda}}(t)\big|_{t=\tau} = \mathbf{0} \\ \dfrac{\partial \bar{\boldsymbol{\lambda}}(t)}{\partial t} - \mathbf{J}(t)\bar{\boldsymbol{\lambda}}(t) = \mathbf{0}\end{cases} \tag{6}$$

where $\mathbf{J}(t) = \partial \mathbf{f}(\mathbf{x}_n,\tau)/\partial \mathbf{x}_n$. Considering that, Eq. (5) is simplified as

$$\frac{d\mathbf{x}_{n+1}}{dt} - \mathbf{J}(\mathbf{x}_n,t)\mathbf{x}_{n+1} = \mathbf{f}(\mathbf{x}_n,t) - \mathbf{J}(\mathbf{x}_n,t)\mathbf{x}_n \tag{7}$$

In this corrective formula, it can be seen that the Lagrange multipliers $\boldsymbol{\lambda}(\tau)$ of the variational iteration method are completely eliminated. This formula can thus be regarded as a generalized version of Newton-Raphson iteration in function space.

Another modification of the variational iteration method is obtained by approximating $\boldsymbol{\lambda}(\tau)$ in the Taylor series. Using Eq. (4), the Taylor series approximation of $\boldsymbol{\lambda}(\tau)$ can be readily obtained in the following form

$$\boldsymbol{\lambda}(\tau) \approx \mathbf{T}_0[\boldsymbol{\lambda}] + \mathbf{T}_1[\boldsymbol{\lambda}](\tau - t) + ... + \mathbf{T}_K[\boldsymbol{\lambda}](\tau - t)^K \tag{8}$$

where $\mathbf{T}_k[\boldsymbol{\lambda}]$ is the k th order differential transformation [8] of $\boldsymbol{\lambda}(\tau)$, *i.e.*

$$\mathbf{T}_k[\boldsymbol{\lambda}] = \frac{1}{k!}\frac{d^k\boldsymbol{\lambda}(\tau)\big|_{\tau=t}}{d\tau^k} \tag{9}$$

Using Eq. (4), $\mathbf{T}_k[\boldsymbol{\lambda}]$ can be determined in an iterative way:

$$\mathbf{T}_0[\boldsymbol{\lambda}] = diag[-1,-1,...], \ \mathbf{T}_{k+1}[\boldsymbol{\lambda}] = -\frac{\mathbf{T}_k[\boldsymbol{\lambda}\mathbf{J}]}{k+1}, \ 0 \le k \le K+1 \tag{10}$$

Let $\mathbf{G} = \dot{\mathbf{x}}_n(\tau) - \mathbf{f}[\mathbf{x}_n(\tau), \tau]$, by substituting Eq. (8) into Eq. (2), we have

$$\mathbf{x}_{n+1}(t) = \mathbf{x}_n(t) + \int_{t_0}^{t} \{\mathbf{T}_0[\boldsymbol{\lambda}] + \mathbf{T}_1[\boldsymbol{\lambda}](\tau - t) + ... + \mathbf{T}_K[\boldsymbol{\lambda}](\tau - t)^K\} \mathbf{G} d\tau \tag{11}$$

Considering that $\mathbf{T}_k[\boldsymbol{\lambda}]$, $k = 0,1,...K$ are functions of $\mathbf{x}_n(t)$ and t, this expression can be rewritten in another form,

$$\mathbf{x}_{n+1}(t) = \mathbf{x}_n(t) + \mathbf{A}_0(t)\int_{t_0}^{t} \mathbf{G} d\tau + \mathbf{A}_1(t)\int_{t_0}^{t} \tau \mathbf{G} d\tau + ... + \mathbf{A}_K(t)\int_{t_0}^{t} \tau^K \mathbf{G} d\tau \tag{12}$$

where the coefficient matrices $\mathbf{A}_k(t)$, $k = 0,1,...K$ are combinations of $\mathbf{T}_k[\boldsymbol{\lambda}]$ and t.

Obviously, if $\boldsymbol{\lambda}(\tau)$ is approximated roughly with $\mathbf{T}_0[\boldsymbol{\lambda}]$, the correctional formula of VIM will degrade to the Picard iteration formula, which is written as

$$\mathbf{x}_{n+1}(t) = \mathbf{x}_n(t) + \int_{t_0}^{t} \mathbf{f}[\mathbf{x}_n(\tau), \tau] d\tau \tag{13}$$

Quasi-Linearization Method

The proposed LVIM is both efficient and accurate for solving IVPs. In order to take this advantage, we designed a Quasi-Linearization & LVIM (QLVIM) to solve TPBVPs. The derivation of QLVIM involves three stages: the quasi-linearization of TPBVP, the transformation of linear TPBVPs into IVPs, and the integration of IVPs using the Local Variational Iteration Method [7].

Quasi Linearization of TPBVP

Generally, we consider nonlinear second order differential equations

$$\boldsymbol{r}'' = \boldsymbol{f}(t, \boldsymbol{r}, \boldsymbol{r}') \tag{14}$$

subjected to the boundary conditions

$$\boldsymbol{r}(t_0) = \boldsymbol{r}_0, \boldsymbol{r}(t_f) = \boldsymbol{r}_f \tag{15}$$

where $\boldsymbol{r}'$ and $\boldsymbol{r}''$ represent $d\boldsymbol{r}/dt$ and $d^2\boldsymbol{r}/dt^2$ respectively.

Eq. (14) can be rewritten as

$$\phi(t,\boldsymbol{r},\boldsymbol{r}',\boldsymbol{r}'') = \boldsymbol{r}'' - \boldsymbol{f}(t,\boldsymbol{r},\boldsymbol{r}') = 0 \tag{16}$$

Denote the nth and (n+1)th iterations by $\boldsymbol{r}_n$ and $\boldsymbol{r}_{n+1}$ respectively and require that, for both iterations, $\phi = 0$. For the nth iteration, this gives

$$\boldsymbol{r}_n'' = \boldsymbol{f}(t,\boldsymbol{r}_n,\boldsymbol{r}_n') \tag{17}$$

For the (n+1)th iteration, we get

$$\begin{aligned}&\phi(t,\boldsymbol{r}_{n+1},\boldsymbol{r}_{n+1}',\ \boldsymbol{r}_{n+1}'') = \phi(t,\boldsymbol{r}_n,\boldsymbol{r}_n',\ \boldsymbol{r}_n'') + \\ &\left(\frac{\partial \phi}{\partial \boldsymbol{r}}\right)_n (\boldsymbol{r}_{n+1} - \boldsymbol{r}_n) + \left(\frac{\partial \phi}{\partial \boldsymbol{r}'}\right)_n (\boldsymbol{r}_{n+1}' - \boldsymbol{r}_n') + \left(\frac{\partial \phi}{\partial \boldsymbol{r}''}\right)_n (\boldsymbol{r}_{n+1}'' - \boldsymbol{r}_n'') + \ldots = 0\end{aligned} \tag{18}$$

$$-\left(\frac{\partial \boldsymbol{f}}{\partial \boldsymbol{r}}\right)_n (\boldsymbol{r}_{n+1} - \boldsymbol{r}_n) - \left(\frac{\partial \boldsymbol{f}}{\partial \boldsymbol{r}'}\right)_n (\boldsymbol{r}_{n+1}' - \boldsymbol{r}_n') + (\boldsymbol{r}_{n+1}'' - \boldsymbol{f}(t,\boldsymbol{r}_n,\boldsymbol{r}_n')) + \ldots = 0 \tag{19}$$

Substituting $\boldsymbol{r}_n''$ from Eq. (17) into Eq. (19), and omitting second- and higher-order terms, we get

$$\boldsymbol{r}_{n+1}'' - \boldsymbol{K}_n \boldsymbol{r}_{n+1}' - \boldsymbol{J}_n \boldsymbol{r}_{n+1} = \boldsymbol{f}_n(t) - \boldsymbol{K}_n \boldsymbol{r}_n' - \boldsymbol{J}_n \boldsymbol{r}_n \tag{20}$$

where $\boldsymbol{K}_n = \left(\frac{\partial \boldsymbol{f}}{\partial \boldsymbol{r}'}\right)_n, \boldsymbol{J}_n = \left(\frac{\partial \boldsymbol{f}}{\partial \boldsymbol{r}}\right)_n, \boldsymbol{f}_n(t) = \boldsymbol{f}(t,\boldsymbol{r}_n,\boldsymbol{r}_n',\boldsymbol{r}_n'')$. The boundary conditions for Eq. (20) are

$$\boldsymbol{r}_{n+1}(t_0) = \boldsymbol{r}_0 \tag{21a}$$

$$\boldsymbol{r}_{n+1}(t_f) = \boldsymbol{r}_f \tag{21b}$$

Thus, the nonlinear TPBVP in Eqs. (14, 15) degenerate to an iterative sequence of linear TPBVPs in Eqs. (20, 21), which are much easier to solve.

Transformation of Linear TPBVP to IVPs

To solve the linear TPBVPs in Eqs. (20, 21), we use the method of superposition to transform linear TPBVP into IVPs. Assume

$$\boldsymbol{r}_{n+1} = \boldsymbol{V} + s_x \cdot \boldsymbol{W}_x + s_y \cdot \boldsymbol{W}_y + s_z \cdot \boldsymbol{W}_z = \boldsymbol{V} + \boldsymbol{W} \cdot \boldsymbol{s} \tag{22}$$

where $\boldsymbol{W}=[\boldsymbol{W}_x,\boldsymbol{W}_y,\boldsymbol{W}_z]$, and

$$\boldsymbol{s}=(s_x,s_y,s_z)^T=\boldsymbol{r}'(t_0) \tag{23}$$

The unknown constant $\boldsymbol{s}$ is identified as the missing initial slope. $\boldsymbol{V},\boldsymbol{W}_x,\boldsymbol{W}_y,\boldsymbol{W}_z$ are column vectors of 3×1. Their physical meanings will be discussed later. Substituting Eq. (22) into Eq. (20), we obtain

$$(\boldsymbol{V}''+\boldsymbol{W}''\cdot\boldsymbol{s})-\boldsymbol{K}_n(\boldsymbol{V}'+\boldsymbol{W}'\cdot\boldsymbol{s})-\boldsymbol{J}_n(\boldsymbol{V}+\boldsymbol{W}\cdot\boldsymbol{s})=\boldsymbol{f}_n(t)-\boldsymbol{K}_n\boldsymbol{r}_n'-\boldsymbol{J}_n\boldsymbol{r}_n \tag{24}$$

Particularly, Eq. (24) can be separated into two differential equations

$$\boldsymbol{V}''-\boldsymbol{K}_n\boldsymbol{V}'-\boldsymbol{J}_n\boldsymbol{V}=\boldsymbol{f}_n(t)-\boldsymbol{K}_n\boldsymbol{r}_n'-\boldsymbol{J}_n\boldsymbol{r}_n \tag{25}$$

$$\boldsymbol{W}''-\boldsymbol{K}_n\boldsymbol{W}'-\boldsymbol{J}_n\boldsymbol{W}=\boldsymbol{0} \tag{26}$$

Similarly, substituting Eq. (22) into Eqs. (21a, 23), we obtain

$$\boldsymbol{r}(t_0)=\boldsymbol{V}(t_0)+s_x\cdot\boldsymbol{W}_x(t_0)+s_y\cdot\boldsymbol{W}_y(t_0)+s_z\cdot\boldsymbol{W}_z(t_0)=\boldsymbol{r}_0 \tag{27}$$

$$\boldsymbol{r}'(t_0)=\boldsymbol{V}'(t_0)+s_x\cdot\boldsymbol{W}_x'(t_0)+s_y\cdot\boldsymbol{W}_y'(t_0)+s_z\cdot\boldsymbol{W}_z'(t_0)=(s_x,s_y,s_z)^T \tag{28}$$

As a special case, Eqs. (27, 28) can be decomposed as

$$\boldsymbol{V}(t_0)=\boldsymbol{r}_0,\boldsymbol{W}_x(t_0)=\boldsymbol{W}_y(t_0)=\boldsymbol{W}_z(t_0)=(0,0,0)^T \tag{29}$$

$$\boldsymbol{V}'(t_0)=(0,0,0)^T,\boldsymbol{W}_x'(t_0)=(1,0,0)^T,\boldsymbol{W}_y'(t_0)=(0,1,0)^T,\boldsymbol{W}_z'(t_0)=(0,0,1)^T \tag{30}$$

Thus, Eqs. (20), (21a) and (23) are separated into four initial value problems

$$\text{IVP1:}\ \boldsymbol{V}''-\boldsymbol{K}_n\boldsymbol{V}'-\boldsymbol{J}_n\boldsymbol{V}=\boldsymbol{f}_n(t)-\boldsymbol{K}_n\boldsymbol{r}_n'-\boldsymbol{J}_n\boldsymbol{r}_n,\ \boldsymbol{V}(t_0)=\boldsymbol{r}_0,\boldsymbol{V}'(t_0)=(0,0,0)^T \tag{31}$$

$$\text{IVP2:}\ \boldsymbol{W}_x''-\boldsymbol{K}_n\boldsymbol{W}_x'-\boldsymbol{J}_n\boldsymbol{W}_x=\boldsymbol{0},\ \boldsymbol{W}_x(t_0)=(0,0,0)^T,\boldsymbol{W}_x'(t_0)=(1,0,0)^T \tag{32}$$

$$\text{IVP3:}\ \boldsymbol{W}_y''-\boldsymbol{K}_n\boldsymbol{W}_y'-\boldsymbol{J}_n\boldsymbol{W}_y=\boldsymbol{0},\ \boldsymbol{W}_y(t_0)=(0,0,0)^T,\boldsymbol{W}_y'(t_0)=(0,1,0)^T \tag{33}$$

$$\text{IVP4:}\ \boldsymbol{W}_z''-\boldsymbol{K}_n\boldsymbol{W}_z'-\boldsymbol{J}_n\boldsymbol{W}_z=\boldsymbol{0},\ \boldsymbol{W}_z(t_0)=(0,0,0)^T,\boldsymbol{W}_z'(t_0)=(0,0,1)^T \tag{34}$$

Now, if we examine the form of Eq. (22) again, its physical meaning will be obvious. $\boldsymbol{V}$ is the state vector of the linear system Eq. (25) with initial values $\boldsymbol{r}_0$ and zero initial velocity. $\boldsymbol{W}_x, \boldsymbol{W}_y, \boldsymbol{W}_z$ are the state vectors of the homogeneous system Eq. (26) starting from the original point with unit velocity in the x direction, y direction, and z direction, respectively. Since we are solving a linear system in this step, the real trajectory of the system is the linear combination of these four component trajectories, and the coefficients (s_x, s_y, s_z) are components of initial velocity in three directions. On the other hand, from the perspective of linear differential equation theory, the form of Eq. (22) is also very easy to understand. As a matter of fact, $\boldsymbol{V}$ is a particular solution of Eq. (25), $\boldsymbol{W}_x, \boldsymbol{W}_y, \boldsymbol{W}_z$ are three nontrivial and distinct solutions obtained from the homogeneous form, Eq. (26), of the original Eq. (25). According to the superposition principle, any solution of Eq. (25) can be represented as the linear combination of $\boldsymbol{V}, \boldsymbol{W}_x, \boldsymbol{W}_y, \boldsymbol{W}_z$, *i.e.*, Eq. (22). However, the value of $\boldsymbol{s}$ is still unknown up to now. By substituting Eq. (22) into Eq. (23), we obtain

$$\boldsymbol{s}_{new} = \boldsymbol{W}(t_f)^{-1} \cdot [\boldsymbol{r}_f - \boldsymbol{V}(t_f)] \tag{35}$$

where $\boldsymbol{W}(t_f) = [\boldsymbol{W}_x(t_f), \boldsymbol{W}_y(t_f), \boldsymbol{W}_z(t_f)]$. Eq. (35) gives the iterative formula of initial slope $\boldsymbol{s}$. A flowchart of QLVIM is attached at the end of this chapter.

PERTURBED ORBIT PROPAGATION

A low Earth orbit (LEO), a high eccentric orbit (HEO) and a geosynchronous Earth orbit (GEO) are used as examples, to test the proposed method. Since these three orbit types are mostly seen in orbit problems, it is possible to evaluate the proposed method comprehensively with them. Table **1** lists the parameters and initial conditions of the three orbit regimes. For comparison, we adopted the fourth-order Runge-Kutta and the Gauss-Jackson method with the same configurations used by [9].

Table 1. Parameters and initial conditions.

Orbit Type	$\mathbf{r}_0$ (m)	$\dot{\mathbf{r}}_0$ (m/s)	Ecc.	Inc. (rad)
LEO	-0.3889e6 7.7388e6 0.6736e6	-3.5794e3 0 6.1997e3	0.1	$\pi/3$
HEO	4.05e6 0	0 9.1464e3	0.7	$\pi/3$

(Table 1) cont.....

	-7.0148e6	0		
GEO	4.2164172e7 0 0	0 3.074660237e3 0	0	0

COMPARISON OF LVIM WITH MCPI

A spherical harmonic gravity field of 40 degrees (EGM2008) is adopted in the orbital model. Air drag force is not considered herein so that we can compare the computational error of Hamiltonian of these two methods. As pointed out in [9], Hamiltonian provides some insights into the highest accuracy that these two methods may achieve, though it cannot capture all the computational errors. Relative error of Hamiltonian that we obtained in numerical results is calculated *via* $\varepsilon = \Delta H / H_0$, where H_0 is the Hamiltonian of the system at an initial instant. ΔH is the discrepancy of Hamiltonian in computation.

The performance of these two methods is demonstrated by using the same number of Chebyshev-Gauss-Lobatto nodes M and the tolerance criterion $tol = norm(\mathbf{U}_{n+1}) - norm(\mathbf{U}_n)$. These configurations have been tuned carefully to ensure the modified Chebyshev-Picard iteration method works optimally, and then the same tuned parameters are used in the proposed method. The step size Δt adopted in these two methods is selected as one orbit period T_p, meaning that the integration during one orbit revolution can be completed within a single step for all three orbit regimes. Much smaller step sizes are used later to give a further comparison. The accuracy and convergence speed of these two methods are evaluated and presented in Figs. (**1 & 2**) for both large and small step sizes respectively. Throughout the simulation, the initial approximation of the methods in each step is selected as a straight line, *i.e.*, the so-called "cold-start".

Table 2. Tuning Parameters ($\Delta t \approx 1$ *orbit period*).

Case	Number of Nodes	Step Size (1 Orbit Period) (s)	Error Tolerance *tol*
LEO	251	8000	1e-5
HEO	1001	5e4	2e-5
GEO	1001	9e4	4e-5

The accuracy of the Feedback-Accelerated Picard Iteration method and the modified Chebyshev-Picard iteration method are almost the same in the case where the step size $\Delta t \approx 1\, orbit\ period$. The results are shown in Fig. (**1**), where these two methods are both of very high accuracy. In these three orbits, the Hamiltonian's relative computational errors are kept near/very close to the machine precision. For HEO, the accuracy deteriorates a little bit. The main reason for it is that the high eccentric orbit is affected more by the gravity force near perigee than near apogee, which means much more nodes are needed to precisely approximate the trajectory near perigee. Against this problem, it is suggested in [10] to use true anomaly segmentations to improve computational accuracy and efficiency.

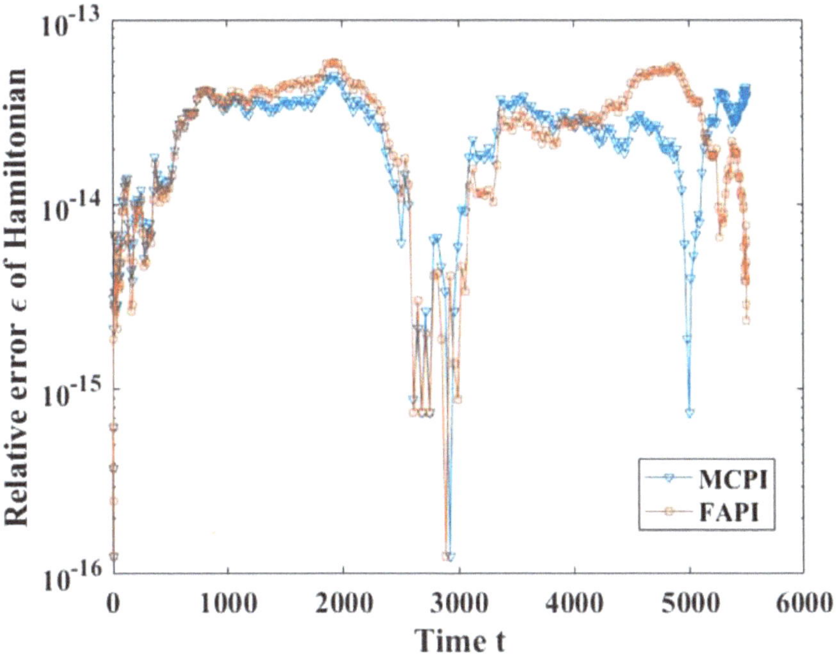

(a)

(Fig. 1) contd.....

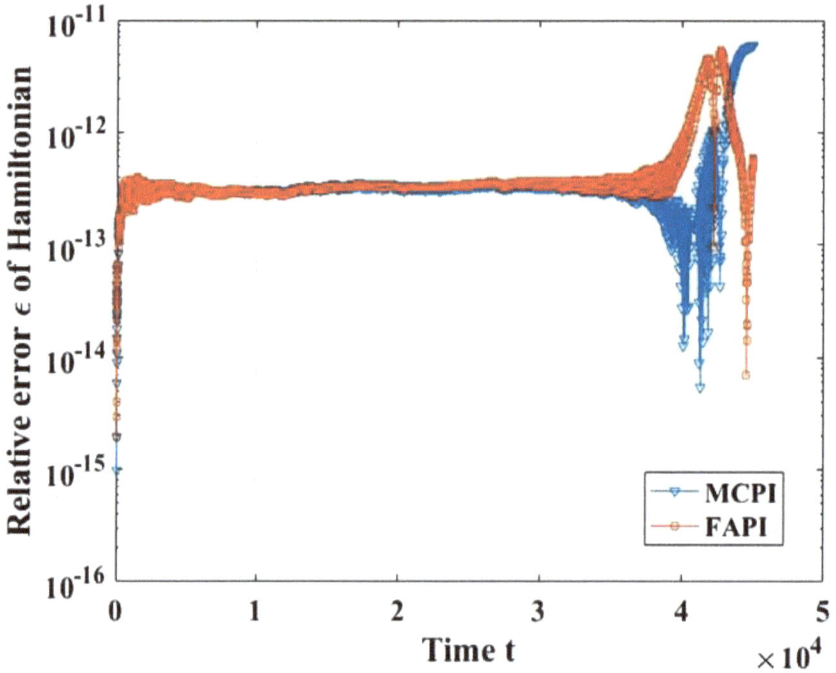

(b)

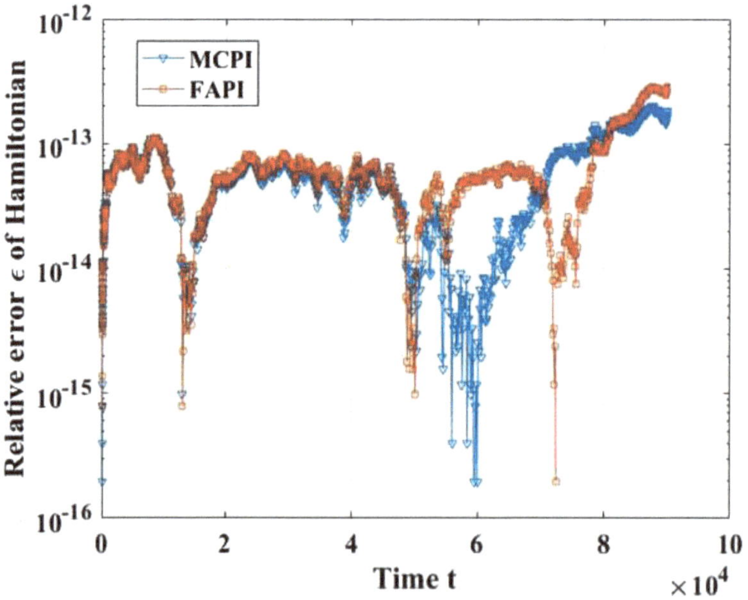

(c)

(Fig. 1) contd.....

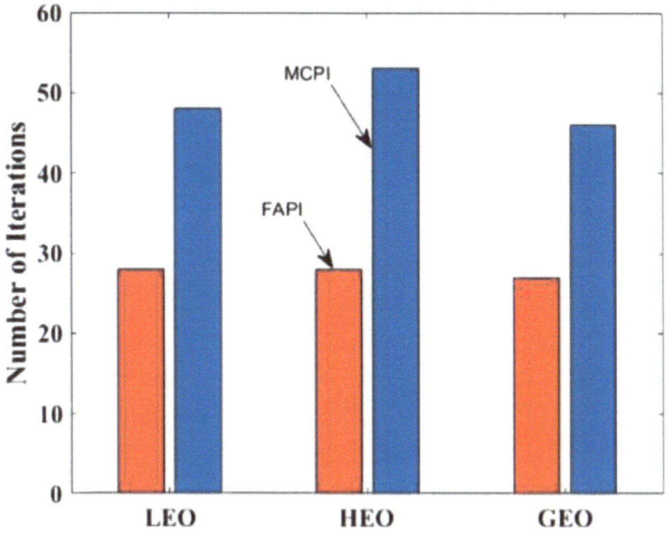

(d)

Fig. (1). Performance of LVIM and MCPI with $\Delta t \approx 1$ *orbit period*.

Fig. (**1**) shows that the convergence speed is the main difference between the proposed method and the modified Chebyshev-Picard iteration method. As shown, the number of iterations needed to converge is halved by using the proposed method. In practical applications, the evaluations of the force model consume most of the computational time, thus the reduction of iterations is significant in reducing the computational cost. Since the computation time in each iteration is roughly the same for these two methods, the convergence speed can be used to evaluate the computational efficiency.

A large number of nodes could occupy a lot of memory and cause a heavy burden to the computer processor, which will in turn, slow down the computation. Table **2** indicates that lots of nodes are needed to accurately approximate the solution in each step by using a very large time step size. Thus, the number of nodes and the iteration steps in the Feedback-Accelerated Picard Iteration method and the modified Chebyshev-Picard iteration method can be much reduced by using a smaller time step size. Small time intervals also guarantee higher stability and faster convergence.

Table 3. Tuning parameters.

Case	Number of Nodes	Step Size (s)	Error Tolerance
LEO	19	500	1e-7
HEO (3 orbits)	31	500	1e-7
GEO	25	1000	1e-7

Table **3** shows that the relative computational error of Hamiltonian can be further reduced in the proposed Feedback-Accelerated Picard Iteration method with a smaller time step size. Fig. (**2**) shows that the error ε is kept well under 10^{-13} in the first orbit revolution for the LEO, HEO (3 orbit revolutions are presented in Fig. (**2b**), and GEO. In Fig. (**2b**), it is observed that the computational error of both methods accumulates relatively fast in the steps near the perigee. The accuracy of the two methods concerned herein is still almost the same except for the HEO case presented in Fig. (**2b**), mainly due to the computational error in the first few steps.

It can be seen in Fig. (**2d**) that the total number of iterations is halved by the proposed method, consistent with the previous result in Fig. (**1d**). In each time step, the iteration numbers of the Feedback-Accelerated Picard Iteration method and the modified Chebyshev-Picard iteration method are about 5-7 and 10-15, respectively.

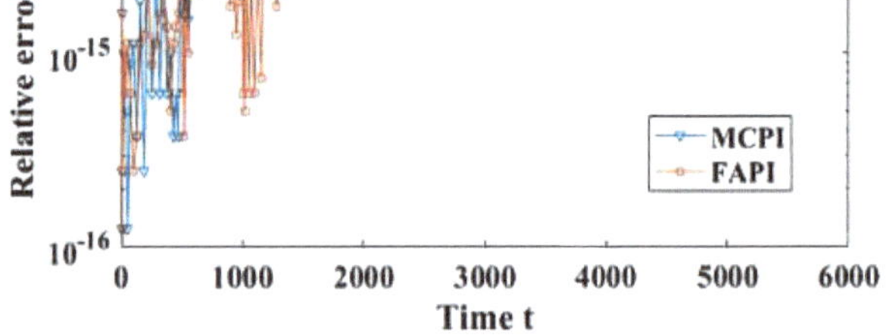

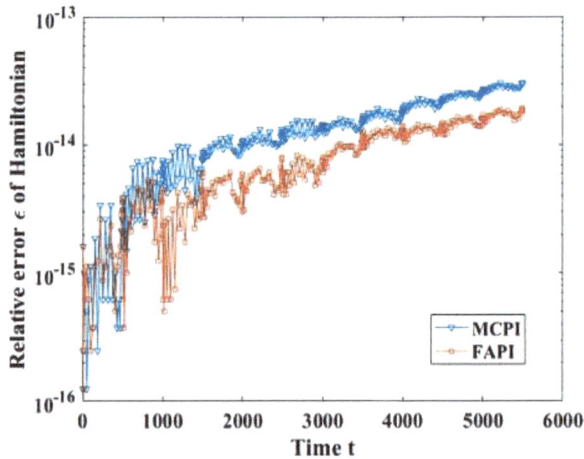

(a)

(Fig. 2) contd.....

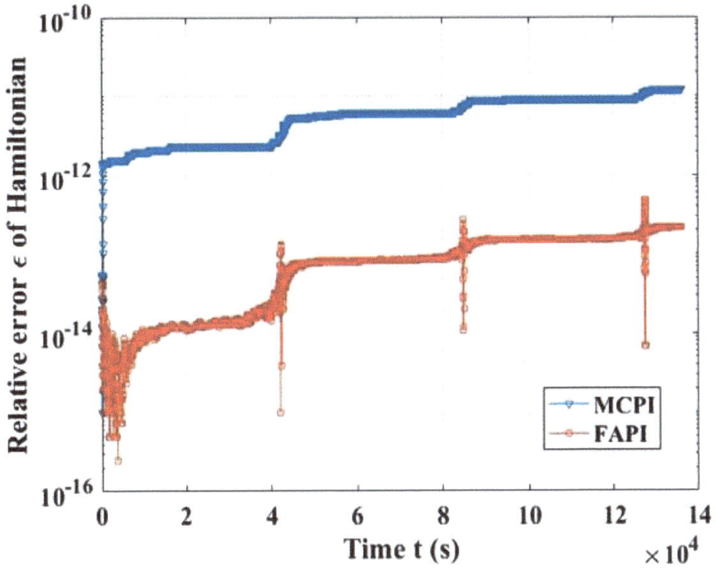

(b)

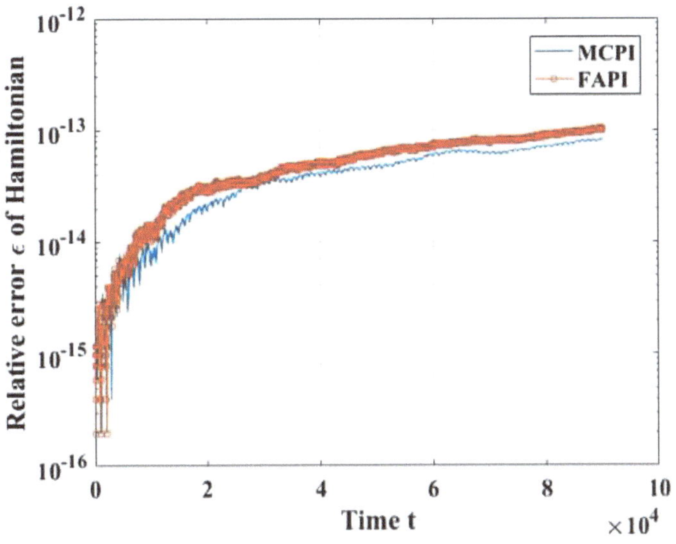

(c)

(Fig. 2) contd.....

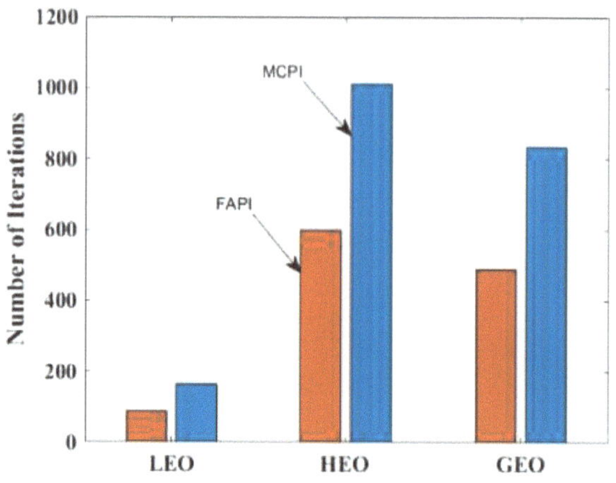

(d)

Fig. (2). Performance of LVIM and MCPI by using small integration steps.

COMPARISON OF LVIM WITH RUNGE-KUTTA 12(10)

The air drag force is included in the perturbed forced model in comparison with Runge-Kutta (RK) 12 (10) method [11], so as to investigate the performance of the proposed method in solving real-life problems. The drag force is evaluated with a basic cannonball drag model with the ballistic coefficient set as $0.01 m^2/kg$. By including the air drag, the force evaluation in LEO and HEO becomes more difficult, and the atmospheric re-entry in the case of HEO may further stress the integrators. The MSIS-E-90 atmosphere model, supported online by NASA, is used herein to provide information on air density.

We used an error ratio ρ in terms of the root-mean-square (RMS) error of the integration as defined in [9] to get reasonable assessments of the integration accuracy. The position error ρ is calculated by

$$\rho = \frac{\Delta r_{RMS}}{r_A N_{orbits}}, \Delta r_{RMS} = \sqrt{\frac{1}{N}\sum_{i=1}^{N}(\Delta r_i)^2}$$

where r_A is the distance of apogee, Δr_i is the position error on each sampling point, N_{orbits} is the number of orbits. The error ratio of velocity can be defined similarly. The exact value of the computational error in the proposed method and the RK 12

(10) method is unlikely to be obtained because there is no accurate solution for reference. Other integration methods are unreliable because these two methods are of relatively high accuracy compared with other methods in the literature. So, the position errors of these two methods are evaluated using the step-size halving technique as verified in [9]. Table **4** lists the parameters of the Feedback-Accelerated Picard Iteration method. The step sizes of RK 12 (10) are set as 50s, 50s, and 100s, respectively, for the three tested cases of LEO, HEO, and GEO.

Table 4. Parameters for LVIM.

Case	Number of Nodes Per Step	Step Size (s)	Error Tolerance
LEO (3orbits)	25	500	1e-7
HEO (3 orbits)	501	1.5e4	2e-6
GEO (3orbits)	301	3e4	2e-6

Table 5. Evaluation of error ratios.

Case	RK12(10)		LVIM	
	Position	Velocity	Position	Velocity
LEO (3orbits)	3e-14	3e-14	3e-14	3e-14
HEO (3 orbits)	2e-9	2e-9	6e-12	6e-12
GEO (3orbits)	5e-13	5e-13	1e-13	1e-13

Table **5** presents the evaluations of error ratios for the RK12 (10) and the proposed method. The accuracy of these two methods are very high in the two cases of LEO and GEO. However, the accuracy of RK 12 (10) deteriorates significantly while the proposed method still achieves relatively high accuracy in the integration of HEO. We made a more concrete comparison in Fig. (**3**). The position errors in three orbit revolutions of LEO, HEO and GEO are plotted separately.

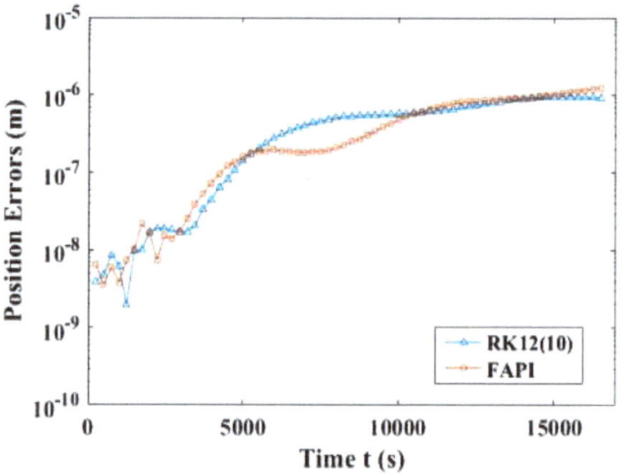

(a)

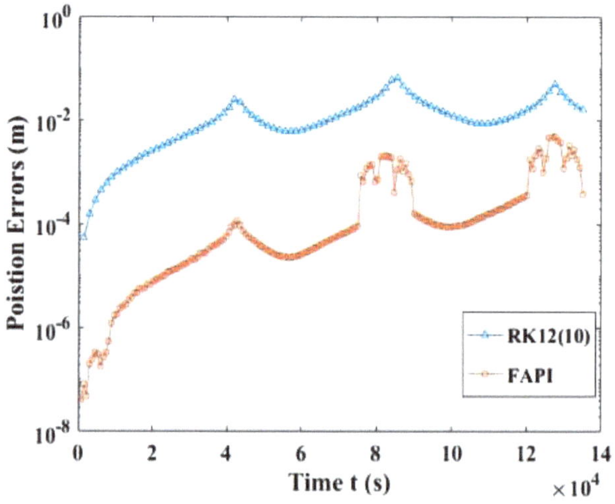

(b)

(Fig. 3) contd.....

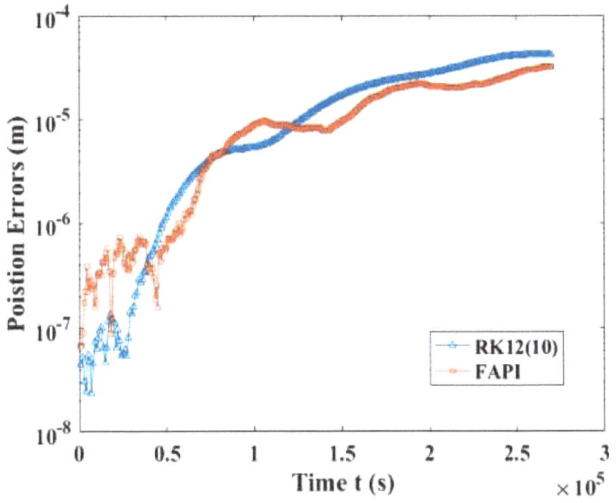

(c)

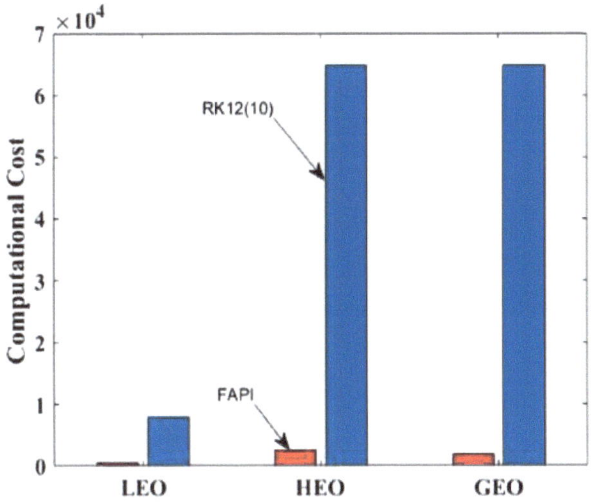

(d)

Fig. (3). Comparison of LVIM and RK 12 (10).

The computational velocity error in these two methods is similar to the position error, except that the value is three magnitudes smaller. The results in Figs. **(3a-c)** and Table **5** agree with each other. The maximum position error in the proposed methods is about 10^{-6} m, 10^{-3} m, and 10^{-4} m in the cases of LEO, HEO and GEO, respectively. Fig. (**3d**) plots the computational cost that reveals the computational time consumed in evaluating the force model.

The computational cost for the LVIM method is calculated as $Cost_{FAPI} = N_{it} t_{FAPI} / t_u$ where N_{it} is the total number of iterations. The time of evaluation of the perturbed force on one node is denoted as the unit time t_u. The computational time for one iteration in the Feedback-Accelerated Picard Iteration method is determined as t_{FAPI}. The computational time for one step of the RK 12 (10) method involves 25 successive iterations, *i.e.*, $25t_u$. The corresponding computational cost is then $Cost_{RK} = 25N_{st}$, where N_{st} is the number of integration steps. Fig. (**4d**) shows that the proposed method is far more efficient than the RK 12 (10) method. The proposed method's cost saving is about 95% regarding the three typical orbit regimes. The efficiency of the proposed method could be further improved if the parallel computing technique is adopted.

The step size of the Feedback-Accelerated Picard Iteration method is fixed in the above simulations. Some steps are over-calculated and others may not be calculated very well as observed in Fig. (**2b**). It means the computational resource is wasted, and errors could accumulate quickly. Thus, adaptive steps should be used in practice to improve the computational efficiency and accuracy when using the proposed method and the MCPI method. A possible approach is to divide the orbit into multiple true anomaly segments and make approximations in each segment, as proposed by [10].

ERTURBED LAMBERT'S PROBLEM

The orbit targeting and determinations are based on the solution to Lambert's problem. There has already been a variety of elegant methods for solving this problem [11, 12] for non-perturbed Keplerian motion, but the solutions of classical Lambert problem can be useless in more general situations, such as the perturbed two-body motion, three-body motion, and relative motion. In contrast, the Feedback-Accelerated Picard Iteration method is not limited to unperturbed motion or any specific two-point boundary value problem. The effectiveness of the proposed method is verified using the perturbed Lambert's problem as an example.

A novel technique named Fish-Scales-Growing Method (FSGM) is depicted to further extend the convergence area of the LVIM. With FSGM, limited computational resources can solve two-point boundary value problems in conservative systems. A high eccentricity orbit transfer problem and a multi-revolution Lambert transfer problem are used to demonstrate this technique.

Using LVIM

The two Lambert transfer problems listed in Table **6** are solved using the proposed Feedback-Accelerated Picard Iteration method. A simple J_2 perturbed gravity force model is used here. The number of collocation points and Chebyshev polynomials are both selected as $N = M = 64$. The initial guess for the solution is selected as a uniform, straight line connecting the initial and final position.

Table 6. Initial and final positions for the Lambert transfer problems.

Cases	Initial Position $\mathbf{r}_0$ (m)	Final Position $\mathbf{r}_f$ (m)	Transfer Time(s)
LEO transfer	-0.3889e6 7.7388e6 0.6736e6	-3.6515e6 -4.2152e6 6.3103e6	2,500
HEO transfer	-1.4e7 2.1e7 2.4249e7	-3.1497e7 -0.0462e7 5.4554e7	25,000

The transfer orbits obtained by the proposed Feedback-Accelerated Picard Iteration method are plotted in Fig. (**4**). The velocities at the boundaries are also determined by the proposed method and listed in Table **7**. As a reference, the MATLAB built-in ODE45 function is used to numerically integrate the transfer orbit with the initial velocity in Table **7**. The ODE45 function uses an explicit Runge-Kutta 4(5) method. An introduction to this function can be found in the literature [13]. The relative and absolute tolerance of ODE45 are set to 10^{-15}. The discrepancies in final positions and velocities between the results obtained by ODE45 and the proposed method are then measured and listed in Table **7**. As is shown, with 64 CGL nodes as collocation points, Lambert's problems considering J_2 perturbation are solved with very high accuracy. It is noted that the modified Chebyshev-Picard iteration method is also used, but somehow it appears to be not convergent in these two cases.

Table 7. Two general transfer orbits obtained using the LVIM method.

	Boundary velocities, m/s		Errors	
Cases	Initial	Final	Position, m	Velocity, m/s
Low Eccentricity	-3579.396550	1798.099253	-3.2143e-05	-2.5436e-07
	0.008964	-5510.306049	1.1292e-05	-7.5450e-06
	6199.705320	-3124.253368	9.1111e-06	-2.4637e-05
High Eccentricity	-1687.308996	20.559111	-2.6535e-05	-1.2676e-06
	-0.025078	-1124.689391	1.0044e-05	1.0028e-06
	2922.606386	-35.693697	-1.2435e-05	-2.5208e-07

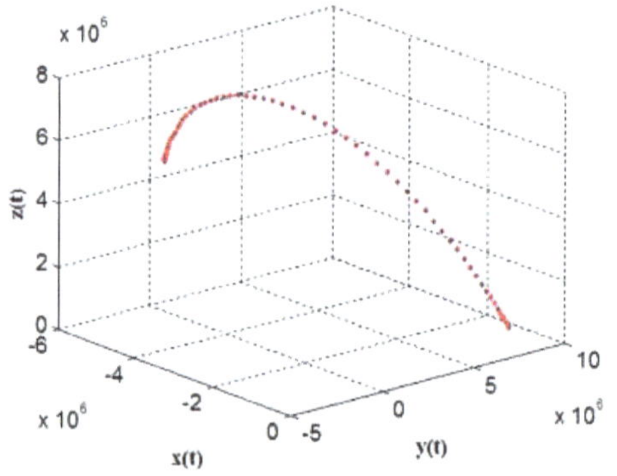

(a)

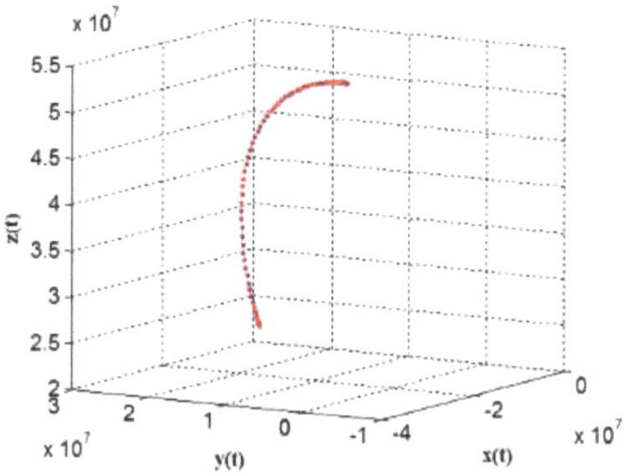

(b)

Fig. (4) **(a)** Low earth transfer orbit and **(b)** high eccentricity transfer orbit.

Using Fish-Scale-Growing Method

In the area of solving two-point boundary value problems, all the methods, including various shooting methods, finite difference methods and collocation methods, are complicated by the long duration or large interval between the initial and final state, especially when the dynamical system is complex and strongly nonlinear. In the simple shooting method, for example, as the shooting distance increases, the method becomes more and more sensitive to the initial guess, and at the same time, the selection of a proper initial guess become more and more difficult. This led to the development of multiple shooting methods and the modified simple shooting method, both of which have larger convergence domain than the simple shooting method. However, these two methods are much more laborious than the simple shooting method, and can still become hard to implement if an accurate initial guess cannot be obtained.

Here we propose a Fish-Scales-Growing Method for solving long-duration two-point boundary value problems. As will be shown, the initialization and the implementation of this method is rarely simple. Moreover, as an iterative method, the Fish-Scales-Growing Method is robust and converges fast. It can also be conveniently coded for parallel computation. The iterative procedure of this method is as follows.

An initial approximation is provided by a reference trajectory, which could be a nominal solution obtained from the linearized problem or unperturbed problem. Then divide the domain (t_0, t_f) of the two-point boundary value problem (TPBVP) into multiple isometric intervals $[t_i, t_{i+1}]$, $0 \le i < N$, $t_{i+1} - t_i = (t_f - t_0)/N$. For each interval, the points $\mathbf{x}(t_i)$ and $\mathbf{x}(t_{i+1})$ on the reference trajectory are set as boundaries. In each iteration, we need to solve the N TPBVPs defined by the boundaries $\mathbf{x}(t_i)$ and $\mathbf{x}(t_{i+1})$ in the corresponding interval $[t_i, t_{i+1}]$. The points $\tilde{\mathbf{x}}(t_j)$, $t_j = (t_i + t_{i+1})/2$ on the solutions of these $N+1$ TPBVPs are collected for the next step. Then we solve the $N-1$ TPBVPs defined by the boundaries $\tilde{\mathbf{x}}(t_j)$ and $\tilde{\mathbf{x}}(t_{j+1})$, $0 \le j < N-1$ in the corresponding interval $[t_j, t_{j+1}]$. After that, the points $\overline{\mathbf{x}}(t_i)$ on the solutions of these $N-1$ TPBVPs are used to replace $\mathbf{x}(t_i)$ and $\mathbf{x}(t_{i+1})$. If $\|\overline{\mathbf{x}}(t_i) - \mathbf{x}(t_i)\| \le \varepsilon$, the iteration ends. Otherwise, one should replace $\mathbf{x}(t_i)$ with $\overline{\mathbf{x}}(t_i)$ and restart the iteration.

According to the principle of least action, the solution of a TPBVP in a conservative system is the trajectory that has the least action among any nearby trajectories that connect the initial and final position. Thus, it can be concluded that in each iteration, the action along the corrected piecewise solution connecting $\mathbf{x}(t_0)$, $\tilde{\mathbf{x}}(t_{j=0})$, ..., $\tilde{\mathbf{x}}(t_{j=N-1})$, $\mathbf{x}(t_f)$ is less than that along the trajectory connecting $\mathbf{x}(t_0)$, $\mathbf{x}(t_{i=1})$, ..., $\mathbf{x}(t_{i=N-1})$, $\mathbf{x}(t_f)$. Therefore, it is guaranteed for a conservative system that the total action along the piecewise trajectory is monotonically decreasing as the iteration goes on. So, this method is absolutely convergent for conservative systems, providing that the sub-TPBVPs are solvable. Although there is no mathematical method to estimate the convergence speed, the numerical examples show that the Fish-Scales-Growing Method converges fast within the first 15 iterations.

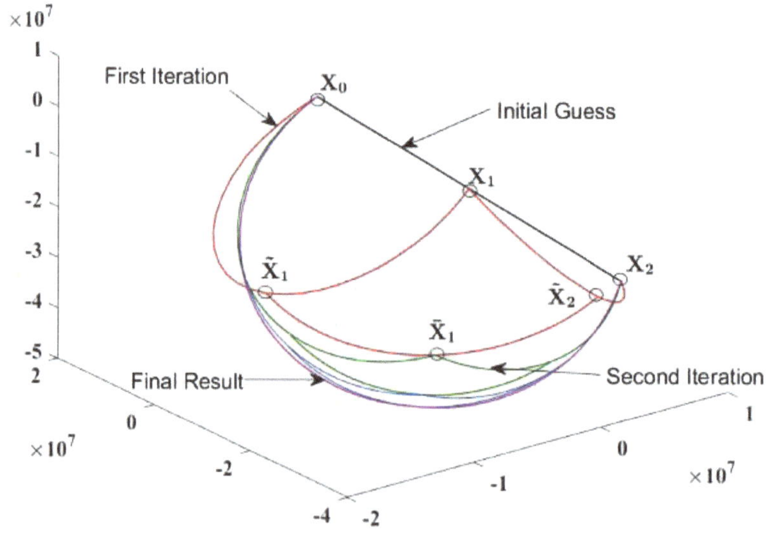

Fig. (5). Illustration of the Fish-Scales-Growing Method.

A schematic of the Fish-Scales-Growing Method is illustrated in Fig. (**5**). The domain $[t_0,t_f]$ is divided into two smaller intervals $[t_0,t_1]$ and $[t_0,t_f]$, where $t_1=(t_0+t_f)/2$. In each iteration, three boundary value problems are defined and need to be solved to correct the position values $\mathbf{x}_1$ at the time instant t_1. As the iteration goes on, $\mathbf{x}_1$ approaches the real value, thus, the piecewise trajectory evolves to the true solution of the original TPBVP.

Table 8. Boundary conditions for orbit transfer problems.

Cases	Initial Position $\mathbf{r}_0$ (m)	Final Position $\mathbf{r}_f$ (m)	Transfer Time (s)
HEO Transfer	-1.4e7 2.1e7 2.4249e7	-1.2572e7 -2.0930e7 2.1772e7	50,000
Multi-Revolution Transfer	4.2164172e7 0 0	3.5737110e7 2.2375813e7 0	180,000

First, a high eccentricity orbit transfer problem is solved using the Fish-Scales-Growing Method. The boundary conditions are listed in Table **8**. A very rough initial guess is selected as

$$\mathbf{r}(t) = \mathbf{r}_0 + (\mathbf{r}_f - \mathbf{r}_0)/(t_f - t_0),$$

Where the initial and final time of orbit transfer are set as $t_0 = 0s$, $t_f = 50{,}000s$. The entire domain is divided into two equal time intervals. The TPBVPs defined in the smaller intervals are solved using the Feedback-Accelerated Picard Iteration method. The number of collocation points are selected as $M = 13$. The results are plotted in Fig. (**6**). It can be seen that the piecewise solution of the Fish-Scales-Growing Method converges to the true solution in a few iterations. This example shows the original problem is divided into multiple sub-TPBVPs, each of which can be solved easily with 13 collocation points. Fig. (**6d**) indicates that the rate of convergence is logarithmic for the first 15 iterations.

In practice, the Fish-Scales-Growing Method can also be combined with other TPBVP solvers to enlarge the application area of these solvers in long-duration transfer problems and strongly nonlinear boundary value problems.

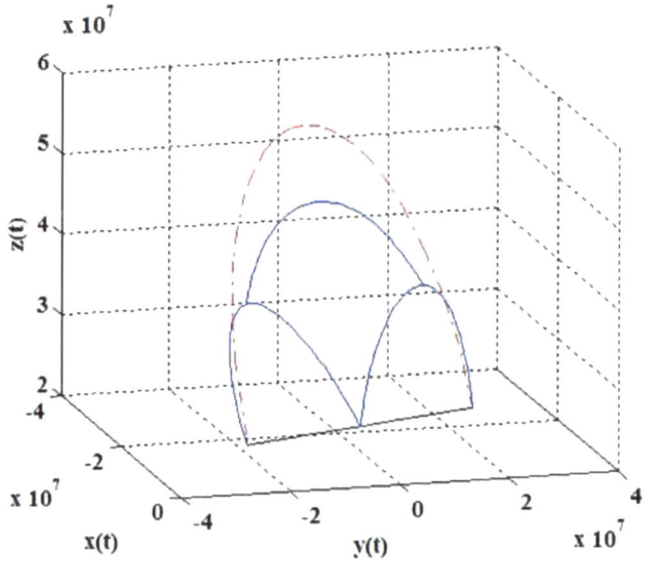

(a)

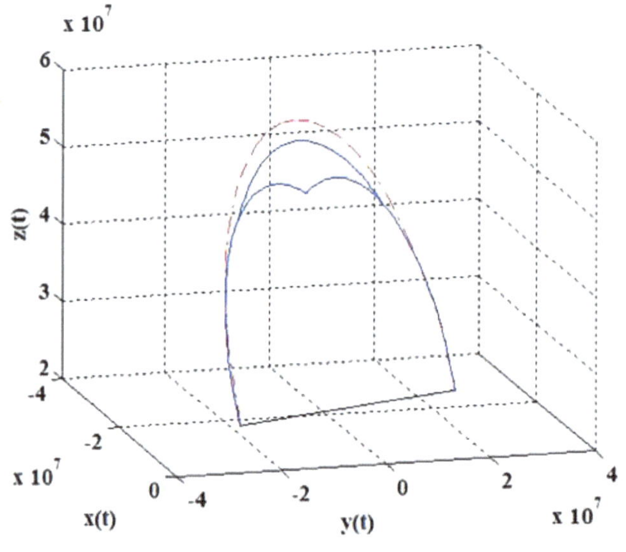

(b)

(Fig. 6) contd.....

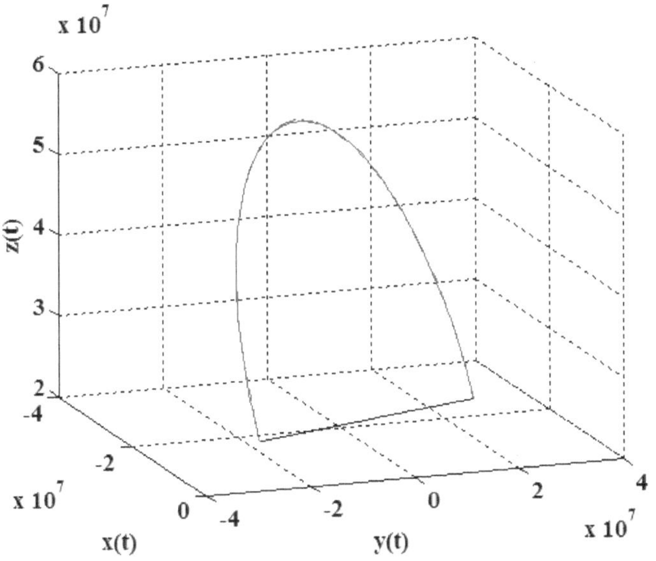

(c)

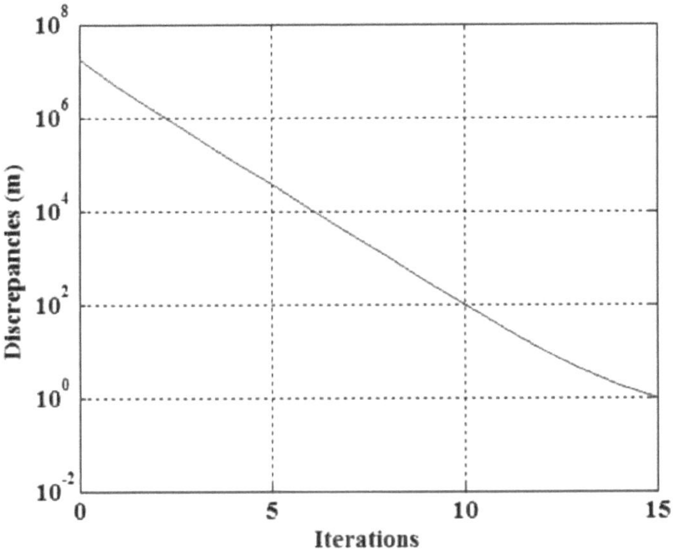

(d)

Fig. (6). Piecewise solution obtained with FSGM.

In solving the multi-revolution orbit transfer problem, the whole time interval $[0s, 180,000s]$ is divided into 10 segments, while in each segment the solution is approximated using 13 collocation points. Unlike the preceding case, the Keplerian orbit is used as an initial approximation. To evaluate the improvement of the solution obtained using the proposed method, the discrepancies δr of position and δv of velocity between the exact solution and the iteratively corrected solution is recorded and plotted in Fig. (**6**). It is shown that the proposed method can further improve the Keplerian solution, but the convergence speed is relatively low.

According to the numerical results in solving the HEO transfer and multi-revolution orbit transfer problems, it is found that the proposed Fish-Scales-Growing-Method is quite efficient in solving Lambert problems within one orbit revolution, while for the multi-revolution problem, the convergence speed could become very slow. This drawback is not addressed in this work, but it is one of our future research to improve the performance of the Fish-Scales-Growing-Method in solving multi-revolution orbit transfer problems (Fig. **7**).

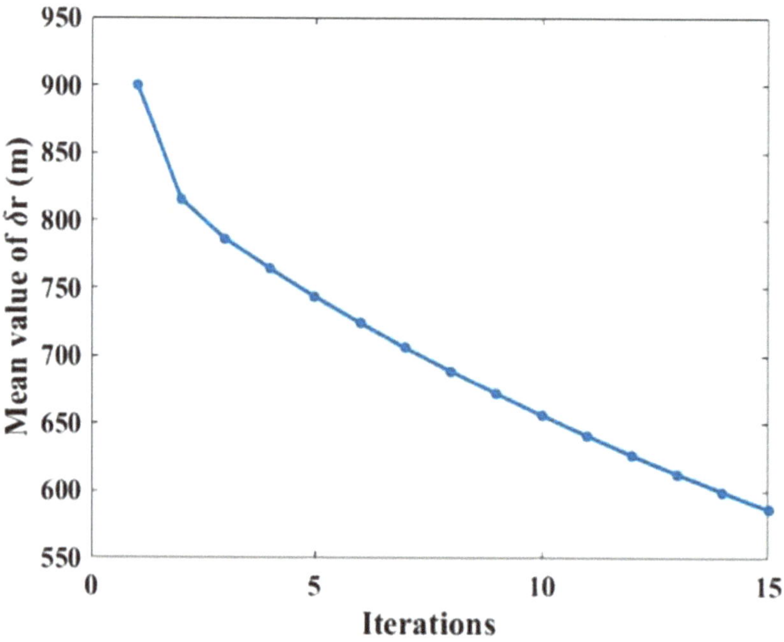

(a)

(Fig. 7) contd.....

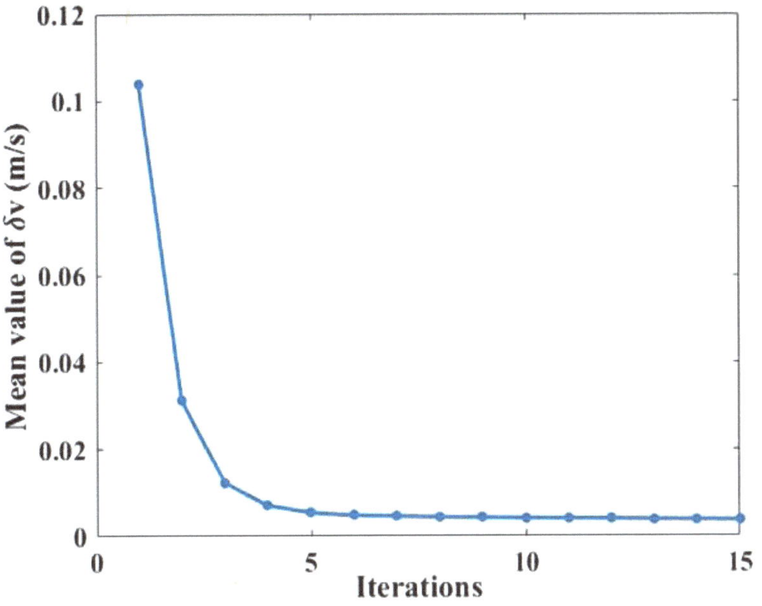

(b)

Fig. (7). Test of FSGM in multi-revolution lambert's problem.

Using QLVIM

While the spacecraft orbiting around the earth, is disturbed by non-spherical earth perturbation, atmospheric drag perturbation, lunisolar gravitational perturbation, and solar radiation pressure perturbation. In low earth orbit (LEO) and high eccentric orbit (HEO) cases, the first two kinds of perturbations are the main perturbations to spacecraft. So we only consider non-spherical earth perturbation and atmospheric drag perturbations in the dynamic equation here. In the equatorial inertial coordinate system, the orbital dynamic equation of the spacecraft is:

$$\ddot{\boldsymbol{r}} = -\frac{\mu}{r^3}\boldsymbol{r} + \boldsymbol{a}_J + \boldsymbol{a}_d \tag{72}$$

where $\boldsymbol{r}=[x, y, z]^T$ is the radius vector, $\mu = GM = 3986004.418 \times 10^8 \mathrm{m}^3/\mathrm{s}^2$ is the geocentric gravitational constant, $r=\|\boldsymbol{r}\|$ is the geocentric distance. $\boldsymbol{a}_J$ is the acceleration caused by non-spherical earth perturbation, here we consider the spherical harmonic gravity field of 10 deg (EGM2008). $\boldsymbol{a}_d$ is the acceleration caused by atmospheric drag force. A basic cannonball drag modal is used, with the ballistic coefficient set as $0.01\text{m}^2/\text{kg}$, to evaluate the drag force. The MSIS-E-90 atmosphere model is used to provide information on air density, which is supported online by NASA. Based on the radius and velocity vectors, $\boldsymbol{a}_J, \boldsymbol{a}_d$ and their Jacobian matrices can be determined by the two models above. Besides the proposed QLVIM, two other solvers are also applied here to solve Lambert's problem. The first is the Newton-RK4 method, which combines the classical Newton-type shooting method with the fourth-order Runge-Kutta method. The other is the FAPI mentioned above method, which is highly precise and efficient. In practice, these three solvers all need initial guesses. For QLVIM and FAPI methods, the initial guess is a trajectory, here, we choose the straight line connecting the two boundary points as the initial approximation, which is also called "cold start". For the Newton-RK4 method, the initial guess is a velocity, here, we choose s=[-3100, 100, 5900] as an initial guess for the LEO case and s=[-1200, 10, 2500] for the HEO case. The boundary conditions for Lambert's problem are listed in Table **6**. The parameters for these methods are listed in Table **9**. M is the number of time nodes in one-time step. N is the degree of basis function. Δt is the time step. $tol_{\tilde{x}}$ is the iteration tolerance of $\tilde{x}_n$. In Eq. (20), denote $\varepsilon_{\tilde{x}} = norm(\tilde{x}_{n+1} - \tilde{x}_n)$, when $\varepsilon_{\tilde{x}} < tol_{\tilde{x}}$ is satisfied, the iteration stops. Similarly, tol_s is the iteration tolerance of s.

The transfer orbits obtained by the proposed QLVIM are plotted in Fig. (**8**). The MATLAB built-in ODE45 function is used to numerically integrate the transfer orbit with the initial velocity determined by the proposed method. The ODE45 function uses an explicit RK 4 (5) method. The relative and absolute tolerance are set to 1e-13, 1e-15, respectively. Fig. (**8**) shows that the results of QLVIM and ODE45 are in good agreement.

The initial velocities calculated by these methods are listed in Table **10**, and the computational errors and computational times are listed in Table **11**. Before the analysis, we give some explanations about the computational errors here. As we know, the key to solving the TPBVP is to find the accurate initial velocity of the transfer orbit. The precision of the initial velocity manifests the precision of the algorithm. The more accurate the calculated initial velocity is, the final position propagated with it is closer to the boundary value $\boldsymbol{r}_f$, so we define the difference between them as computational error.

Table 9. Tuning Parameters.

Algorithm	Cases	M	N	Δt	$tol_{\tilde{x}}$	tol_s
QLVIM	LEO	10	10	250	1e-7	1e-5
	HEO	10	10	2500	1e-7	1e-5
Newton-RK4	LEO	--	--	3.125	--	1e-5
	HEO	--	--	12.5	--	1e-5
FAPI	LEO	32	32	2500	1e-5	--
	HEO	32	32	25000	4e-6	--

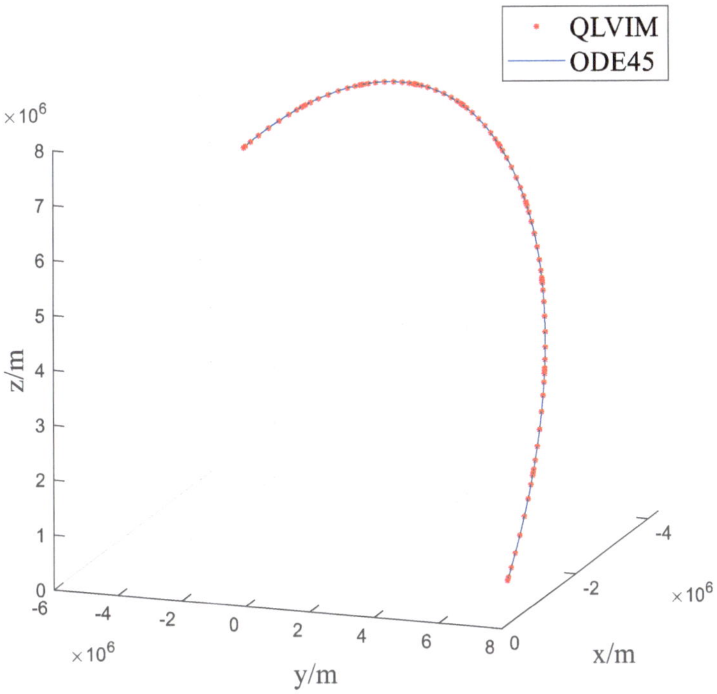

(a)

(Fig. 8) contd.....

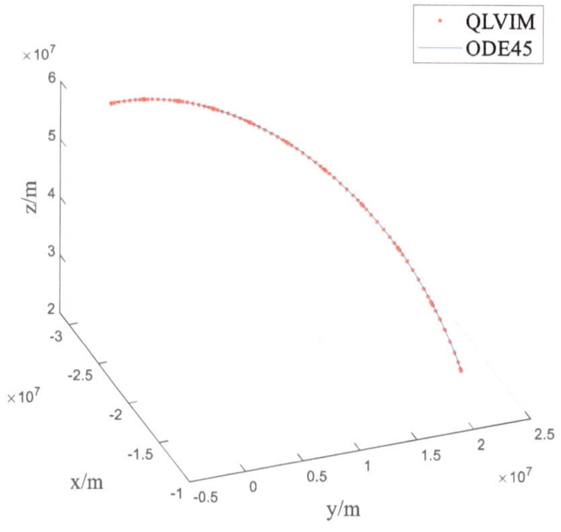

(b)

Fig. (8). (a) Low Earth transfer Orbit and **(b)** High-Eccentricity transfer Orbit.

Table 10. Initial velocities Calculated by three Algorithms.

Algorithm	LEO, m/s	HEO, m/s
QLVIM	-3583.94021700243; 0.428692017154153; 6194.3206904754	-1687.37254745742; 0.033017745670880; 2922.5938559553
Newton-RK4	-3583.94021718111; 0.428692011011873; 6194.3206903777	-1687.37254745767; 0.033017746529212; 2922.5938559549
FAPI	-3583.94021700365; 0.428692017691296; 6194.3206904743	-1687.37254745773; 0.033017745704072; 2922.5938559552

Table 11. Calculation Precision and Efficiency of Three Algorithms.

Algoritm	LEO		HEO	
	Position Error, m	Computing Time, s	Position Error, m	Computing Time, s
QLVIM	-1.50e-6	0.81	-1.68e-6	0.65

(Table 11) cont.....

	6.04e-7		-4.85e-7	
	2.60e-6		2.97e-6	
Newton-RK4	-2.7e-6	45.79	-2.09e-6	119.76
	9.38e-6		2.52e-6	
	4.63e-6		3.48e-6	
FAPI	-2.63e-6	1.97	-8.74e-6	1.65
	1.44e-6		7.60e-7	
	1.22e-6		3.40e-6	

Table **10** shows that, in LEO and HEO cases, the six decimal places of the velocities calculated by the three algorithms are exactly the same, so the validity of the algorithms and the accuracy of the initial velocities are proved. Table **11** shows that QLVIM can solve these two problems accurately with very low computational costs. While all the computational errors reach the magnitude of 1e-6, the computational speed of QLVIM is more than 10 times higher than the Newton-RK4 method and 2 times higher than the FAPI method.

Both QLVIM and FAPI methods apply the variational iteration method, but the computational speed of the former is faster. The reason is that when solving the TPBVP with the FAPI method, the iteration must be implemented over the entire time interval. Since the time interval is relatively large, we have to set more basis functions and collocation nodes for the method, which increases the computational amounts and the number of iterations. While using LVIM for solving IVPs, the entire time interval is divided into several subintervals, and the iteration is implemented within each subinterval. The solutions could reach the same precision with much less collocation nodes and basis functions.

Comparing with the traditional Newton-RK4 method, the QLVIM spends much less time to derive the solution. Generally, both the QL and the Newton's methods converge quadratically to the solution of the original equation. So, the high computational efficiency of the QLVIM mainly originates from the advantage of LVIM. While the RK4 method propagates the solution from one node to another, the LVIM approximates the solution by iterating a series of collocation points. And the time step of the LVIM can be chosen much larger than the RK4 method without losing computational accuracy. This comparison also proves that the LVIM is more efficient than RK4 for solving IVPs (Fig. **9**).

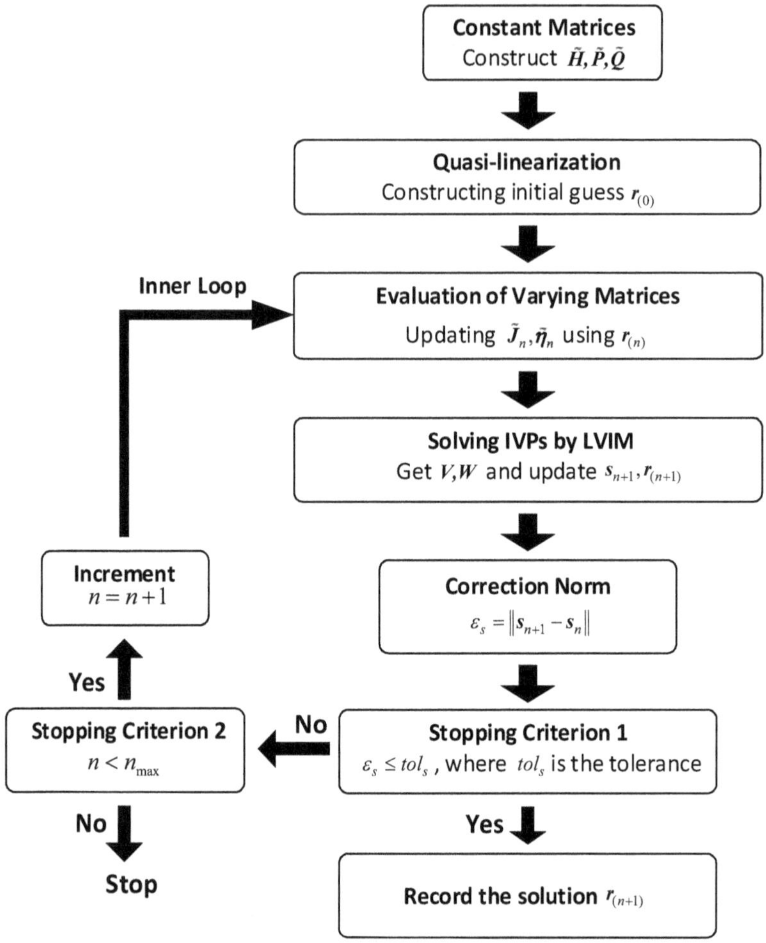

Fig. (9). The flow chart of QLVIM for solving TPBVP.

CONCLUSION

The orbit propagation and Lambert's problem for a perturbed two-body system are studied using the LVIM method. With the use of Chebyshev polynomials of the first kind as the basis functions, a time integrator is developed to solve the initial value problems in orbit propagation. The examples show that the proposed LVIM method is very efficient and highly accurate for the long-term integration of the perturbed orbital equations. It proves that the proposed method is theoretically correct and practically effective. Then Lambert's problem is also solved for both low earth orbit transfer and high eccentricity orbit transfer using the proposed method. The solution achieves very high accuracy in both the final position and final velocity.

A Fish-Scales-Growing Method is further proposed to solve two-point boundary value problems (TPBVPs) in conservative systems. An example is provided to illustrate that the Fish-Scales-Growing Method converges to an accurate solution in a straightforward approach. Among the merits of the Fish-Scales-Growing method are that it is insensitive to the initial guess and only needs to solve the sub-TPBVPs, which provide much convenience in manipulation.

CONSENT FOR PUBLICATION

Not applicable.

CONFLICT OF INTEREST

The authors declare no conflict of interest, financial or otherwise.

ACKNOWLEDGEMENT

Declared none.

REFERENCES

[1] Dai H, Yue X, Yuan J, et al. A time domain collocation method for studying the aeroelasticity of a two dimensional airfoil with a structural nonlinearity. Journal of Computational Physics, 2014, 270: 214-237.

[2] Woollands R M, Bani Younes A, Junkins J L. New solutions for the perturbed lambert problem using regularization and picard iteration. Journal of Guidance, Control, and Dynamics, 2015, 38(9): 1548-1562.

[3] Bai X, Junkins J L. Modified Chebyshev-Picard iteration methods for solution of boundary value problems. The Journal of the Astronautical Sciences, 2011, 58(4): 615-642.

[4] Bai X, Junkins J L. Modified Chebyshev-Picard iteration methods for orbit propagation. The Journal of the Astronautical Sciences, 2011, 58(4): 583-613.

[5] Bai X. Modified Chebyshev-Picard iteration methods for solution of initial value and boundary value problems. Texas A&M University, 2010.

[6] Wang X, Atluri S N. A novel class of highly efficient and accurate time-integrators in nonlinear computational mechanics. Computational Mechanics, 2017, 59(5): 861-876.

[7] Wang X, Yue X, Dai H, et al. Feedback-accelerated Picard iteration for orbit propagation and lambert's problem. Journal of Guidance, Control, and Dynamics, 2017, 40(10): 2442-2451.

[8] Jang M J, Chen C L, Liy Y C. On solving the initial-value problems using the differential transformation method. Applied Mathematics and Computation, 2000, 115(2-3): 145-160.

[9] Berry M, Healy L. Comparison of accuracy assessment techniques for numerical integration. Univelt, Inc., 2003.

[10] Macomber B, Probe A B, Woollands R, et al. Enhancements to modified Chebyshev-Picard iteration efficiency for perturbed orbit propagation. CMES-Computer Modeling in Engineering and Sciences, 2016, 111(1): 29-64.

[11] Hairer E, Nørsett S P, Wanner G. Solving ordinary differential equations, I: Nonstiff problems, Springer Verlag. Berlin, Germany, 1993.

[12] Shi K, Liu C, Sun Z, et al. Coupled orbit-attitude dynamics and trajectory tracking control for spacecraft electromagnetic docking. Applied Mathematical Modelling, 2022, 101: 553-572.

[13] Moler C. Numerical computing with Matlab. Philadelphia, USA, SIAM (Society for Industrial and Applied Mathematics). 2004.

CHAPTER 3

Adaptive Event-triggered Sliding Mode Control for Spacecraft Attitude Tracking

Jianqiao Zhang[1,*]

[1]*Shanghai Institute of Satellite Engineering, Shanghai, 201109, China*

Abstract: The spacecraft attitude tracking control problem with limited communication is addressed by employing event-triggered based sliding mode control theory. The attitude dynamics are directly developed on the Special Orthogonal Group SO(3), and singularities and ambiguities associated with other attitude representations are avoided successfully, taking model uncertainties and external disturbances into consideration. Based on the developed model, an adaptive event-triggered sliding mode controller is designed to ensure the closed-loop system that is uniformly ultimately bounded, by using fuzzy logic theory to deal with the disturbance. Due to the application of the event-triggered theory, the control signal is only updated and transmitted at some discrete instants. Therefore, the communication burden is decreased significantly. Finally, numerical simulations are conducted to demonstrate the effectiveness of the proposed control method.

Keywords: Communication burden, Event-triggered, SO(3), Fuzzy logic theory, Spacecraft attitude tracking control.

INTRODUCTION

Spacecraft attitude tracking is a key technology to accomplish a wide range of orbital missions such as pointing and slewing of spacecraft, Earth observation, and spacecraft rendezvous [1-4]. Therefore, the spacecraft attitude tracking control problems have been studied extensively in the past decades by utilizing nonlinear control techniques such as sliding mode control [1-3], proportional-derivative (PD) control [4], H_∞ control [5], optimal control [6, 7], and back-stepping control [8], *etc.*, where local coordinate representations, the modified Rodriguez parameters (MRPs) or quaternions were utilized to develop the spacecraft attitude model. However, as stated in [9, 10], Euler angles and MRPs suffer from kinematic singularities, and an unwinding problem exists in quaternions. Fortunately, pioneering studies on spacecraft attitude control have been done in [11-13] *via*

[*]**Corresponding author Jianqiao Zhang:** Shanghai Institute of Satellite Engineering, Shanghai, 201109, China; Tel: +86 13352220328; E-mail: zhangjianqiao@hit.edu.cn

Chuang Liu, Honghua Dai, Xiaokui Yue & Yiqing Ma (Eds.)

rotation matrix, which can represent a rigid body's attitude globally and uniquely on the Special Orthogonal Group SO(3). Consequently, the problems mentioned above associated with other attitude representations can be avoided. However, SO(3) is a nonlinear non-Euclidean manifold, and the classical control schemes designed on Euclidean space cannot be applied to this model directly. A common way to deal with this problem is to construct a smooth positive-definite function to measure the attitude error and then design an attitude tracking controller with the error function. For example, the attitude control problems were solved in [9, 14, 15]. However, the spacecraft attitude tracking control problem solved in [15] did not take model uncertainties into consideration.

In spacecraft attitude control, the most commonly used method is proportion-integral-derivative (PID) control. But the properties of slow response and poor control performance to model uncertainties and external disturbances make PID control unsuitable for the problem investigated in this paper [16]. An alternative way is to use sliding mode control (SMC), which has been widely applied to guarantee the system robustness against these drawbacks, where model uncertainties and disturbances are regarded as the system total disturbance with an upper bound, and then controllers are designed such as [2, 17, 18]. However, the disturbance upper bound is not easy to obtain in practice. Moreover, the sign function was used in [18] to reject disturbance, which leads to the chattering phenomenon. Chattering is a serious problem for the practical use of sliding mode control. It not only leads to self-oscillations in the system but also increases the consumption of energy [19]. Though the sign function can be replaced by a tanh function, this technique implies deterioration in accuracy and robustness. An effective way to tackle this problem is using the estimation of the system disturbance and applying adaptive control theory to design a continuous robust adaptive controller, where the control gains can be altered and are as small as possible to mitigate chattering effects, but large enough to ensure that sliding can be maintained in the presence of disturbance, and also leads to a reduction of the system energy cost [19-22]. An alternative estimation method is to use a fuzzy logic system (FLS) to approximate the disturbance with high accuracy and then design SMC based on the estimated information. As a result, the limitation on the disturbance upper bound can be released. Moreover, among various fuzzy control methods, researchers have paid more attention to Takagi-Sugeno (T-S) method for its capability of approximating an unknown smooth nonlinear function over a compact set with an arbitrary accuracy [23], and lots of results have been obtained based on this way [24-29]. In addition, the combination of the fuzzy controller with SMC can guarantee assured tracking accuracy even in the presence of high model uncertainties [30]. For instance, an event-triggered adaptive sliding mode controller was designed in [24] to address the control problem of a class of T-S fuzzy systems with actuator failures. Better results were obtained compared with the existing results. For spacecraft attitude tracking control problems in [31-33], FLSs were applied successfully to approximate the disturbance, and then sliding mode controllers were designed, where a better result was obtained compared with the traditional SMC method.

Furthermore, the concepts of spacecraft architectures, such as cellularized spacecraft [32], plug-and-play spacecraft [2], and fractionated spacecraft [34], are becoming more attractive recently with the potential to progress the Technology Readiness Level, in which the functional components of a spacecraft are connected by wireless networks, and data transmission between different modules is achieved through wireless communication. Therefore, for these kinds of spacecraft, it is essential to design an attitude controller with less communication burden, while the control accuracy required for the tracking system should be guaranteed at the same time. An extraordinary choice is an event-triggered control, where signals are transmitted only when a predesigned event-triggering condition is met. Compared with traditional periodic or time-triggered control, where signals are transmitted periodically, and may occupy the onboard communication channel unnecessarily, the transmission time and control performance of event-triggered control can be adjusted by designing the triggering strategy properly. Recently, great progress has been made on event-triggered control for many kinds of nonlinear systems, such as multiagent systems in [35-38], T-S fuzzy systems in [24, 39], and nonlinear dynamic systems in [40, 41]. And the state-of-the-art event-triggered control has been extended to spacecraft attitude control [2, 32, 34]. However, model uncertainties were not considered in these studies and the dynamics were developed based on quaternions. Moreover, the input-to-state stability (ISS) assumption is required for the attitude control system, which is hard to satisfy in practice [41]. Though the ISS assumption on a simple nonlinear system has been released successfully in [41] by designing the triggering strategy based on the control signal measurement error, extending the result to spacecraft attitude control on SO(3) directly is difficult since the nonlinearity of the spacecraft dynamics on SO(3) is high. Therefore, based on the above discussions, the main purposes of this paper are to develop the attitude dynamics on SO(3) in a unique and singular-free way in the presence of model uncertainties and external disturbances and to extend the state-of-the-art in fuzzy controller design with SMC in [31–33] and event-triggered control without ISS assumption in [41] to solve spacecraft attitude tracking control problem, such that the results obtained in this paper will be suitable for the on-orbit application.

The spacecraft attitude tracking control problem is solved by proposing a fuzzy-logic-based adaptive event-triggered SMC, further reducing the communication burden. The main contributions are: 1) Considering model uncertainties and disturbances, the spacecraft attitude tracking problem is described on SO(3) directly with the help of Lie group and Lie algebra. Defining a smooth, positive error function to measure the attitude tracking error, an SMC is proposed to achieve the control objective of this paper, where FLS is added to the controller to deal with the total system disturbance, and a better control result is obtained compared with the PD controller in [42]; 2) An event-triggering strategy is designed based on the control signal measurement error, where the requirement for the ISS assumption in [2] is released. Consequently, the communication burden is reduced significantly; 3) As analyzed in [43-45], it is essential to finish a task in real-time, and the shortcoming of FLS is that the system suffers training time and data length to optimize the model structure. However, in this study, the control signal is only updated when the triggering strategy is violated, which gives the FLS enough training time. 4) The relative threshold strategy is designed based on the sliding surface and a constant number. The existence of the constant number guarantees that the Zeno behavior is always successfully avoided even the sliding surface reaches zero.

This chapter is organized as follows. In Section 2, the mathematical model of the attitude tracking system is developed on SO(3), and the control objective is given. The main results of this study are given in Section 3, including the adaptive controller design and its stability analysis. Then numerical simulations are conducted in Section 4 to show the effectiveness of the proposed controller. Finally, the conclusion of this study is given in Section 5.

PRELIMINARIES

Spacecraft Attitude Motion Modeling *Via* Rotation Matrix

In this chapter, we study the attitude motion of a fully-actuated rigid spacecraft. A standard Earth-centered inertial frame FI and a body-fixed frame Fb are introduced to model the spacecraft attitude motion. $\mathbf{F}_b$ is defined with its origin located in the mass center of the spacecraft, and its three axes coincide with the spacecraft's principal body axes. Then, the attitude of the spacecraft is the rotation matrix from $\mathbf{F}_b$ to $\mathbf{F}_I$. Here, the rotation matrix $R \in SO(3)$ is applied to represent the attitude. The Special Orthogonal Group $SO(3)$ is a Lie group, and can be described by $SO(3) = \{R \in \mathbb{R}^{3\times 3} : R^T R = I_{3\times 3}, \det(R) = 1\}$, where $I_{n\times n}$ is an identity matrix with proper dimension, and $\det(\cdot)$ represents the determinant of a matrix. The attitude kinematics and dynamics of a rigid spacecraft can be described as [10]

$$\begin{cases} \dot{\boldsymbol{R}} = \boldsymbol{R}\Omega^{\times} \\ \boldsymbol{J}\dot{\Omega} = -\Omega^{\times}\boldsymbol{J}\Omega + \boldsymbol{u}(t) + \boldsymbol{d}(t) \end{cases} \tag{1}$$

where $\Omega \in \mathbb{R}^{3\times1}$ denotes the spacecraft's angular velocity expressed in $\mathbf{F}_{\mathrm{b}}$, $\boldsymbol{J} \in \mathbb{R}^{3\times3}$ is the spacecraft's inertia matrix, $\boldsymbol{u}(\boldsymbol{t}) \in \mathbb{R}^{3\times1}$ is the control torque, and $\boldsymbol{d}(\boldsymbol{t}) \in \mathbb{R}^{3\times1}$ is the external disturbance.

The symbol $(\cdot)^{\times} : \mathbb{R}^{3\times1} \to \varsigma o(3)$ can convert a 3 × 1 vector into a 3 × 3 skew-symmetric matrix such that $\boldsymbol{x}^{\times}\boldsymbol{y} = \boldsymbol{x} \times \boldsymbol{y}$ for any $\boldsymbol{x}, \boldsymbol{y} \in \mathbb{R}^{3\times1}$. More specifically, for a vector $\boldsymbol{x} = [\mathrm{x}_1, \mathrm{x}_2, \mathrm{x}_3]^{\mathrm{T}} \in \mathbb{R}^{3\times1}$, we have

$$\boldsymbol{x}^{\times} = \begin{bmatrix} 0 & -x_3 & x_2 \\ x_3 & 0 & -x_1 \\ -x_2 & x_1 & 0 \end{bmatrix} \tag{2}$$

$\varsigma o(3) = \left\{ \boldsymbol{A} \in \mathbb{R}^{3\times3} \middle| \boldsymbol{A}^{\mathrm{T}} = -\boldsymbol{A} \right\}$ is the Lie algebra of $SO(3)$. The inverse of $(\cdot)^{\times} : \varsigma o(3) \to \mathbb{R}^3$ is represented by the *vee* map $\left(\boldsymbol{x}^{\times}\right)^{\vee} = \boldsymbol{x}$, whose properties can be described by the following equations [14]

$$\begin{cases} tr\left[\boldsymbol{A}\boldsymbol{x}^{\times}\right] = \frac{1}{2} tr\left[\boldsymbol{x}^{\times}\left(\boldsymbol{A} - \boldsymbol{A}^{\mathrm{T}}\right)\right] = -\boldsymbol{x}^{\mathrm{T}}\left(\boldsymbol{A} - \boldsymbol{A}^{\mathrm{T}}\right)^{\vee} \\ \boldsymbol{x}^{\times}\boldsymbol{A} + \boldsymbol{A}^{\mathrm{T}}\boldsymbol{x}^{\times} = (\{tr[\boldsymbol{A}]\boldsymbol{I}_{3\times3} - \boldsymbol{A}\}\boldsymbol{x})^{\times} \\ \boldsymbol{R}\boldsymbol{x}^{\times}\boldsymbol{R}^{\mathrm{T}} = (\boldsymbol{R}\boldsymbol{x})^{\times} \end{cases} \tag{3}$$

Attitude Error Dynamics of a Spacecraft

The spacecraft attitude tracking control problem studied in this chapter requires the development of a control law to guarantee that the spacecraft attitude can track a given desired smooth attitude $R_d(t) \in SO(3)$, which can be described by [46].

$$\dot{R}_d = R_d \Omega_d^{\times}(t), \forall t \geq 0, \Omega_d \in C^1 \cap L_{\infty} \tag{4}$$

where Ω_d is the desired angular velocity. Then the attitude error can be obtained, which is $R_e = R_d^{\mathrm{T}} R \in SO(3)$. However, $SO(3)$ is a nonlinear non-Euclidean manifold, and it is difficult to construct a control scheme $SO(3)$ through classical nonlinear control theories directly. The following key steps can state the common way to deal with this problem: (1) Design a positive function $\Psi : SO(3) \times SO(3) \rightarrow \mathbb{R}$ to measure the attitude tracking error; (2) Use the derivative of Ψ to obtain an attitude error vector and an angular velocity error vector; (3) Design an attitude tracking controller with the above two vectors [14]. In this study, the modified trace function defined $SO(3)$ in [47] is utilized:

$$\psi(R, R_d) = \frac{1}{2} \mathrm{tr}\left[K(I_{3\times3} - R_d^{\mathrm{T}} R) \right] \tag{5}$$

where $K = \mathrm{diag}(k_1 \; k_2 \; k_3) > 0$, and $\mathrm{tr}(\cdot)$ is the trace of a matrix. For the configuration error function Ψ, the following statements hold [48]

(1) ψ is locally positive definite about $R = R_d$.

(2)The critical points of $\psi = 0$, are $R = \{R_d\} \cup \{R_d \exp(\pi e^{\times}) | e \in \{e_1, e_2, e_3\}\}$, where e_1, e_2, e_3 are unit column vectors of R_d.

Taking the derivative of Ψ, we have

$$\frac{d}{dt}(\Psi(R, R_d)) = -\frac{1}{2} tr(K) \frac{d}{dt}(R_d^R R)) = -\frac{1}{2} tr\left(K R_d^R R e_{\Omega}^{\times} \right) \tag{6}$$

where $e_{\Omega} = \Omega - R^{\mathrm{T}} R_d \Omega_d$ is the angular velocity tracking error. By the properties in Eq.(3), we have

$$\frac{d}{dt}(\psi(R, R_d)) = e_R^T e_{\Omega} \tag{7}$$

where $e_R : SO(3) \times SO(3) \rightarrow \mathbb{R}^3$ is the attitude error vector.

$$e_R = \frac{1}{2}(KR_d^T R - R^T R_d K)^{\vee} \tag{8}$$

Taking the derivatives of e_R and e_Ω, the attitude error dynamics studied in this paper can be obtained with the aid of Eqs. (1) and (3).

$$\dot{e}_R = G(R, R_d)e_\Omega \tag{9}$$

$$\dot{e}_\Omega = J^{-1}(-\Omega^\times J\Omega + u + d) + e_\Omega^\times R_e^T \Omega_d - R_e^T \dot{\Omega}_d \tag{10}$$

where $G(R, R_d) = \frac{1}{2}(tr(R_e^T K)I_{3\times3} - R_e^T K)$.

In addition, due to the effects of fuel consumption or onboard motion, the inertia matrix of a spacecraft cannot be known exactly [49]. Hence, the actual inertia matrix of the spacecraft should be $J = J_0 + \Delta J$, where J_0 is the nominal part and ΔJ the uncertain part. As a result, Eq.(10) can be rewritten as

$$\dot{e}_\Omega = (J_0 + \Delta J)^{-1}(-\Omega^\times (J_0 + \Delta J)\Omega + u + d) + e_\Omega^\times R_e^T \Omega_d - R_e^T \dot{\Omega}_d \tag{11}$$

Furthermore, $(J_0 + \Delta J)^{-1}$ can be denoted by $(J_0 + \Delta J)^{-1} = J_0^{-1} + \Delta\hat{J}$, where $\Delta\hat{J}$ is also an uncertain part. Regarding the inertia uncertainties and external disturbances as the total disturbance of the spacecraft, Eq.(11) can be rewritten as

$$\dot{e}_\Omega = J_0^{-1}(-\Omega^\times J_0\Omega + u) + e_\Omega^\times R_e^T \Omega_d - R_e^T \dot{\Omega}_d + \Delta\tilde{d} \tag{12}$$

where $\Delta\tilde{d} = J_0^{-1}(-\Omega^\times J\Omega) + \Delta\hat{J}(-\Omega^\times J\Omega + u) + J^{-1}d$ is the total disturbance. For the developed attitude error dynamics, the following assumptions hold.

Assumption 1. e_R and e_Ω can be measured by the spacecraft sensors directly, which means they are available in the attitude tracking control law design.

Assumption 2. The total disturbance of the spacecraft $\Delta\tilde{d}$ is unknown but bounded.

Remark 1. The motion of the spacecraft attitude tracking system can be described by Eqs. (9) and (12), and we can find that the model is developed in a coordinate-free way and the controller $\boldsymbol{u}$ can be designed byusing the defined attitude error function in a singular-free way [12]. When $\Psi = 0$, the desired attitude has been tracked. Moreover, the uncertainties in inertia and external disturbances due to solar radiation pressure, air drag, and gravity gradient perturbation, cannot be obtained precisely, but they are all bounded. Consequently, Assumption 2 is reasonable. In addition, it should be noted that compared with [10, 48], inertia uncertainties and external disturbances are considered in this study, which increases the nonlinearity of the developed model. To some extent, each of these two issues can severely degrade the control performance and even make the system unstable, and especially it will be more difficult to meet the control accuracy requirement when both of them are considered simultaneously. Therefore, the designed attitude controller $\boldsymbol{u}$ should have strong robustness against these disturbances.

Control Objective

This study investigates the attitude tracking control problem of a rigid spacecraft. The control objective can be stated as follows: Based on the relative attitude dynamics described by Eqs. (9) and (12), and under the above reasonable assumptions, for any given desired attitude $\boldsymbol{R}_d$, design a control law $\boldsymbol{u}$ such that the spacecraft attitude can track $\boldsymbol{R}_d$ with acceptable accuracy and the closed-loop system should be uniformly ultimately bounded, and $\lim_{t\to\infty} \boldsymbol{e}_\Omega = \lim_{t\to\infty} \boldsymbol{e}_R = 0$. Meanwhile,the controller should be robust against model uncertainties and external disturbances.

GEOMETRIC TRACKING CONTROL DESIGN ON SO(3) AND STABILITY ANALYSIS

In this section, the spacecraft attitude tracking problem will be addressed based on the relative attitude dynamics described by Eqs. (9) and (12). The key point is to deal with the system disturbance, which has large uncertainty and strong nonlinearity, to obtain fine control perfomance. Here, the fuzzy-logic-based estimation scheme is adopted, which can approximate any smooth function over a compact set to any degree of accuracy to estimate the system disturbance [50]. Then, an adaptive SMC is designed with the estimation information based on the event-triggered mechanism to guarantee that the spacecraft attitude $\boldsymbol{R}$ can track $\boldsymbol{R}_d$.

Fuzzy Logic System

For a vector $\boldsymbol{X} = [x_1, ..., x_n]^T \in \mathbb{R}^n$ with n input variables, the FLS can be obtained by a collection of if-then fuzzy rules with the following expression

$$IF\ x_1\ is\ A_1^l\ and\ ...\ and\ x_n\ is\ A_n^l, THEN\ z\ is\ B_l \tag{13}$$

where $l = 1, 2, ..., m$ is the fuzzy rules number, z is the system output, and A_1^l, $k = (1, ..., n)$ and B_l are fuzzy sets. For the convenience of the following analysis, the FLS can be normalized by using some standard tools (a center-average defuzzifier, singleton function, and product inference) as follows [25]

$$z = \sum_{l=1}^{m} \frac{\bar{z}^l \prod_{k=1}^{n} \mu_{A_k^l}(x_k)}{\sum_{l=1}^{m} \prod_{k=1}^{n} \mu_{A_k^l}(x_k)} \tag{14}$$

Generally, $\mu_{A_k^l}(x_k)$ is obtained by the Gaussian membership function

$$\boldsymbol{\mu}_{A_k^l}(x_k) = a_k^l \exp\left(-\frac{1}{2}\left(\frac{x_k - \bar{x}_k^l}{\varsigma_k^l}\right)^2\right) \tag{15}$$

where $0 < a_k^l < 1, \varsigma_k^l > 0, \bar{x}_k^l > 0$, and $\mu_{A_k^l}(x_k)$ gets its maximum value at $\bar{x}_k^l$.

The fuzzy basis functions are defined as

$$\boldsymbol{h}_l(X) = \frac{\prod_{k=1}^{n} \boldsymbol{\mu}_{A_k^l}(x_k)}{\sum_{l=1}^{m} \prod_{k=1}^{n} \boldsymbol{\mu}_{A_k^l}(x_k)} \tag{16}$$

Then the FLS in Eq.(14) can be rewritten as

$$z = \boldsymbol{W}\boldsymbol{h}(X) \tag{17}$$

where $\boldsymbol{W} = (\bar{z}^1, \bar{z}^2, ..., \bar{z}^m) \in \mathbb{R}^{n \times m}$ is the constant weight of the FLS, and $\boldsymbol{h}(X) = [h_1(X), ..., h_m(X)]^T$ is the basic function vector.

As stated in [50], an unknown smooth nonlinear function $\boldsymbol{f}(X)\in\mathbb{R}^n$ can be approximated by a FLS on a compact set Δ_1, and can be expressed as

$$\boldsymbol{f}(X)=\boldsymbol{W}^*\boldsymbol{h}(X)+\delta^*,\forall X\in\Delta_1 \tag{18}$$

where $W^*\in\mathbb{R}^{n\times m}$ is the optimal constant weight of the FLS, and δ^* is the estimation error, which satisfies $\|\delta^*\|\le\delta^*$, where $\delta^*>0$ is a small constant, and $\|\delta^*\|$ represents the Euclidean norm of δ^*.If the FLS is used to estimate the disturbance, we have $\boldsymbol{J}_0\Delta\tilde{\boldsymbol{d}}=\boldsymbol{f}(X)=\boldsymbol{W}^*\boldsymbol{h}(X)+\delta^*$.

FLS-Based Adaptive Event-Triggered SMC Design

A sliding surface is designed with the following form

$$\boldsymbol{S}=\boldsymbol{e}_\Omega+\vartheta\boldsymbol{e}_R \tag{19}$$

where $\vartheta>0$ is a constant number. Taking the derivative of $\boldsymbol{S}$, and substituting Eqs. (9) and (12) into the result, the relative spacecraft attitude dynamics can be rewritten as

$$\dot{\boldsymbol{S}}=\dot{\boldsymbol{e}}_\Omega+\vartheta\dot{\boldsymbol{e}}_R=\boldsymbol{J}_0^{-1}(\Omega^\times\boldsymbol{J}_0\Omega+\boldsymbol{u})+\boldsymbol{L}+\Delta\tilde{\boldsymbol{d}} \tag{20}$$

where $\boldsymbol{L}=\boldsymbol{e}_\Omega^\times\boldsymbol{R}_e^T\Omega_d-\boldsymbol{R}_e^T\dot{\Omega}_d+\vartheta G(\boldsymbol{R},\boldsymbol{R}_d)\boldsymbol{e}_\Omega$.

When the event-triggered mechanism is applied, the control signal $\boldsymbol{u}$ will be updated at t_k when the condition of the designed triggering strategy is violated, and then it will be held as a constant through a zero-holder (ZOH) when $t\in[t_k,t_{k+1})$, which can be expressed by

$$\boldsymbol{u}(t)=\boldsymbol{u}_{out}(t),t\in[t_k,t_{k+1}) \tag{21}$$

where $\boldsymbol{u}_{out}$ is the designed control law, $\boldsymbol{u}(t)\{t_k\}_{=0}^{\infty},t_0=0$ is the triggering instants sequence generated by the designed event-triggering strategy, and $T_k=t_{k+1}-t_k$ is called the inter-event time. In this study, the triggering strategy is designed based on the control signal measurement error $\boldsymbol{e}_c = \boldsymbol{u}(t) - \boldsymbol{u}_{out}(t)$ and the sliding surface $\boldsymbol{S}$. The triggering instants are determined by

$$t_{k+1} = \inf\left\{t > t_k \left| \|\boldsymbol{e}_c\| \geq a\|\boldsymbol{S}\| + \gamma \right.\right\} \tag{22}$$

where $a > 0$ and $\gamma > 0$ are constant numbers. To this end, based on the above analysis, the main results of this study can be presented by the following theorems.

Theorem 1. *For the relative spacecraft attitude dynamics described by Eqs. (9) and (12), the fuzzy-logic based adaptive event-triggered sliding mode controller is designed as follows*

$$\boldsymbol{u}_{out}(t) = \Omega^{\times}\boldsymbol{J}_0\Omega - \boldsymbol{W}\boldsymbol{h}(X) - \boldsymbol{J}_0\boldsymbol{L} - (K_1 + K_2(\|\boldsymbol{S}\|))\boldsymbol{S} - \boldsymbol{e}_R \tag{23}$$

where $\boldsymbol{W}$ is the estimation of $\boldsymbol{W}^*$, and it is updated by

$$\dot{\boldsymbol{W}} = \lambda\boldsymbol{S}\boldsymbol{h}(\boldsymbol{X})^{\mathrm{T}} \tag{24}$$

where $\lambda > 0$, $\boldsymbol{h}(\boldsymbol{X}) = \boldsymbol{g}(\boldsymbol{S})$ is a function of $\boldsymbol{S}$, and $\tilde{\boldsymbol{W}} = \boldsymbol{W}^* - \boldsymbol{W}$ is the estimation error of $\boldsymbol{W}^*$. $K_1 > a$, and $K_2(\|\boldsymbol{S}\|)$ is a positive adaptive gain parameter

$$K_2(\|\boldsymbol{S}\|) = \frac{\hat{\varphi} + \ell\int_{t_k}^{t}\hat{\varphi}(s)ds}{\|\boldsymbol{S}\| + \varepsilon\exp(-\sigma t)} \tag{25}$$

where $t \in [t_k, t_{k+1})$, φ is the upper bound of $\|\delta^*\| + \gamma$, and $\hat{\varphi}$ is the estimation of φ, which is updated by:

$$\dot{\hat{\varphi}} = -\ell\hat{\varphi} + \frac{\|\boldsymbol{S}\|^2}{\|\boldsymbol{S}\| + \varepsilon\exp(-\sigma t)} \tag{26}$$

where $\varepsilon > 0$, $\ell > 0$, and $\sigma > 0$ are constant numbers. The control objective can be realized, and the Zeno behaviour can be avoided.

Proof. Considering the following Lyapunov function candidate

$$V = \Psi(\boldsymbol{R}, \boldsymbol{R}_d) + \frac{1}{2}\boldsymbol{S}^{\mathrm{T}}\boldsymbol{J}_0\boldsymbol{S} + \frac{1}{2}\tilde{\varphi}^2 + \frac{1}{2\lambda}\tilde{\boldsymbol{W}}^{\mathrm{T}}\tilde{\boldsymbol{W}} \tag{27}$$

where $\tilde{\varphi} = \varphi - \hat{\varphi} - \ell\int_{t_k}^{t}\hat{\varphi}(s)ds$. There is no doubt that V is positive definite. Taking the derivative of V with respect to time $t \in [t_k, t_{k+1})$ and taking Eq.(20) into the result, we have

$$\begin{aligned} \dot{V} &= \boldsymbol{e}_R \cdot \boldsymbol{e}_\Omega + \boldsymbol{S}^{\mathrm{T}}\boldsymbol{J}_0\dot{\boldsymbol{S}} + \dot{\tilde{\varphi}}\tilde{\varphi} + \frac{1}{\lambda}\tilde{\boldsymbol{W}}^{\mathrm{T}}\dot{\tilde{\boldsymbol{W}}} \\ &= \boldsymbol{e}_R \cdot \boldsymbol{e}_\Omega + \boldsymbol{S}^{\mathrm{T}}(-\Omega^{\times}\boldsymbol{J}_0\Omega + \boldsymbol{u}) + \boldsymbol{S}^{\mathrm{T}}\boldsymbol{J}_0\boldsymbol{L} \\ &\quad + \boldsymbol{S}^{\mathrm{T}}\boldsymbol{J}_0\Delta\tilde{\boldsymbol{d}} + \dot{\tilde{\varphi}}\tilde{\varphi} + \frac{1}{\lambda}\tilde{\boldsymbol{W}}^{\mathrm{T}}\dot{\tilde{\boldsymbol{W}}} \end{aligned} \tag{28}$$

Substituting (23) into (28), we can get

$$\begin{aligned} \dot{V} &= \boldsymbol{e}_R \cdot \boldsymbol{e}_\Omega + \boldsymbol{S}^{\mathrm{T}}(-\Omega^{\times}J_0\Omega + \boldsymbol{u}_{out}(t) + \boldsymbol{e}_c) + \boldsymbol{S}^{\mathrm{T}}\boldsymbol{J}_0\boldsymbol{L} \\ &\quad + \boldsymbol{S}^{\mathrm{T}}\boldsymbol{J}_0\Delta\tilde{\boldsymbol{d}} + \dot{\tilde{\varphi}}\tilde{\varphi} + \frac{1}{\lambda}\tilde{\boldsymbol{W}}^{\mathrm{T}}\dot{\tilde{\boldsymbol{W}}} \\ &= \boldsymbol{e}_R \cdot \boldsymbol{e}_\Omega + \boldsymbol{S}^{\mathrm{T}}(-\Omega^{\times}\boldsymbol{J}_0\Omega + \Omega^{\times}\boldsymbol{J}_0\Omega - \boldsymbol{W}\boldsymbol{h}(X) - \boldsymbol{J}_0\boldsymbol{L} \\ &\quad - (K_1 + K_2(\|\boldsymbol{S}\|))\boldsymbol{S} - \boldsymbol{e}_R + \boldsymbol{e}_c) + \boldsymbol{S}^{\mathrm{T}}\boldsymbol{J}_0\boldsymbol{L} \\ &\quad + \boldsymbol{S}^{\mathrm{T}}\boldsymbol{J}_0\Delta\tilde{\boldsymbol{d}} + \dot{\tilde{\varphi}}\tilde{\varphi} + \frac{1}{\lambda}\tilde{\boldsymbol{W}}^{\mathrm{T}}\dot{\tilde{\boldsymbol{W}}} \\ &= \boldsymbol{e}_R \cdot \boldsymbol{e}_\Omega + \boldsymbol{S}^{\mathrm{T}}(-\boldsymbol{W}\boldsymbol{h}(X) - (K_1 + K_2(\|\boldsymbol{S}\|))\boldsymbol{S} - \boldsymbol{e}_R + \boldsymbol{e}_c) \\ &\quad + \boldsymbol{S}^{\mathrm{T}}\boldsymbol{J}_0\Delta\tilde{\boldsymbol{d}} + \dot{\tilde{\varphi}}\tilde{\varphi} + \frac{1}{\lambda}\tilde{\boldsymbol{W}}^{\mathrm{T}}\dot{\tilde{\boldsymbol{W}}} \end{aligned} \tag{29}$$

Substituting (24), (25) and (26) into (29), one can obtain

$$\begin{aligned}\dot{V} &= \boldsymbol{e}_R \cdot \boldsymbol{e}_\Omega - \boldsymbol{S}^{\mathrm{T}}\boldsymbol{e}_R + \boldsymbol{S}^{\mathrm{T}}(-\boldsymbol{W}\boldsymbol{h}(X) + \boldsymbol{J}_0\Delta\tilde{\boldsymbol{d}} - \tilde{\boldsymbol{W}}\boldsymbol{h}(X)) \\ &\quad + \boldsymbol{S}^{\mathrm{T}}(-(K_1 + K_2(\|\boldsymbol{S}\|))\boldsymbol{S} + \boldsymbol{e}_c) \\ &\quad + (\boldsymbol{\varphi} - \hat{\boldsymbol{\varphi}} - \ell\int_{t_k}^{t}\hat{\varphi}(s)ds)(\ell\hat{\boldsymbol{\varphi}} - \frac{\|\boldsymbol{S}\|^2}{\|\boldsymbol{S}\| + \varepsilon\exp(-\sigma t)} - \ell\hat{\boldsymbol{\varphi}}) \\ &= \boldsymbol{e}_R \cdot \boldsymbol{e}_\Omega - \boldsymbol{S}^{\mathrm{T}}\boldsymbol{e}_R + \boldsymbol{S}^{\mathrm{T}}\delta^* + \boldsymbol{S}^{\mathrm{T}}(-(K_1 + K_2(\|\boldsymbol{S}\|))\boldsymbol{S} + \boldsymbol{e}_c) \\ &\quad + (\boldsymbol{\varphi} - \hat{\boldsymbol{\varphi}} - \ell\int_{t_k}^{t}\hat{\boldsymbol{\varphi}}(s)ds)(-\frac{\|\boldsymbol{S}\|^2}{\|\boldsymbol{S}\| + \varepsilon\exp(-\sigma t)})\end{aligned} \tag{30}$$

Utilizing the triggering strategy in Eq. (22), we have

$$\begin{aligned}\dot{V} &\le \boldsymbol{e}_R \cdot \boldsymbol{e}_\Omega - \boldsymbol{S}^{\mathrm{T}}\boldsymbol{e}_R - (K_1 - a)\|\boldsymbol{S}\|^2 + (\|\delta^*\| + \gamma)\|\boldsymbol{S}\| \\ &\quad - K_2(\|\boldsymbol{S}\|)\|\boldsymbol{S}\|^2 + (\boldsymbol{\varphi} - \hat{\boldsymbol{\varphi}} - \ell\int_{t_k}^{t}\hat{\boldsymbol{\varphi}}(s)ds)(-\frac{\|\boldsymbol{S}\|^2}{\|\boldsymbol{S}\| + \varepsilon\exp(-\sigma t)})\end{aligned} \tag{31}$$

Taking $\boldsymbol{K}_2(\|\boldsymbol{S}\|)$ in (25) into (31), and using the equality that $\boldsymbol{e}_\Omega = \boldsymbol{S} - \vartheta\boldsymbol{e}_R$, yield

$$\begin{aligned}\dot{V} &\le \boldsymbol{e}_R \cdot (\boldsymbol{S} - \vartheta\boldsymbol{e}_R) - \boldsymbol{S}^{\mathrm{T}}\boldsymbol{e}_R - (K_1 - a)\|\boldsymbol{S}\|^2 + \boldsymbol{\varphi}\|\boldsymbol{S}\| \\ &\quad - \frac{\hat{\boldsymbol{\varphi}} + \ell\int_{t_k}^{t}\hat{\boldsymbol{\varphi}}(s)ds}{\|\boldsymbol{S}\| + \varepsilon\exp(-\sigma t)}\|\boldsymbol{S}\|^2 + (\boldsymbol{\varphi} - \hat{\boldsymbol{\varphi}} - \ell\int_{t_k}^{t}\hat{\boldsymbol{\varphi}}(s)ds)(-\frac{\|\boldsymbol{S}\|^2}{\|\boldsymbol{S}\| + \varepsilon\exp(-\sigma t)}) \\ &\le -\vartheta\boldsymbol{e}_R^{\mathrm{T}}\boldsymbol{e}_R - (K_1 - a)\|\boldsymbol{S}\|^2 - \frac{(\boldsymbol{\varphi} - \hat{\boldsymbol{\varphi}} - \ell\int_{t_k}^{t}\hat{\boldsymbol{\varphi}}(s)ds)\|\boldsymbol{S}\|^2}{\|\boldsymbol{S}\| + \varepsilon\exp(-\sigma t)} \\ &\quad + \frac{(\boldsymbol{\varphi} - \hat{\boldsymbol{\varphi}} - \ell\int_{t_k}^{t}\hat{\boldsymbol{\varphi}}(s)ds)\|\boldsymbol{S}\|^2 + \varepsilon\boldsymbol{\varphi}\|\boldsymbol{S}\|\exp(-\sigma t)}{\|\boldsymbol{S}\| + \varepsilon\exp(-\sigma t)} \\ &\le -\vartheta\boldsymbol{e}_R^{\mathrm{T}}\boldsymbol{e}_R - c\|\boldsymbol{S}\|^2 + \frac{\varepsilon\boldsymbol{\varphi}\|\boldsymbol{S}\|\exp(-\sigma t)}{\|\boldsymbol{S}\| + \varepsilon\exp(-\sigma t)}\end{aligned} \tag{32}$$

where $c = K_1 - a > 0$. Since $\varepsilon\exp(-\sigma t) > 0$, thus the following inequality equation holds

$$\frac{\varepsilon\varphi\|\boldsymbol{S}\|\exp(-\sigma t)}{\|\boldsymbol{S}\|+\varepsilon\exp(-\sigma t)}<\varepsilon\exp(-\sigma t) \tag{33}$$

Then (32) can be rewritten as

$$\dot{V}\leq -c\|\boldsymbol{S}\|^2+\varepsilon\varphi\exp(-\sigma t) \tag{34}$$

Therefore, the closed-loop system is uniformly ultimately bounded, and $\boldsymbol{S}$ will converge to Δ_2:

$$\Delta_2=\left\{\boldsymbol{S}\in\mathbb{R}^3\middle|\|\boldsymbol{S}\|\leq\sqrt{\frac{\varepsilon\varphi\exp(-\sigma t)}{c}}\right\} \tag{35}$$

Since $\sigma>0$, as time increases, Δ_2 will decrease. Thus, it can be concluded that $\lim_{t\to\infty}\Delta_2=0$, which means that when $t\to\infty$, the following equation holds

$$\boldsymbol{S}=\boldsymbol{e}_\Omega+\vartheta\boldsymbol{e}_R=0 \tag{36}$$

Then, another Lyapunov function $V_1=\Psi(\boldsymbol{R},\boldsymbol{R}_d)$ is considered to prove the convergences of $\boldsymbol{e}_\Omega$ and $\boldsymbol{e}_R$. Taking the time derivative of $\Psi(\boldsymbol{R},\boldsymbol{R}_d)$, and taking $\boldsymbol{e}_\Omega=-\vartheta\boldsymbol{e}_R$ into the result, we have

$$\dot{V}_1=\frac{d}{dt}(\Psi(\boldsymbol{R},\boldsymbol{R}_d))=\boldsymbol{e}_R\cdot\boldsymbol{e}_\Omega=-\vartheta\boldsymbol{e}_R^{\mathrm{T}}\boldsymbol{e}_R=-\vartheta\|\boldsymbol{e}_R\|^2\,dt \tag{37}$$

Integrating both sides of Eq. (37), one can obtain

$$V_1(t)\leq V_1(0)-\vartheta\int_0^t\|\boldsymbol{e}_R\|^2\,dt \tag{38}$$

Then, it is obvious that $\boldsymbol{e}_\Omega,\boldsymbol{e}_R\in L_2$. According to Barbalat's lemma in [51], it can be concluded that $\lim_{t\to\infty}\boldsymbol{e}_\Omega=\lim_{t\to\infty}\boldsymbol{e}_R=0$.

Moreover, for event-triggered control, the triggering instant t_k is reasonable if the inter-event time $t_{k+1}-t_k$ is lower bounded by a positive value to avoid the Zeno behavior. Zeno behavior is defined that the triggering strategy that is motivated infinitely in a finite-time interval $[t_k, t_{k+1})$ [35]. Avoiding Zeno behavior is critical for the application of attitude controllers in practice since the actuator cannot be executed continuously [2]. To prove that for $\forall k$, there exists $t^* > 0$ such that $t_{k+1} - t_k > t^*$, we recall that at the beginning of each triggering instant $t = t_k$, $\|\boldsymbol{e}_c = \boldsymbol{u}(t) - \boldsymbol{u}_{out}(t)\| = 0$, and $u(t)$ is a constant for a certain time interval $t \in [t_k, t_{k+1})$. For $t \in [t_k, t_{k+1})$, taking the time derivative of $\|\boldsymbol{e}_c\|$, we have

$$\frac{d}{dt}\|\boldsymbol{e}_c\| = \frac{d}{dt}\sqrt{\boldsymbol{e}_c^{\mathrm{T}}\boldsymbol{e}_c} = \frac{\boldsymbol{e}_c^{\mathrm{T}}}{\|\boldsymbol{e}_c\|}\dot{\boldsymbol{e}}_c \le \|\dot{\boldsymbol{e}}_c\| = \|\dot{\boldsymbol{u}}_{out}(t)\| \tag{39}$$

The components of $\boldsymbol{u}_{out}(t)$, namely Ω, $\boldsymbol{J}_0$, $\boldsymbol{Wh}(X)$, $\boldsymbol{S}$, and $\boldsymbol{e}_R$ are all differentiable and bounded. Thus, it is reasonable to assume that there exists a positive constant number δ, such that $\|\dot{\boldsymbol{u}}_{out}(t)\| \le \delta$. Integrating both sides of (39) yields

$$\|\boldsymbol{e}_c\| = \int_{t_k}^{t} \|\dot{\boldsymbol{u}}_{out}(t)\| ds \le (t - t_k)\delta \tag{40}$$

which means that $\lim\limits_{t \to t_{k+1}}(t - t_k) = t^* \ge \dfrac{\|\boldsymbol{e}_c\|}{\delta}$. Since $\|\boldsymbol{e}_c\| > 0$ and $\delta > 0$, thus $t^* > 0$, which means that the Zeno behavior has been avoided successfully. Thereby, the above analysis completes the proof.

Remark 2. The designed controller (23) consists of three parts: $\Omega^{\times}\boldsymbol{J}_0\Omega - \boldsymbol{J}_0\boldsymbol{L} - \boldsymbol{e}_R$ is the equivalent control to ensure $\dot{\boldsymbol{S}} \to 0$ for the nominal system without model uncertainties and external disturbances, $-\boldsymbol{Wh}(X)$ is designed to reject the system disturbance, and $-(K_1 + K_2(\|\boldsymbol{S}\|))\boldsymbol{S}$ ensures the stability of the control system after considering the triggering strategy. In addition, from the properties of Ψ, it can be seen that the designed controller can guarantee the attitude tracking error $\boldsymbol{e}_R$ converges to zero, but R can converge to the desired attitude $\boldsymbol{R}_d$ or the other three unstable critical points. This problem cannot be avoided for any continuous attitude controller for the topological property of $SO(3)$ [48]. Therefore, if the initial spacecraft attitude is in a neighborhood of these undesired points, the hybrid controller in [42] can be adopted.

Remark 3. We have to pay attention to actuator saturation since the physical limitations on the actuator cannot be ignored. The actual control torque generated by the actuator can be expressed as

$$sat(u_{outi}(t)) = \begin{cases} u_{\max}, & if\ u_{outi}(t) > u_{\max} \\ u_{outi}(t), & if\ |u_{outi}(t)| \le u_{\max} \\ -u_{\max}, & if\ u_{outi}(t) < -u_{\max} \end{cases} \quad i = 1,2,3 \tag{41}$$

where $u_{\max} > 0$ is the maximum actuator output.

Remark 4. A state-dependent relative threshold triggering strategy is designed *via* the sliding surface $\boldsymbol{S}$ and a constant number γ, while the measurement error of the strategy relies on the control signal. When $\boldsymbol{e}_c$ reaches the threshold, the triggering instant t_k will be generated, and $\boldsymbol{u}(t)$ will be updated. Then $\boldsymbol{u}(t)$ will be held constant through a ZOH until the next transmission command is received. $-(K_1 + K_2(\|\boldsymbol{S}\|))\boldsymbol{S}$ is designed to compensate for the measurement error. The larger the values of a and γ, the longer the inter-event times can be achieved. Consequently, the communication burden can be reduced significantly. But, it also means K_1 and K_2 should be large enough to guarantee that the closed-loop system remains stable, which requires more control effort and will deteriorate the control performance. Hence, a trade-off should be made between the control cost and the communication burden. Moreover, γ is an extraordinarily important term to guarantee that t^* is always lower bounded because $\boldsymbol{S}$ may converge to zero. From the proof of the avoidance of Zeno behavior in (40), it can be seen that when $\boldsymbol{S}$ reaches zero, t^* is determined by γ.

SIMULATIONS

In this section, numerical simulations are conducted to show the effectiveness of the designed controller (23) with an application to a spacecraft attitude tracking scenario, in which the spacecraft operates in a low Earth orbit (LEO), and the nominal part of the spacecraft's inertia matrix is

$$\boldsymbol{J}_0 = \begin{bmatrix} 22 & 1.5 & 1 \\ 1.5 & 24 & -1.2 \\ 1 & -1.2 & 23 \end{bmatrix} kg.m^2 \tag{42}$$

The uncertain inertia matrix of the spacecraft is $\Delta \boldsymbol{J} = 0.15\boldsymbol{J}_0$. In addition, the upper bound of the control torque is $u_{\max} = 0.2N \cdot m$. Initially, the attitude and angular velocity of the spacecraft are

$$\boldsymbol{R}(0) = \begin{bmatrix} 1 & 0 & 0 \\ 0 & 1 & 0 \\ 0 & 0 & 1 \end{bmatrix}, \Omega(0) = [0, 0, 0.001]^T \text{ deg/ } s \tag{43}$$

The desired attitude and angular velocity are

$$R_d(0) = \begin{bmatrix} -0.766044359753917 & -0.6427877090337114 & 0.000000171074392 \\ 0.454519500236212 & -0.541675292832748 & -0.707106711210961 \\ 0,454519595610833 & -0.541675030110651 & 0.707106851162107 \end{bmatrix} \tag{44}$$

$$\Omega_d(0) = [0.00, -0.002, 0]^T \text{ deg/ } s$$

Then, the trajectory of the desired time-varying attitude can be obtained by (1) with $\dot{\Omega}_d = 0$. In addition, since the spacecraft operates in LEO, the external disturbances are mainly generated by atmospheric drag and solar radial pressure, whose magnitudes are $10^{-3} N \cdot m$ and can be calculated as in [52]. Therefore, in the simulation, the external disturbances that the spacecraft suffers are assumed to be

$$\boldsymbol{d} = [\sin(1 + 0.12t), \cos(1 + 0.15t), \sin(1 + 0.18t)] \times 10^{-3} N \cdot m \tag{45}$$

Two cases are considered in this section: Case 1 is done to demonstrate that, the control accuracy of the controller Eq. (23) is higher than the PD-like controller in [42], which is designed as follows

$$\boldsymbol{u} = -\boldsymbol{e}_R - K_\Omega \boldsymbol{e}_\Omega + (\boldsymbol{R}^{\mathrm{T}} \boldsymbol{R}_d \Omega_d)^\times \boldsymbol{J}_0 \boldsymbol{R}^{\mathrm{T}} \boldsymbol{R}_d \Omega_d + \boldsymbol{J}_0 \boldsymbol{R}^{\mathrm{T}} \boldsymbol{R}_d \dot{\Omega}_d \tag{46}$$

Case 2 is used to show the importance of different values of a and γ in the trade-off of the control effort and the communication burden. In Case 1, the parameters of the controller (23) are designed to be: $\vartheta = 0.2$, $K_1 = 2.5$, $a = 0.6$, $\gamma = 0.001$,

$\hat{\varphi}(0)=1, \ell=0.4, \varepsilon=1, \sigma=0.6, \lambda=0.05, W(0)=zeros(3,7)$ and the Gaussian basis functions contain 7 nodes with centers $\bar{x}_k^l$ evenly spaced in the interval [−1, 1]. For controller (46), $K_\Omega=2.5$. In Case 2, the event-triggering parameters designed for comparison are: $a_1=0.6$ $\gamma_1=0.001$, $a_2=0.6, \gamma_2=0.01, a_3=1.6$, and $\gamma_3=0.01$.

Figs. (**1-7**) presents the simulation results of Case 1. The time responses of the attitude and angular velocity tracking errors in terms of $\boldsymbol{e}_R$ and $\boldsymbol{e}_\Omega$ under the controller (23) are depicted in Figs. (**1** and **2**). We can find that the spacecraft can track its target attitude in the 80s, and the attitude tracking errors are $|\boldsymbol{e}_{Ri}|\le 0.05$ (deg) and $|\boldsymbol{e}_{\Omega i}|\le 2\times 10^{-3}$ (deg/s). The plots of the corresponding control torques are shown in Fig. (**3**), and we can see that the control limit in Eq. (41) is ensured. The plots of the triggering strategy in (22) are given in Fig. (**4**), and it can be found that $\|\boldsymbol{e}_c\|$ is always no larger than $a\|\boldsymbol{S}\|+\gamma$. The plots of the event-triggering release instants and inter-event times generated by the triggering strategy are shown in Fig. (**5**).

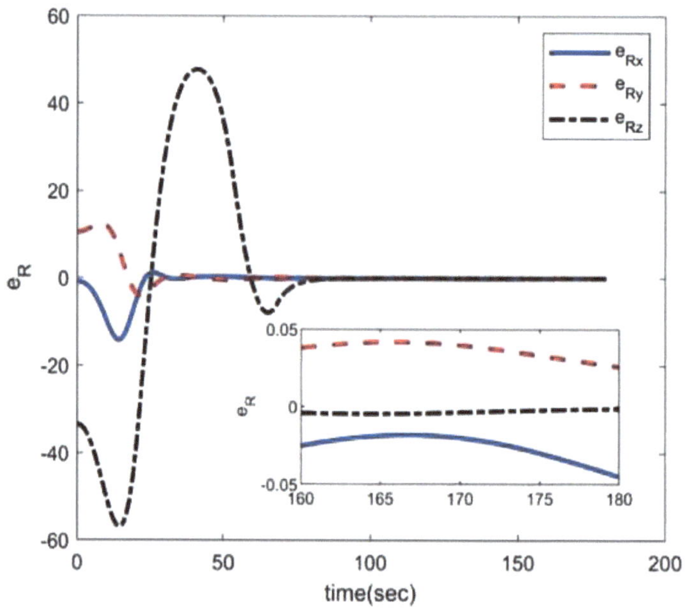

Fig. (1). Time responses of $\boldsymbol{e}_R$.

It can be seen that the Zeno behavior is avoided. During the whole simulation process, the number of the triggering instants is 156, and the mean value of the inter-event time is about 1.154s. For a spacecraft attitude control system, the frequency is usually 4 Hz or higher [32], which means the communication burden has been reduced by more than 78.3% under the triggering strategy. In summary, the results of Case 1 indicate the control objective is achieved successfully.

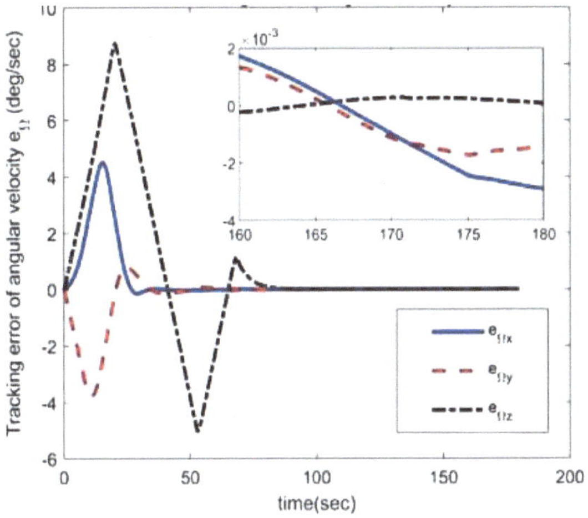

Fig. (2). Time responses of $\boldsymbol{e}_{\Omega}$.

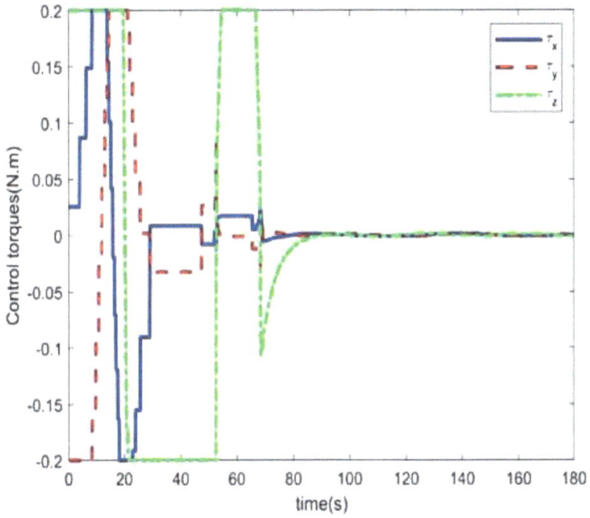

Fig. (3). Time responses of control torques.

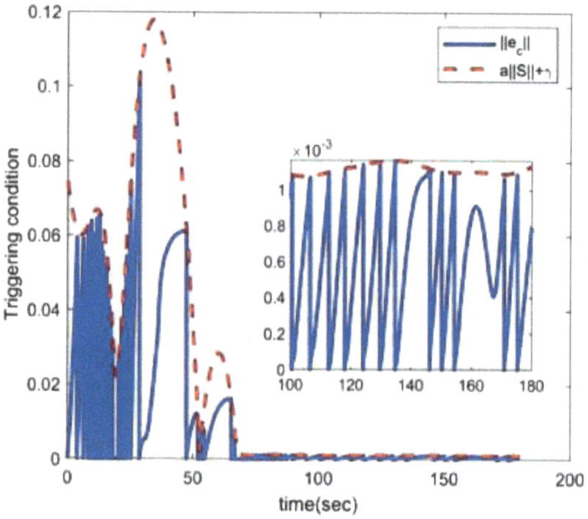

Fig. (4). Time responses of $\|\boldsymbol{e}_c\|$ and instants and $a\|\boldsymbol{S}\|+\gamma$.

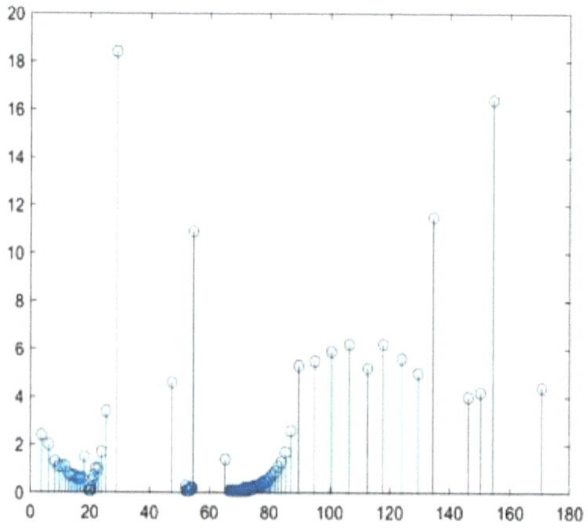

Fig. (5). Event-triggering release inter-event times.

Additionally, the time responses of the attitude and angular velocity tracking errors under the controller of [42] without triggering strategy are shown in Figs. (**6** and **7**), from which it can be observed that the attitude tracking errors fall to tolerance in 200s and the steady-state attitude tracking errors are $|\boldsymbol{e}_{Ri}| \le 0.2$ (deg) and $|\boldsymbol{e}_{\Omega i}| \le 0.05$ (deg/s). Comparing the simulation results of the proposed controller (23) with the controller (46) in [42], it is obvious that not only the control accuracy and convergence time have been improved, but also the communication burden has been reduced significantly.

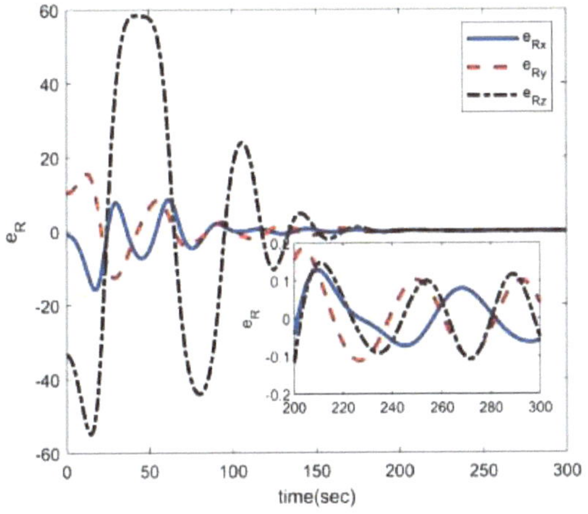

Fig. (6). Time responses of e_R (Ref.[42]).

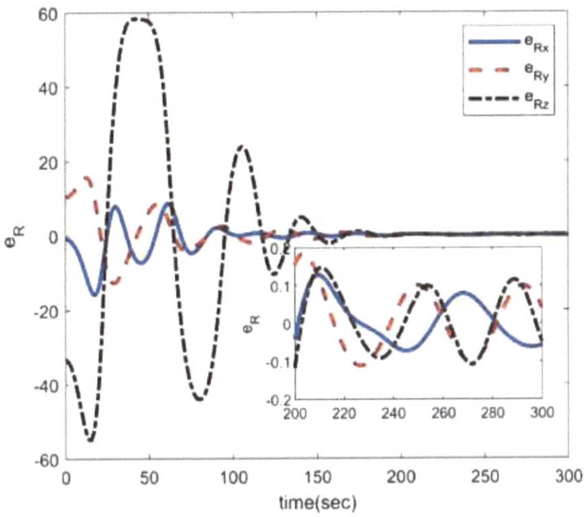

Fig. (7). Time responses of e_Ω (Ref.[42]).

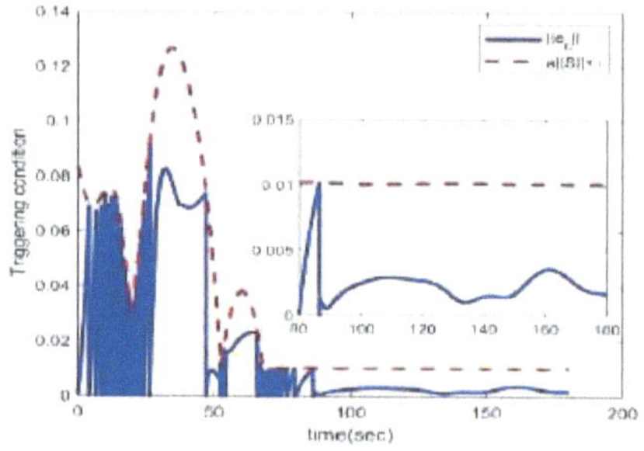

(a)

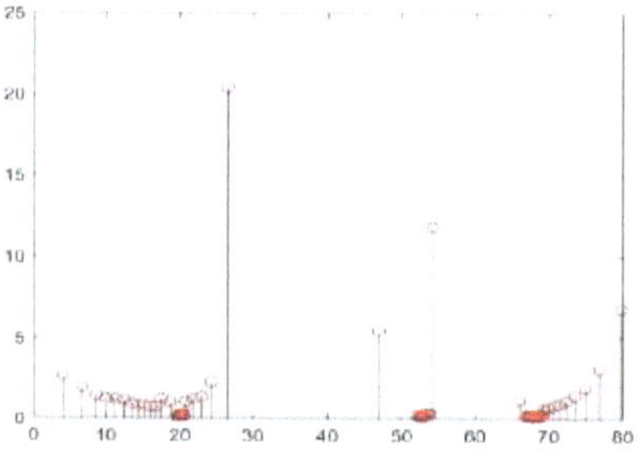

(b)

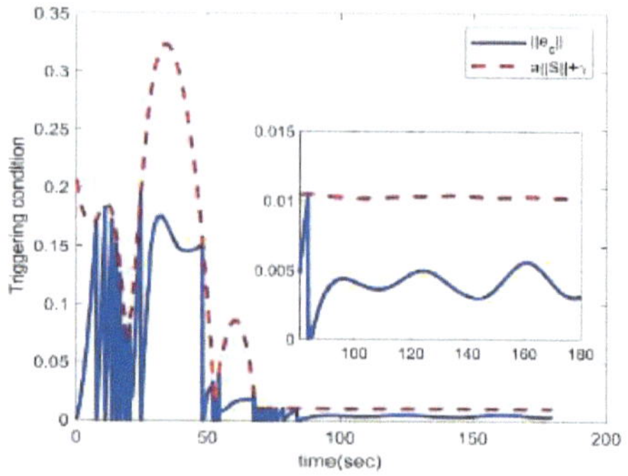

(c)

(Fig. 8) contd.....

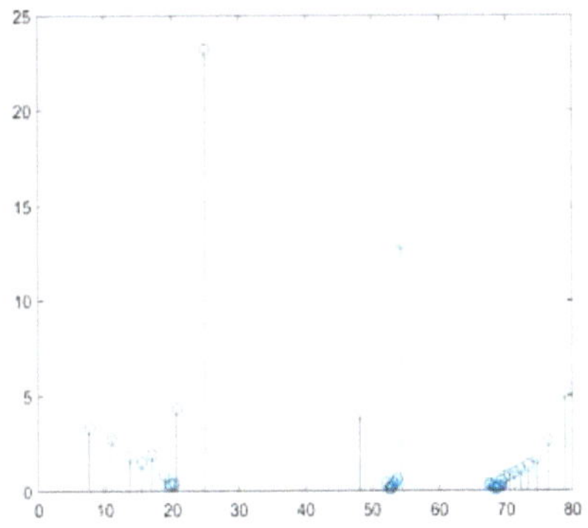

(d)

Fig. (8). (a) Time responses of $\|\boldsymbol{e}_c\|$ and $a_2\|\boldsymbol{S}\|+\gamma_2$, (b) Event-triggering release instants and inter-event times, (c) Time responses of $\|\boldsymbol{e}_c\|$ and $a_3\|\boldsymbol{S}\|+\gamma_3$, (d) Event-triggering release instants and inter-event times.

Table 1. Effects of the parameters a and γ .

a	γ	NTI	$\int_0^T \|u\| dt$	DCF
0.6	0.001	156	16.78	78.3%
0.6	0.01	60	16.78	91.67%
1.6	0.01	40	17.93	94.44%

[a] NTI is the number of the triggering instants
[b] DCF is the decrement of communication frequency

As analyzed in Remark 4, the way to design a and γ is to make a trade-off between the control cost and the communication burden. Since $-(K_1+K_2(\|\boldsymbol{S}\|))\boldsymbol{S}$ is designed to deal with the event-triggering strategy, different values of a and γ will not affect the control accuracy. The only concern is the balance between the control cost and the decrement of communication frequency. Therefore, Case 2 is presented to illustrate how different values of a and γ will affect the performance of the control system. The simulation results are given in Fig. (**8**) and Table **1**. From Figs. (**4, 5,** and **8**), it can be seen that initially, the value of the sliding surface $\boldsymbol{S}$ is large, and the triggering condition is mainly determined by $\boldsymbol{S}$. Consequently, a larger a will lead to a longer inter-event time. After the system converges, $\boldsymbol{S}$ will be close to the origin, and the triggering condition will be mainly determined by γ. A larger γ will lead to a longer inter-event time. Therefore, the number of the triggering instants of $a_3 = 1.6$, $\gamma_3 = 0.01$ is minimum, and the communication burden can be reduced by more than 94.44%. However, the largest control torque consumption is required. In addition, from Fig. (**8**) it can be found that when t > 80s, if $\gamma = 0.01$ the triggering strategy will no more be triggered, which means a larger control torque will be held as a constant and never decrease. When a is the same, a larger γ requires more control cost, which can be observed from Table **1**. Based on the above analysis, it can be summarized that employing the adaptive event-triggered controller designed in this study can obtain a higher control performance and needs a lower communication source than [42]. Furthermore, Zeno behavior avoidance can be achieved.

CONCLUSION

This study solved the spacecraft attitude tracking control problem with model uncertainties and external disturbances. The attitude dynamics were directly developed on $SO(3)$, where the singularity of MRPs and the unwinding problem of quaternions were avoided. Based on the designed positive configuration error function, an adaptive sliding mode controller was designed to ensure the stability of the closed-loop system, where an adaptive fuzzy logic system was utilized to estimate the disturbance, and high robustness and less chattering of the controller could be guaranteed. In addition, an event-triggering strategy was employed in the controller, which can reduce the usage of the spacecraft communication channel greatly and can also guarantee the Zeno free execution during the whole control process. Finally, simulation results were conducted to show that the control objective can be achieved through the designed control law. Moreover, compared with the results in [42], not only the control accuracy and convergence time have been improved, and the communication burden has been reduced greatly. However, it should be pointed out that the control cost and the communication burden are contradictory, and a trade-off has to be made when we design the parameters of the triggering strategy. Furthermore, flexible appendages are extensively applied in the design of modern spacecraft because they can enhance spacecraft functionality with low launch weight. Therefore, extending the modeling and control methods to flexible spacecraft attitude control could be one of the future studies.

CONSENT FOR PUBLICATION

Not applicable.

CONFLICT OF INTEREST

The authors declare no conflict of interest, financial or otherwise.

ACKNOWLEDGEMENTS

This work was supported by the National Natural Science Foundation of China under Grants 62103284 and 62103336.

REFERENCES

[1] Xiao B, Hu Q, Zhang Y. Finite-time attitude tracking of spacecraft with fault-tolerant capability. IEEE Transactions on Control Systems Technology, 2014, 23(4): 1338-1350.

[2] Wu B, Shen Q, Cao X. Event-triggered attitude control of spacecraft. Advances in Space Research, 2018, 61(3): 927-934.

[3] Li B, Hu Q, Yang Y. Continuous finite-time extended state observer based fault tolerant control for attitude stabilization. Aerospace Science and Technology, 2019, 84: 204-213.

[4] Hu Q, Zhang X, Niu G. Observer-based fault tolerant control and experimental verification for rigid spacecraft. Aerospace Science and Technology, 2019, 92: 373-386.

[5] Liu C, Sun Z, Shi K, et al. Robust dynamic output feedback control for attitude stabilization of spacecraft with nonlinear perturbations. Aerospace Science and Technology, 2017, 64: 102-121.

[6] Huang X, Biggs J D, Duan G. Post-capture attitude control with prescribed performance. Aerospace Science and Technology, 2020, 96: 105572.

[7] Li B, Hu Q, Ma G. Extended state observer based robust attitude control of spacecraft with input saturation. Aerospace Science and Technology, 2016, 50: 173-182.

[8] Shi K, Liu C, Sun Z, *et al.* Coupled orbit-attitude dynamics and trajectory tracking control for spacecraft electromagnetic docking. Applied Mathematical Modelling, 2022, 101: 553-572.

[9] Bohn J, Sanyal A K. Almost global finite-time stabilization of rigid body attitude dynamics using rotation matrices. International Journal of Robust and Nonlinear Control, 2016, 26(9): 2008-2022.

[10] Zhang J, Biggs J D, Ye D, *et al.* Finite-time attitude set-point tracking for thrust-vectoring spacecraft rendezvous. Aerospace Science and Technology, 2020, 96: 105588.

[11] Chen T, Shan J. Continuous constrained attitude regulation of multiple spacecraft on SO (3). Aerospace Science and Technology, 2020, 99: 105769.

[12] Biggs J D, Colley L. Geometric attitude motion planning for spacecraft with pointing and actuator constraints. Journal of Guidance, Control, and Dynamics, 2016, 39(7): 1672-1677.

[13] Shi K, Liu C, Biggs J D, *et al.* Observer-based control for spacecraft electromagnetic docking[J]. Aerospace Science and Technology, 2020, 99: 105759.

[14] Lee T. Exponential stability of an attitude tracking control system on SO (3) for large-angle rotational maneuvers. Systems & Control Letters, 2012, 61(1): 231-237.

[15] Guo Y, Guo J, Song S. Backstepping control for attitude tracking of the spacecraft under

input saturation. Acta Astronautica, 2017, 138: 318-325.

[16] Aslan O, Altan A, Hacioglu R. The control of blast furnace top gas pressure by using fuzzy PID//Proceedings of the fifth international conference on advances in mechanical and robotics engineering–AMRE. 2017: 22-26.

[17] Shi X N, Zhou Z G, Zhou D. Finite-time attitude trajectory tracking control of rigid spacecraft. IEEE Transactions on Aerospace and Electronic Systems, 2017, 53(6): 2913-2923.

[18] Wu S N, Sun X Y, Sun Z W, et al. Robust sliding mode control for spacecraft global fast-tracking manoeuvre. Proceedings of the Institution of Mechanical Engineers, Part G: Journal of Aerospace Engineering, 2011, 225(7): 749-760.

[19] Edwardsl C, Shtessel Y. Adaptive continuous higher order sliding mode control. Automatica, 2016, 65: 183-190.

[20] Shtessel Y, Shkolnikov I, Levant A. Smooth second-order sliding modes: Missile guidance application. Automatica, 2007, 43: 1470-1476.

[21] Besnard L, Shtessel Y, Landrum B. Quadrotor vehicle control *via* sliding mode controller driven by sliding mode disturbance observer, Journal of the Franklin Institute, 2012, 349: 658-684.

[22] Shtessel Y, Tournes C. Integrated higher-order sliding mode guidance and autopilot for dual control missiles. Journal of Guidance, Control, and Dynamics, 2009, 32(1): 79-94.

[23] Tanaka K, Sugeno M. Stability analysis and design of fuzzy control systems. Fuzzy sets and systems, 1992, 45(2): 135-156.

[24] Liu C, Yue X, Shi K, et al. Inertia-free attitude stabilization for flexible spacecraft with active vibration suppression. International Journal of Robust and Nonlinear Control, 2019, 29(18): 6311-6336..

[25] Zhang M, Shi P, Ma L, et al. Quantized feedback control of fuzzy Markov jump systems. IEEE transactions on cybernetics, 2018, 49(9): 3375-3384.

[26] Liu C, Yue X, Yang Z. Are nonfragile controllers always better than fragile controllers in attitude control performance of post-capture flexible spacecraft?[J]. Aerospace Science and Technology, 2021, 118: 107053.

[27] Shi P, Yu J. Dissipativity-based consensus for fuzzy multiagent systems under switching directed topologies. IEEE Transactions on Fuzzy Systems, 2020, 29(5): 1143-1151.

[28] Zhang J, Shi P, Qiu J, et al. A novel observer-based output feedback controller design for discrete-time fuzzy systems. IEEE Transactions on Fuzzy Systems, 2014, 23(1): 223-229.

[29] Wang Z, Sun J, Zhang H. Stability analysis of T–S fuzzy control system with sampled-dropouts based on time-varying Lyapunov function method. IEEE Transactions on Systems, Man, and Cybernetics: Systems, 2018, 50(7): 2566-2577.

[30] Palm R. Robust control by fuzzy sliding mode. Automatica, 1994, 30(9): 1429-1437.

[31] Huo B, Xia Y, Lu K, et al. Adaptive fuzzy finite-time fault-tolerant attitude control of rigid spacecraft. Journal of the Franklin Institute, 2015, 352(10): 4225-4246.

[32] Liu W, Geng Y, Wu B, et al. Neural-network-based adaptive event-triggered control for spacecraft attitude tracking. IEEE transactions on neural networks and learning systems, 2019, 31(10): 4015-4024.

[33] Zou A M, Kumar K D, Hou Z G, et al. Finite-time attitude tracking control for spacecraft

using terminal sliding mode and Chebyshev neural network. IEEE Transactions on Systems, Man, and Cybernetics, Part B (Cybernetics), 2011, 41(4): 950-963.

[34] Wang C, Guo L, Wen C, *et al.* Event-triggered adaptive attitude tracking control for spacecraft with unknown actuator faults. IEEE Transactions on Industrial Electronics, 2019, 67(3): 2241-2250.

[35] Zhang X M, Han Q L, Zhang B L. An overview and deep investigation on sampled-data-based event-triggered control and filtering for networked systems. IEEE Transactions on industrial informatics, 2016, 13(1): 4-16.

[36] Xi C, Dong J. Event-triggered adaptive fuzzy distributed tracking control for uncertain nonlinear multi-agent systems. Fuzzy Sets and Systems, 2021, 402: 35-50.

[37] Yang J, Xiao F, Chen T. Event-triggered formation tracking control of nonholonomic mobile robots without velocity measurements. Automatica, 2020, 112: 108671.

[38] Yang J, Xiao F, Ma J. Model-based edge-event-triggered containment control under directed topologies. IEEE transactions on cybernetics, 2018, 49(7): 2556-2567.

[39] Fan X, Wang Z. Event-triggered sliding-mode control for a class of T–S fuzzy systems. IEEE Transactions on Fuzzy Systems, 2019, 28(10): 2656-2664.

[40] Su X, Wen Y, Shi P, *et al.* Event-triggered fuzzy filtering for nonlinear dynamic systems *via* reduced-order approach. IEEE Transactions on Fuzzy Systems, 2018, 27(6): 1215-1225.

[41] Xing L, Wen C, Liu Z, *et al.* Event-triggered adaptive control for a class of uncertain nonlinear systems. IEEE transactions on automatic control, 2016, 62(4): 2071-2076.

[42] Lee T, Global exponential attitude tracking controls on SO(3), IEEE Transactions on Automatic Control, 2015, 60(10): 2837-2842.

[43] Liu C, Shi K, Yue X, *et al.* Inertia-free saturated output feedback attitude stabilization for uncertain spacecraft. International Journal of Robust and Nonlinear Control, 2020, 30(13): 5101-5121.

[44] Altan A, Aslan Ö, Hacıoğlu R. Real-time control based on NARX neural network of hexarotor UAV with load transporting system for path tracking//2018 6th international conference on control engineering & information technology (CEIT). IEEE, 2018: 1-6.

[45] Altan A, Hacıoğlu R. Model predictive control of three-axis gimbal system mounted on UAV for real-time target tracking under external disturbances. Mechanical Systems and Signal Processing, 2020, 138: 106548.

[46] Invernizzi D, Lovera M, Zaccarian L. Dynamic attitude planning for trajectory tracking in thrust-vectoring UAVs. IEEE Transactions on Automatic Control, 2019, 65(1): 453-460.

[47] Mayhew C G, Teel A R. Synergistic hybrid feedback for global rigid-body attitude tracking on SO(3), IEEE Transactions on Automatic Control, 2013, 58(11): 2730-2742.

[48] Lee T. Robust adaptive attitude tracking on SO(3) with an application to a quadrotor UAV, IEEE Transactions on Control Systems Technology, 2012, 21(5): 1924-1930.

[49] Liu C, Vukovich G, Sun Z, *et al.* Observer-based fault-tolerant attitude control for spacecraft with input delay. Journal of Guidance, Control, and Dynamics, 2018, 41(9): 2041-2053.

[50] Sun R, Wang J, Zhang D, *et al.* Neural-network-based sliding-mode adaptive control for spacecraft formation using aerodynamic forces. Journal of Guidance, Control, and Dynamics, 2018, 41(3): 757-763.

[51] Ioannou P A, Sun J. Robust adaptive control. Courier Corporation, 2012.

[52] Huang X, Yan Y, Zhou Y, *et al.* Dual-quaternion based distributed coordination control of six-DOF spacecraft formation with collision avoidance. Aerospace Science and Technology, 2017, 67: 443-455.

CHAPTER 4

Robust Finite-time Adaptive Control Algorithm for Satellite Attitude Maneuver

Li You[1,*]

[1]Xidian University, Xi'an 710126, China

Abstract: A robust adaptive finite-time controller for satellite fast attitude maneuver is proposed in this paper. The standard sliding mode is robust to some typical disturbances, but the convergence speed is slow and often could not meet the system requirements. The finite-time sliding mode not only has the robustness of the classical sliding mode, but also could greatly improve the terminal convergence speed. In order to deal with inertia matrix uncertainty, a finite-time adaptive law for inertia matrix estimation variables is proposed. A new method to deal with the singularity problem is proposed,based on the properties of Euler rotations. Considering that the variable estimation system has no direct feedback, an auxiliary state that converges slower than the system is designed to achieve finite-time stability. The Lyapunov method is used to demonstrate the global finite-time stability of the ensemble, and the numerical simulation results demonstrate the performance of the controller.

Keywords: Adaptive control, Fast attitude maneuver, Finite-time control, Robust control

INTRODUCTION

As a mature control method, sliding mode control has been widely used in satellite attitude control issues. The standard sliding mode has a clearl physical meaning that angular velocity is reversed to attitude quaternion, and this property could suppress some typical perturbations caused by disturbance, model uncertainty, actuator error and so on; hence standard sliding mode has its inherent robustness. However, the convergence rate of the system near the equilibrium point is relatively low, and the system control capability is not fully utilized. Therefore, it is necessary

[*]**Corresponding author Li You:** Xidian University, Xi'an 710126, China; Tel: +86 18392070397; E-mail: liyou@xidian.edu.cn

Chuang Liu, Honghua Dai, Xiaokui Yue & Yiqing Ma (Eds.)

to improve the convergence speed when dealing with fast attitude maneuvering problems for the classical sliding mode.

In order to improve the convergence speed of the system, many people have done a lot of research on this. Refs. [1-4] pointed out that the key to improving the convergence speed of the system is to design the angular velocity trajectory reasonably, so they designed a constant angular velocity maneuvering stage, which could maintain the convergence speed. However, most of the work does not solve the exponential convergence problem, and the terminal convergence speed needs to be improved [5,6], Ref. [7] designed the angular velocity "braking curve" of satellite attitude maneuver, optimized the angular velocity trajectory, and designed the deceleration process trajectory. However, the problem of exponential convergence speed still exists.

The finite-time control method is also an effective method to improve the convergence rate of the system, so this method is suitable for the fast maneuvering of satellites. The improvement of the convergence speed near the equilibrium point and the improvement of the torque control efficiency make the system performance near the equilibrium point the main focus of this method, so many researchers have done a lot of research on time-finite controllers for satellite attitude control. Ye, Refs. [8,9] focused on the control torque distribution algorithm and fault-tolerant algorithm, and designed a finite-time controller for satellite control. Refs. [10,11] proposed finite-time stabilization methods such as the Lyapunov and terminal sliding mode methods based on the finite-time controller structure and terminal sliding-mode standard structure of classical nonlinear systems. Refs. [12-14] designed a finite-time controller for satellite formation attitude control and analyzed the finite-time stability using the Lyapunov method. Existing approaches to deal with finite-time stability could be generalized as designing fractional state feedback and singularities near equilibrium points. Furthermore, existing finite-time control methods often require some special modifications to deal with unknown disturbance torque and model uncertainty.

Since the precise inertia matrix and some other unknown disturbances could not be known in the satellite attitude control problem, many researchers will design a controller with a symbolic function to suppress the disturbance for the problem uncertainty mentioned above. Refs. [15-17] designed a fault-tolerant controller to estimate the uncertainty of the system. Although this method is suitable for several typical uncertainty models, it is not suitable for random noise models, so there are still some shortcomings. Refs. [18-20] designed some robust controllers for system uncertainty and added symbolic function items to the controller, but the symbolic function items would bring high-frequency vibration and cause damage to actuators and physical systems. To deal with uncertainty without causing high-frequency vibrations, some researchers have delved into adaptive control. Ref. [21] designed an output feedback attitude stabilization controller for the control problem of rigid body satellites. Ref. [22] proposed a disturbance observer-based attitude controller for spacecraft based on a time-varying inertial matrix model. Ref. [23] designed a sliding mode controller for coupled orbit-attitude tracking of spacecraft electromagnetic docking. Ref. [24] and Ref. [25] designed an adaptive law for the quadrotor UAV control problem, taking into account the problems of fuzzy sensor data and model uncertainty. Ref. [26] designed a space exploration law that accommodates the uncertainty of the gravity gradient model. Ref. [27] designed a synchronous input and state estimation algorithm for integrated motor drive systems in a controller area network environment via an adaptive unscented Kalman filter. Ref. [28] designed a warp-based finite-time consensus for Euler-Lagrangian systems with an event-triggered strategy. Researchers have done other work focusing on adaptive and finite-time controllers [29-35], and the main idea of designing an adaptive law is to use system state feedback. Since the error state of the model is difficult to get feedback directly, this paper proposes another method to design the adaptive law and design the auxiliary state to achieve the purpose of finite time stabilization. This paper proposes a finite-time controller based on a standard sliding mode for the satellite control problem. This method could greatly improve the convergence speed and achieve the purpose of finite-time convergence under the premise of maintaining the advantages of standard sliding mode. Then, the adaptive law of inertia matrix uncertainty is given, and an auxiliary state that converges slower than the system state is designed to overcome the difficulty of estimating the system without direct feedback. The Lyapunov method is used to analyze the overall finite-time stability under the adaptive law, and the controller performance is discussed. The paper is organized as follows: Section 1 introduces the paper, and Section 2 describes the mathematical model used in this paper. Section 3 presents an adaptive finite-time controller for the satellite attitude stabilization problem, and Section 4 discusses the satellite attitude tracking problem. The numerical simulation results will be presented in Section 5, and Section 6 will conclude this paper.

DYNAMIC AND KINETIC MODEL

The dynamic model of the rigid satellite could be modeled as follows [1-4]

$$\boldsymbol{J}\dot{\boldsymbol{\omega}} + \boldsymbol{\omega}^{\times}\boldsymbol{J}\boldsymbol{\omega} = \boldsymbol{u} + \boldsymbol{d} \tag{1}$$

where $\boldsymbol{\omega}$ is the angular velocity, $\boldsymbol{J}$ is the inertia matrix of a satellite which is a symmetric matrix, $\boldsymbol{d}$ is unknown disturbance torque with norm upper bound $\|\boldsymbol{d}\| < \bar{d}$. The product matrix $\boldsymbol{r}^{\times}$ of the vector $\boldsymbol{r}$ is defined as

$$\boldsymbol{r}^{\times} = \begin{bmatrix} 0 & -r_3 & r_2 \\ r_3 & 0 & -r_1 \\ -r_2 & r_1 & 0 \end{bmatrix} \tag{2}$$

where r_i is the ith component of the vector $\boldsymbol{r}$.

Generally, it is very complicated to model all disturbances, such as gravitational gradient moments, atmospheric moments, solar pressure, space radiation, and payload moments. These perturbations are assumed to be Gauss white noise and the norm upper bound $\bar{d}$ could be obtained through experience and $\bar{d}$ is assumed to be $10^{-3}\,\text{Nm}$ in this paper through other works [1-4].

Also, the inertia matrix $\boldsymbol{J}$ could not be accurately known, and it is assumed that

$$\boldsymbol{J} = \hat{\boldsymbol{J}} + \tilde{\boldsymbol{J}} \tag{3}$$

where $\hat{\boldsymbol{J}}$ is the inertia matrix estimation and $\tilde{\boldsymbol{J}}$ is the error matrix. In this paper, the error matrix $\tilde{\boldsymbol{J}}$ could be treated as a disturbance in the control system and one of the main goals is to design an adaptive law to suppress this disturbance. Considering that $\tilde{\boldsymbol{J}}$ has six independent variables, define a six-dimensional vector $\boldsymbol{\psi}$ to describe $\boldsymbol{J}$ as follows

$$\begin{gathered} \boldsymbol{\psi} = \begin{bmatrix} J_{11} & J_{12} & J_{13} & J_{22} & J_{23} & J_{33} \end{bmatrix}^T \\ \hat{\boldsymbol{\psi}} = \begin{bmatrix} \hat{J}_{11} & \hat{J}_{12} & \hat{J}_{13} & \hat{J}_{22} & \hat{J}_{23} & \hat{J}_{33} \end{bmatrix}^T \\ \tilde{\boldsymbol{\psi}} = \begin{bmatrix} \tilde{J}_{11} & \tilde{J}_{12} & \tilde{J}_{13} & \tilde{J}_{22} & \tilde{J}_{23} & \tilde{J}_{33} \end{bmatrix}^T \\ \boldsymbol{\psi} = \hat{\boldsymbol{\psi}} + \tilde{\boldsymbol{\psi}} \end{gathered} \tag{4}$$

In order to simplify the text, define manipulator $\boldsymbol{L}(\bullet)$ for three-dimensional vector $\boldsymbol{r}$ as follows

$$L(r)=\begin{bmatrix} r_1 & r_2 & r_3 & 0 & 0 & 0 \\ 0 & r_1 & 0 & r_2 & r_3 & 0 \\ 0 & 0 & r_1 & 0 & r_2 & r_3 \end{bmatrix} \tag{5}$$

The essence of the manipulator $L(\bullet)$ is that it satisfies the following transformation and this property will be used in later text.

$$Jr = L(r)\psi \tag{6}$$

The kinetic model based on attitude quaternion could be written as follows [1-4]:

$$\begin{cases} \dot{q}_0 = -\dfrac{1}{2} q_v^T \omega \\ \dot{q}_v = \dfrac{1}{2}\left(q_0 I_3 + q_v^{\times}\right)\omega = \dfrac{1}{2} F\omega \end{cases} \tag{7}$$

Considering that q and $-q$ describes the same attitude, the scalar part of attitude quaternion is assumed to be non-negative in this paper *i.e.* $q_0 \geq 0$.

In satellite attitude tracking issue, error quaternion and error angular velocity are defined as follows [1-4]:

$$\begin{gathered} q_e = q_d^* \otimes q = \begin{bmatrix} q_{d0} q_0 + q_{dv}^T q_v \\ q_{d0} q_v - q_0 q_{dv} - q_{dv} \times q_v \end{bmatrix} \\ R(q_e) = \left(q_{e0}^2 - q_{ev}^T q_{ev}\right) I_3 + 2 q_{ev} q_{ev}^T - 2 q_{e0} q_{ev}^{\times} \\ \omega_e = \omega - R(q_e)\omega_d \end{gathered} \tag{8}$$

where q_d is desired attitude quaternion, ω_d is desired angular velocity and q_d^* is the conjugated quaternion defined as follows:

$$q_d^* = \begin{bmatrix} q_{d0} & -q_{dv}^T \end{bmatrix}^T \tag{9}$$

Substitute (8) in a dynamic model (1), the dynamic model of error angular velocity could be written as (10).

$$J\dot{\omega}_e + JR\dot{\omega}_d - J\omega_e^{\times} R\omega_d + \left(\omega_e + R\omega_d\right)^{\times} J\left(\omega_e + R\omega_d\right) = u + d \tag{10}$$

The kinetic model of error quaternion could be written as follows [1-4].

$$\begin{cases} \dot{q}_{e0} = -\dfrac{1}{2} q_{ev}^T \omega_e \\ \dot{q}_{ev} = \dfrac{1}{2}\left(q_{e0} I_3 + q_{ev}^{\times}\right)\omega_e = \dfrac{1}{2} F_e \omega_e \end{cases} \tag{11}$$

In this paper, the tracking target is assumed to be cooperative, which means $q_d, \omega_d, \dot{\omega}_d$ are all accurate and known.

It is also worth noticing that in this paper, the finite-time stability refers to system states that could reach the neighborhood near the equilibrium point within finite time *i.e.* there exists a small positive scalar ε to satisfy $\|x\| \le \varepsilon$ for any $t \ge T_0$.

ATTITUDE STABILIZATION ISSUE

In this section, the satellite attitude stabilization issue is discussed. First, the finite-time sliding mode is given and the finite-time controller without inertia matrix uncertainty is given. Then the adaptive law is proposed and a finite-time adaptive controller is given to deal with the system uncertainty issue.

Finite-time Controller

In satellite attitude control issue, the standard sliding mode could be written as follows

$$s = \omega + kq_v, (k > 0) \tag{12}$$

The standard sliding mode (12) has a simple structure, and the controller based on (12) is easy to design. Furthermore, the convergence speed is exponential when the system state is far from the equilibrium point. Although the system performance is good, the convergence time will tend to infinity, thus, the exponential convergence speed is also the main disadvantage of the sliding mode (12). In order to obtain the finite-time sliding mode for the fast maneuvering issue, the sliding mode (12) is modified as follows

$$\boldsymbol{s}=\begin{cases}\boldsymbol{\omega}+k_1\boldsymbol{q}_v,\|\boldsymbol{q}_v\|>\alpha\\ \boldsymbol{\omega}+k_2\boldsymbol{sig}^r(\boldsymbol{q}_v),\|\boldsymbol{q}_v\|\le\alpha\end{cases}\tag{13}$$

where α,k_1,k_2 are positive scalars, r is a positive scalar which satisfies $0<r<1/2$, and vector function $\boldsymbol{sig}^r(\boldsymbol{q}_v)$ is defined as follows

$$\boldsymbol{sig}^r(\boldsymbol{q}_v)=\boldsymbol{q}_v/\|\boldsymbol{q}_v\|^r\tag{14}$$

Based on the definition of $\boldsymbol{sig}^r(\boldsymbol{q}_v)$ it could be found that when $0<r<1$ is satisfied, there would be no singularity issue when $\boldsymbol{q}_v\to\boldsymbol{0}$.

In order to get a continuous sliding mode *i.e.* sliding mode (13) is continuous at the switching point, control parameters k_1,k_2 should satisfy following equation.

$$\alpha^r k_1=k_2\tag{15}$$

Based on the definition of sliding mode (13) it could be found that it is constructed as two stages: the first stage is totally the same as standard sliding mode (12) and during this stage, the system convergence rate is exponential, and the system state would converge into the neighborhood of the equilibrium $\|\boldsymbol{q}_v\|\le\alpha$ within finite-time; the second stage is finite-time converge stage, when maneuvering along this trajectory, it could be got that

$$\boldsymbol{\omega}=-k_2\boldsymbol{sig}^r(\boldsymbol{q}_v)\tag{16}$$

$$\begin{aligned}\frac{d(1-q_0)}{dt}&=-\dot{q}_0=\frac{1}{2}\boldsymbol{q}_v^T\boldsymbol{\omega}=-\frac{k_2}{2}\boldsymbol{q}_v^T\boldsymbol{sig}^r(\boldsymbol{q}_v)=-\frac{k_2}{2}\|\boldsymbol{q}_v\|^{2-r}=-\frac{k_2}{2}\sqrt{1-q_0^2}^{\,2-r}\\&=-\frac{k_2}{2}(1+q_0)^{1-\frac{r}{2}}(1-q_0)^{1-\frac{r}{2}}\\&\le-\frac{k_2}{2}(1-q_0)^{1-\frac{r}{2}}\end{aligned}\tag{17}$$

Therefore, the system will converge to its equilibrium point in a finite time when maneuvering along the second stage. Most importantly, the system has finite time stability when maneuvering along the sliding mode (13).

In order to avoid the singularity issue in finite time control, it is necessary to discuss the derivative of the sliding mode state $\boldsymbol{s}$. Obviously, there is no singularity issue in the first stage, and the second stage could be

$$\begin{aligned}\dot{\boldsymbol{s}} &= \dot{\boldsymbol{\omega}} + k_2 \frac{d\boldsymbol{sig}^r(\boldsymbol{q}_v)}{dt} \\ &= \boldsymbol{J}^{-1}\left(\boldsymbol{u}+\boldsymbol{d}-\boldsymbol{\omega}^\times \boldsymbol{J}\boldsymbol{\omega}\right) + k_2 \frac{\left\|\boldsymbol{q}_v\right\|^r \dot{\boldsymbol{q}}_v}{\left\|\boldsymbol{q}_v\right\|^{2r}} - k_2 \frac{\left\|\boldsymbol{q}_v\right\|^{r-2} \boldsymbol{q}_v^T \dot{\boldsymbol{q}}_v \boldsymbol{q}_v}{\left\|\boldsymbol{q}_v\right\|^{2r}} \\ &= \boldsymbol{J}^{-1}\left(\boldsymbol{u}+\boldsymbol{d}-\boldsymbol{\omega}^\times \boldsymbol{J}\boldsymbol{\omega}\right) + \frac{1}{2}k_2 \left\|\boldsymbol{q}_v\right\|^{-r}\left(q_0 \boldsymbol{I}_3 + \boldsymbol{q}_v^\times\right)\boldsymbol{\omega} - \frac{1}{2}k_2\left\|\boldsymbol{q}_v\right\|^{-r-2} q_0 \boldsymbol{q}_v^T \boldsymbol{\omega}\boldsymbol{q}_v\end{aligned} \quad \textbf{(18)}$$

In order to make system sates converge along the sliding mode (13) *i.e.* $\boldsymbol{s}=\boldsymbol{0}$, it could be found that

$$\boldsymbol{u} = -k_s \boldsymbol{sig}^r(\boldsymbol{s}) + \boldsymbol{\omega}^\times \boldsymbol{J}\boldsymbol{\omega} - \frac{1}{2}k_2 \boldsymbol{J}\left[\left\|\boldsymbol{q}_v\right\|^{-r}\left(q_0\boldsymbol{I}_3 + \boldsymbol{q}_v^\times\right)\boldsymbol{\omega} - \left\|\boldsymbol{q}_v\right\|^{-r-2} q_0 \boldsymbol{q}_v^T\boldsymbol{\omega}\boldsymbol{q}_v\right] \quad \textbf{(19)}$$

Based on (19) it could be found that there are two singularity terms caused by $\left\|\boldsymbol{q}_v\right\|^{-r}$ and $\left\|\boldsymbol{q}_v\right\|^{-r-2}$, and if the system state approaches the equilibrium point but does not reach the sliding mode, *i.e.* $\boldsymbol{q}_v = \boldsymbol{0}, \boldsymbol{s} \neq \boldsymbol{0}$, the singularity issue occurs. Therefore, to solve the singularity problem, the angular velocity vector and the attitude quaternion vector could be partially reversed. When the system state reaches the sliding mode, it could be obtained that

$$\boldsymbol{\omega} = -k_2 \boldsymbol{sig}^r(\boldsymbol{q}_v) \quad \textbf{(20)}$$

$$\boldsymbol{u} = \boldsymbol{\omega}^\times \boldsymbol{J}\boldsymbol{\omega} + \frac{1}{2}k_2^2 q_0 \boldsymbol{J}(1-r)\boldsymbol{sig}^{2r}(\boldsymbol{q}_v) \quad \textbf{(21)}$$

According to the definition of the function $\boldsymbol{sig}(\bullet)$ and the value range of r, it could be found that when the system state reaches the sliding mode, there is no singularity problem in the controller. It is worth noting that for large-angle maneuvers, the system reaches the first stage of the sliding mode (13) before the condition $\left\|\boldsymbol{q}_v\right\| \leq \alpha$ is satisfied and α is a proper parameter. In other words, this article needs to make an assumption:

Assumption: when the system reaches the second stage of sliding mode (13), the angular velocity vector and the vector part of the attitude quaternion have been reversed.

After the finite time sliding mode (13) is given, to ensure that the system states could move along the sliding mode under any initial conditions. A controller without system inertial uncertainty could be written as follows

$$\boldsymbol{u}=\begin{cases}-k_s\boldsymbol{sig}^{r_s}(\boldsymbol{s})+\boldsymbol{\omega}^{\times}\boldsymbol{J\omega}-\frac{1}{2}k_1\boldsymbol{JF\omega}-\bar{d}\boldsymbol{sign}(\boldsymbol{s}),\|\boldsymbol{q}_v\|>\alpha\\ -k_s\boldsymbol{sig}^{r_s}(\boldsymbol{s})+\boldsymbol{\omega}^{\times}\boldsymbol{J\omega}+\frac{1}{2}k_2^2q_0(1-r)\boldsymbol{Jsig}^{2r}(\boldsymbol{q}_v)-\bar{d}\boldsymbol{sign}(\boldsymbol{s}),\|\boldsymbol{q}_v\|\le\alpha\end{cases} \tag{22}$$

where r_s is a positive scalar which satisfies $0<r_s<1$, $\bar{d}$ is a positive scalar which satisfies $\bar{d}\ge\|\boldsymbol{d}\|$, k_s is a positive scalar, $\boldsymbol{sign}(\boldsymbol{x})$ is the sign function of the vector $\boldsymbol{x}$.

In order to prove system finite-time stability, select the Lyapunov function as follows

$$V_s=\frac{1}{2}\boldsymbol{s}^T\boldsymbol{Js} \tag{23}$$

When $\|\boldsymbol{q}_v\|>\alpha$, calculate the derivative of V_s it could be got that

$$\begin{aligned}\dot{V}_s&=\boldsymbol{s}^T\boldsymbol{J}\dot{\boldsymbol{s}}=\boldsymbol{s}^T\boldsymbol{J}\left(\dot{\boldsymbol{\omega}}+k_1\dot{\boldsymbol{q}}_v\right)\\&=\boldsymbol{s}^T\left(-k_s\boldsymbol{sig}^{r_s}(\boldsymbol{s})+\boldsymbol{\omega}^{\times}\boldsymbol{J\omega}-\frac{1}{2}k_1\boldsymbol{JF\omega}-\bar{d}\boldsymbol{sign}(\boldsymbol{s})\right)\\&\quad+\boldsymbol{s}^T\boldsymbol{d}-\boldsymbol{s}^T\boldsymbol{\omega}^{\times}\boldsymbol{J\omega}+\frac{1}{2}k_1\boldsymbol{s}^T\boldsymbol{JF\omega}\\&=\boldsymbol{s}^T\left(-k_s\boldsymbol{sig}^{r_s}(\boldsymbol{s})-\bar{d}\boldsymbol{sign}(\boldsymbol{s})\right)+\boldsymbol{s}^T\boldsymbol{d}\\&\le-k_s\boldsymbol{s}^T\boldsymbol{sig}^{r_s}(\boldsymbol{s})\\&=-k_s\|\boldsymbol{s}\|^{2-r_s}\\&\le-k_s\left(\frac{V_s}{\lambda_m(\boldsymbol{J})}\right)^{1-\frac{r_s}{2}}\end{aligned} \tag{24}$$

When $\|\boldsymbol{q}_v\| \le \alpha$, *i.e.* the angular velocity and the vector part of attitude quaternion have been reversed, calculate the derivative of V_s it could be got that

$$\begin{aligned}
\dot{V}_s &= \boldsymbol{s}^T \boldsymbol{J}\dot{\boldsymbol{s}} = \boldsymbol{s}^T \boldsymbol{J}\left(\dot{\boldsymbol{\omega}} + k_2 \frac{d\boldsymbol{sig}^r(\boldsymbol{q}_v)}{dt}\right) \\
&= \boldsymbol{s}^T\left(\boldsymbol{u} + \boldsymbol{d} - \boldsymbol{\omega}^\times \boldsymbol{J}\boldsymbol{\omega}\right) - \frac{1}{2}k_2^2 q_0 \boldsymbol{s}^T \boldsymbol{J}(1-r)\boldsymbol{sig}^{2r}(\boldsymbol{q}_v) \\
&= \boldsymbol{s}^T\left(-k_s \boldsymbol{sig}^{r_s}(\boldsymbol{s}) - \bar{d}\boldsymbol{sign}(\boldsymbol{s})\right) + \boldsymbol{s}^T\boldsymbol{d} \\
&\le -k_s \boldsymbol{s}^T \boldsymbol{sig}^{r_s}(\boldsymbol{s}) \\
&= -k_s \|\boldsymbol{s}\|^{2-r_s} \\
&\le -k_s \left(\frac{V_s}{\lambda_m(\boldsymbol{J})}\right)^{1-\frac{r_s}{2}}
\end{aligned} \tag{25}$$

In (24) and (25), $\lambda_m(\boldsymbol{J})$ is the minimum eigenvalue of the inertia matrix $\boldsymbol{J}$.

It could be found from (24) and (25) that systems (1) and (7) governed by controller (22) would reach the sliding mode (13) within finite time, system (1) and (7) would converge to the equilibrium point within finite-time.

Adaptive Finite-Time Controller

In the previous section, the finite-time controller has some shortcomings, such as ignoring the uncertainty of the system model. Therefore, this section discusses the finite-time controller under the uncertainty of the inertia matrix based on the adaptive update law.

The adaptive controller could be written as follows

$$\boldsymbol{u} = \begin{cases} -k_s \boldsymbol{sig}^{r_s}(\boldsymbol{s}) + \boldsymbol{\omega}^\times \hat{\boldsymbol{J}}\boldsymbol{\omega} - \frac{1}{2}k_1 \hat{\boldsymbol{J}}\boldsymbol{F}\boldsymbol{\omega} - (\bar{d}+\gamma)\boldsymbol{sign}(\boldsymbol{s}), \|\boldsymbol{q}_v\| > \alpha \\ -k_s \boldsymbol{sig}^{r_s}(\boldsymbol{s}) + \boldsymbol{\omega}^\times \hat{\boldsymbol{J}}\boldsymbol{\omega} + \frac{1}{2}k_2^2 q_0 (1-r)\hat{\boldsymbol{J}}\boldsymbol{sig}^{2r}(\boldsymbol{q}_v) - (\bar{d}+\gamma)\boldsymbol{sign}(\boldsymbol{s}), \|\boldsymbol{q}_v\| \le \alpha \end{cases} \tag{26}$$

$$\dot{\hat{\psi}} = \begin{cases} -\boldsymbol{s}^T\left(\omega^{\times}\boldsymbol{L}(\omega)-\frac{k_1}{2}\boldsymbol{L}(\boldsymbol{F}\omega)\right), \|\boldsymbol{q}_v\| > \alpha \\ -\boldsymbol{s}^T\left(\omega^{\times}\boldsymbol{L}(\omega)+\frac{1}{2}k_2^2 q_0(1-r)\boldsymbol{L}\left(\boldsymbol{sig}^{2r}(\boldsymbol{q}_v)\right)\right), \|\boldsymbol{q}_v\| \le \alpha \end{cases} \tag{27}$$

$$\dot{\gamma} = -k_\gamma \gamma, \gamma(0) \ge \|\tilde{\psi}(0)\| \tag{28}$$

where k_γ is a positive scalar. Based on (28) it could be found that γ is a positive scalar and would converge to zero. The initial value of γ should be selected larger than the norm of the initial value of the error estimation state $\tilde{\psi}(0)$.

It could be found from equations (26)-(28) that the update rule of the model estimation state $\hat{\psi}$ plays a decisive role in suppressing the uncertainty disturbance of the system. By appropriately designing the update law, the system could reach the neighborhood of sliding mode state *i.e.* within a finite time, $\|\boldsymbol{s}\| \le \varepsilon$ could be satisfied where ε is a small positive scalar. Moreover, the sign function term in (26) consists of $\bar{d}$ and γ, the former term is a small term and the later term approaches to zero along the convergence process. Larger sign function terms are harmful to the physical system and cause high-frequency vibrations, but the sign function terms in Eq. (26) are easier to implement in engineering practice and could mitigate vibrations.

In order to analyze system stability, it is necessary to propose a lemma.

Lemma: for positive scalar $x > 0, y > 0$, and $1 > p > 0$, following inequality holds [10,11].

$$x^p + y^p > (x+y)^p \tag{29}$$

Select Lyapunov function as follows

$$V_s = \frac{1}{2}\boldsymbol{s}^T\boldsymbol{J}\boldsymbol{s} + \frac{1}{2}\tilde{\psi}^T\tilde{\psi} \tag{30}$$

When $\|\boldsymbol{q}_v\| > \alpha$, calculate the derivative of V_s it could be got that

$$\begin{aligned}
\dot{V}_s &= s^T J\dot{s} + \tilde{\psi}^T \dot{\tilde{\psi}} \\
&= s^T \left(-k_s sig^{r_s}(s) + \omega^{\times} \hat{J}\omega - \frac{1}{2} k_1 \hat{J} F\omega - (\bar{d} + \gamma) sign(s) \right) \\
&\quad + s^T d - s^T \omega^{\times} J\omega + \frac{1}{2} k_1 s^T JF\omega - \tilde{\psi}^T \dot{\hat{\psi}} \\
&\le -k_s s^T sig^{r_s}(s) - \gamma s^T sign(s) - s^T \omega^{\times} \tilde{J}\omega + \frac{1}{2} k_1 s^T \tilde{J} F\omega - \dot{\hat{\psi}}^T \tilde{\psi} \\
&= -k_s s^T sig^{r_s}(s) - \gamma s^T sign(s) - s^T \omega^{\times} L(\omega)\tilde{\psi} + \frac{1}{2} k_1 s^T L(F\omega)\tilde{\psi} - \dot{\hat{\psi}}^T \tilde{\psi} \\
&= -k_s s^T sig^{r_s}(s) - \gamma s^T sign(s) - s^T \left(\omega^{\times} L(\omega) - \frac{1}{2} k_1 s^T L(F\omega) \right) \tilde{\psi} \\
&\quad + s^T \left(\omega^{\times} L(\omega) - \frac{k_1}{2} L(F\omega) \right) \tilde{\psi} \\
&= -k_s s^T sig^{r_s}(s) - \gamma s^T sign(s) \\
&= -k_s \|s\|^{2-r_s} - \gamma \|s\|
\end{aligned} \tag{31}$$

When system states do not reach the sliding mode, there exists a small positive scalar ε that satisfies $\|s\| \le \varepsilon$ and (31) could be transformed to

$$\dot{V}_s \le -k_s \|s\|^{2-r_s} - \gamma\varepsilon \tag{32}$$

it could be found from equations (31) and (32) that system states s and $\tilde{\psi}$ have a convergence rate faster than the exponential rate, considering that the coefficient γ has a standard exponential convergence rate and its initial value satisfies $\gamma(0) \ge \|\tilde{\psi}(0)\|$, hence it could be obtained that

$$\gamma \ge \|\tilde{\psi}\| \tag{33}$$

Inequalities (31) and (32) could be transformed to

$$\begin{aligned}
\dot{V}_s &\le -k_s \|s\|^{2-r_s} - \varepsilon \|\tilde{\psi}\| \le -k_s \|s(0)\|^{1-r_s} \|s\| - \varepsilon \|\tilde{\psi}\| \\
&\le -\min\left(\varepsilon, k_s \|s(0)\|^{1-r_s}\right)\left(\|s\| + \|\tilde{\psi}\|\right) \\
&\le -\varepsilon\left(\|s\| + \|\tilde{\psi}\|\right)
\end{aligned} \tag{34}$$

where $\boldsymbol{s}(0)$ is the initial value of the sliding mode state. In (34), the property that ε is a small positive scalar is used *i.e.* it is assumed that $\varepsilon \le k_s \|\boldsymbol{s}(0)\|^{1-r_s}$.

Considering that the eigenvalue of the inertia matrix $\boldsymbol{J}$ is far larger than that of the unit matrix, hence (30) could be transformed to

$$V_s = \frac{1}{2}\boldsymbol{s}^T\boldsymbol{J}\boldsymbol{s} + \frac{1}{2}\tilde{\boldsymbol{\psi}}^T\tilde{\boldsymbol{\psi}} \le \frac{1}{2}\lambda_M(\boldsymbol{J})\left(\|\boldsymbol{s}\|^2 + \|\tilde{\boldsymbol{\psi}}\|^2\right) \tag{35}$$

where $\lambda_M(\boldsymbol{J})$ is the maximum eigenvalue of the inertia matrix $\boldsymbol{J}$.

Based on the lemma and (34) (35), it could be got that

$$\begin{gathered}\frac{\dot{V}_s}{\varepsilon} = -\left(\|\boldsymbol{s}\| + \|\tilde{\boldsymbol{\psi}}\|\right) \le -\left(\|\boldsymbol{s}\|^2 + \|\tilde{\boldsymbol{\psi}}\|^2\right)^{\frac{1}{2}} \le -\left(\frac{2V_s}{\lambda_M(\boldsymbol{J})}\right)^{\frac{1}{2}} \\ \dot{V}_s \le -\varepsilon\left(\frac{2V_s}{\lambda_M(\boldsymbol{J})}\right)^{\frac{1}{2}}\end{gathered} \tag{36}$$

When $\|\boldsymbol{q}_v\| \le \alpha$, calculate the derivative of V_s it could be obtained that

$$\begin{aligned}\dot{V}_s &= \boldsymbol{s}^T\boldsymbol{J}\dot{\boldsymbol{s}} + \tilde{\boldsymbol{\psi}}^T\dot{\tilde{\boldsymbol{\psi}}} \\ &= \boldsymbol{s}^T\left(-k_s\boldsymbol{sig}^{r_s}(\boldsymbol{s}) + \boldsymbol{\omega}^\times\hat{\boldsymbol{J}}\boldsymbol{\omega} + \frac{1}{2}k_2^2 q_0(1-r)\hat{\boldsymbol{J}}\boldsymbol{sig}^{2r}(\boldsymbol{q}_v) - (\bar{d}+\gamma)\boldsymbol{sign}(\boldsymbol{s})\right) \\ &\quad + \boldsymbol{s}^T\boldsymbol{d} - \boldsymbol{s}^T\boldsymbol{\omega}^\times\boldsymbol{J}\boldsymbol{\omega} - \frac{1}{2}k_2^2 q_0(1-r)\boldsymbol{J}\boldsymbol{sig}^{2r}(\boldsymbol{q}_v) - \tilde{\boldsymbol{\psi}}^T\dot{\hat{\boldsymbol{\psi}}} \\ &\le -k_s\boldsymbol{s}^T\boldsymbol{sig}^{r_s}(\boldsymbol{s}) - \gamma\boldsymbol{s}^T\boldsymbol{sign}(\boldsymbol{s}) - \boldsymbol{s}^T\boldsymbol{\omega}^\times\tilde{\boldsymbol{J}}\boldsymbol{\omega} - \frac{1}{2}k_2^2 q_0(1-r)\boldsymbol{s}^T\tilde{\boldsymbol{J}}\boldsymbol{sig}^{2r}(\boldsymbol{q}_v) - \dot{\hat{\boldsymbol{\psi}}}^T\tilde{\boldsymbol{\psi}} \\ &= -k_s\boldsymbol{s}^T\boldsymbol{sig}^{r_s}(\boldsymbol{s}) - \gamma\boldsymbol{s}^T\boldsymbol{sign}(\boldsymbol{s}) - \boldsymbol{s}^T\boldsymbol{\omega}^\times\boldsymbol{L}(\boldsymbol{\omega})\tilde{\boldsymbol{\psi}} - \frac{1}{2}k_2^2 q_0(1-r)\boldsymbol{s}^T\boldsymbol{L}\left(\boldsymbol{sig}^{2r}(\boldsymbol{q}_v)\right)\tilde{\boldsymbol{\psi}} \\ &\quad - \dot{\hat{\boldsymbol{\psi}}}^T\tilde{\boldsymbol{\psi}}\end{aligned}$$

$$
\begin{aligned}
&= -k_s \boldsymbol{s}^T \boldsymbol{sig}^{r_s}(\boldsymbol{s}) - \gamma \boldsymbol{s}^T \boldsymbol{sign}(\boldsymbol{s}) - \boldsymbol{s}^T\left(\omega^\times \boldsymbol{L}(\omega) + \frac{1}{2}k_2^2 q_0 (1-r)\boldsymbol{s}^T \boldsymbol{L}\left(\boldsymbol{sig}^{2r}(\boldsymbol{q}_v)\right)\right)\tilde{\psi} \\
&+\boldsymbol{s}^T\left(\omega^\times \boldsymbol{L}(\omega) + \frac{1}{2}k_2^2 q_0 (1-r)\boldsymbol{L}\left(\boldsymbol{sig}^{2r}(\boldsymbol{q}_v)\right)\right)\tilde{\psi} \\
&= -k_s \boldsymbol{s}^T \boldsymbol{sig}^{r_s}(\boldsymbol{s}) - \gamma \boldsymbol{s}^T \boldsymbol{sign}(\boldsymbol{s}) \\
&= -k_s \|\boldsymbol{s}\|^{2-r_s} - \gamma\|\boldsymbol{s}\| \le -k_s \|\boldsymbol{s}\|^{2-r_s} - \gamma\varepsilon \le -\varepsilon\left(\|\boldsymbol{s}\| + \|\tilde{\psi}\|\right) \\
&\le -\varepsilon\left(\frac{2V_s}{\lambda_M(\boldsymbol{J})}\right)^{\frac{1}{2}}
\end{aligned}
\tag{37}
$$

Based on equations (36) and (37), the system could be considered to have reached a sliding mode when the system has converged to less than a suitable scalar ε. Therefore, the systems (1) and (7) controlled by the controller (26) are stable within a finite time. .

ATTITUDE TRACKING ISSUE

Similar as discussed in the previous section, the finite-time sliding mode for attitude tracking issue could be written as follows

$$
\boldsymbol{s}_e = \begin{cases} \omega_e + k_1 \boldsymbol{q}_{ev}, \|\boldsymbol{q}_{ev}\| > \alpha \\ \omega_e + k_2 \boldsymbol{sig}^r(\boldsymbol{q}_{ev}), \|\boldsymbol{q}_{ev}\| \le \alpha \end{cases} \tag{38}
$$

$$
\alpha^r k_1 = k_2, k_1 > 0, 0 < r < 1 \tag{39}
$$

Based on the discussion in section 3.1, it is easy to find that systems (10) and (11) are finite-time stable on sliding mode (38). The adaptive finite-time controller based on this sliding mode could be written as follows

$$
\boldsymbol{u} = \begin{cases} -k_s \boldsymbol{sig}^{r_s}(\boldsymbol{s}_e) + \hat{\boldsymbol{J}}\boldsymbol{R}\dot{\omega}_d - \hat{\boldsymbol{J}}\omega_e^\times \boldsymbol{R}\omega_d + \omega^\times \hat{\boldsymbol{J}}\omega \\ \quad -\frac{1}{2}k_1 \hat{\boldsymbol{J}}\boldsymbol{F}_e \omega_e - (\bar{d} + \gamma)\boldsymbol{sign}(\boldsymbol{s}_e) & \|\boldsymbol{q}_{ev}\| > \alpha \\ -k_s \boldsymbol{sig}^{r_s}(\boldsymbol{s}_e) + \hat{\boldsymbol{J}}\boldsymbol{R}\dot{\omega}_d - \hat{\boldsymbol{J}}\omega_e^\times \boldsymbol{R}\omega_d + \omega^\times \hat{\boldsymbol{J}}\omega \\ +\frac{1}{2}k_2^2 q_{e0}(1-r)\hat{\boldsymbol{J}}\boldsymbol{sig}^{2r}(\boldsymbol{q}_{ev}) - (\bar{d} + \gamma)\boldsymbol{sign}(\boldsymbol{s}_e) & \|\boldsymbol{q}_{ev}\| \le \alpha \end{cases} \tag{40}
$$

$$\dot{\hat{\psi}} = \begin{cases} -s_e^T\left(\omega^\times L(\omega) - L\left(R\dot{\omega}_d - \omega_e^\times R\omega_d\right) - \dfrac{k_1}{2}L\left(F_e\omega_e\right)\right), \|q_{ev}\| > \alpha \\ -s_e^T\left(\omega^\times L(\omega) - L\left(R\dot{\omega}_d - \omega_e^\times R\omega_d\right) + \dfrac{1}{2}k_2^2 q_{e0}(1-r)L\left(sig^{2r}(q_{ev})\right)\right), \|q_{ev}\| \le \alpha \end{cases} \tag{41}$$

$$\dot{\gamma} = -k_\gamma \gamma, \gamma(0) \ge \|\tilde{\psi}(0)\| \tag{42}$$

where r_s is a positive scalar which satisfies $0 < r_s < 1$, $\bar{d}$ is a positive scalar which satisfies $\bar{d} \ge \|d\|$, k_s is a positive scalar.

In order to analyze the finite-time stability of the controller (40), select the Lyapunov function as follows

$$V_s = \frac{1}{2}s^T Js + \frac{1}{2}\tilde{\psi}^T\tilde{\psi} \tag{43}$$

When $\|q_{ev}\| > \alpha$, calculate the derivative of V_s it could be got that

$$\begin{aligned}
\dot{V}_s &= s_e^T J\dot{s}_e + \tilde{\psi}^T\dot{\tilde{\psi}} \\
&= s_e^T\left(-k_s sig^{r_s}(s_e) + \hat{J}R\dot{\omega}_d - \hat{J}\omega_e^\times R\omega_d + \omega^\times\hat{J}\omega - \frac{1}{2}k_1\hat{J}F_e\omega_e - \left(\bar{d}+\gamma\right)sign(s_e)\right) \\
&\quad + s_e^T d - s_e^T\omega^\times J\omega + s_e^T JR\dot{\omega}_d - s_e^T J\omega_e^\times R\omega_d + \frac{1}{2}k_1 s^T JF_e\omega_e - \tilde{\psi}^T\dot{\hat{\psi}} \\
&\le -k_s s_e^T sig^{r_s}(s_e) - \gamma s_e^T sign(s_e) - s_e^T\left(\omega^\times\tilde{J}\omega - \tilde{J}R\dot{\omega}_d + \tilde{J}\omega_e^\times R\omega_d\right) \\
&\quad + \frac{1}{2}k_1 s_e^T\tilde{J}F_e\omega_e - \dot{\hat{\psi}}^T\tilde{\psi} \\
&= -k_s s_e^T sig^{r_s}(s_e) - \gamma s_e^T sign(s_e) - s_e^T\left(\omega^\times L(\omega) - L\left(R\dot{\omega}_d\right) + L\left(R\omega_e^\times R\omega_d\dot{\omega}_d\right)\right)\tilde{\psi} \\
&\quad + \frac{1}{2}k_1 s_e^T L\left(F_e\omega_e\right)\tilde{\psi} - \dot{\hat{\psi}}^T\tilde{\psi} \\
&= -k_s s_e^T sig^{r_s}(s_e) - \gamma s_e^T sign(s_e) \\
&= -k_s\|s_e\|^{2-r_s} - \gamma\|s_e\| \le -k_s\|s_e\|^{2-r_s} - \gamma\varepsilon \le -\varepsilon\left(\|s_e\| + \|\tilde{\psi}\|\right) \\
&\le -\varepsilon\left(\frac{2V_s}{\lambda_M(J)}\right)^{\frac{1}{2}}
\end{aligned} \tag{44}$$

Hence systems (10) and (11) would converge to the field of $\|s_e\| \le \varepsilon$ within finite time where ε is a small positive scalar.

In the large angle attitude maneuver issue, it could be assumed that the error angular velocity vector and the vector part of the error quaternion have been reversed when the system reaches the second stage of sliding mode (38). Calculate the derivative of V_s when $\|q_{ev}\| \le \alpha$ it could be obtained that

$$
\begin{aligned}
\dot{V}_s &= s_e^T J\dot{s}_e + \tilde{\psi}^T \dot{\tilde{\psi}} \\
&= s_e^T \left(-k_s sig^{r_s}(s_e) + \hat{J}R\dot{\omega}_d - \hat{J}\omega_e^{\times} R\omega_d + \omega^{\times}\hat{J}\omega + \frac{1}{2}k_2^2 q_{e0}(1-r)\hat{J}sig^{2r}(q_{ev}) \right) \\
&\quad -(\bar{d}+\gamma)s_e^T sign(s_e) + s_e^T d - s_e^T \omega^{\times} J\omega + s_e^T JR\dot{\omega}_d - s_e^T J\omega_e^{\times} R\omega_d \\
&\quad -\frac{1}{2}k_2^2 q_{e0}(1-r)s_e^T J sig^{2r}(q_{ev}) - \tilde{\psi}^T \dot{\hat{\psi}} \\
&\le -k_s s_e^T sig^{r_s}(s_e) - \gamma s_e^T sign(s_e) - s_e^T\left(\omega^{\times}\tilde{J}\omega - \tilde{J}R\dot{\omega}_d + \tilde{J}\omega_e^{\times}R\omega_d\right) \\
&\quad -\frac{1}{2}k_2^2 q_{e0}(1-r)s_e^T \tilde{J} sig^{2r}(q_{ev}) - \dot{\hat{\psi}}^T \tilde{\psi} \\
&= -k_s s_e^T sig^{r_s}(s_e) - \gamma s_e^T sign(s_e) \\
&= -k_s \|s_e\|^{2-r_s} - \gamma\|s_e\| \le -k_s\|s_e\|^{2-r_s} - \gamma\varepsilon \le -\varepsilon\left(\|s_e\| + \|\tilde{\psi}\|\right) \\
&\le -\varepsilon\left(\frac{2V_s}{\lambda_M(J)}\right)^{\frac{1}{2}}
\end{aligned}
\tag{45}
$$

According to Equations (44) and (45), it could be considered that the system has reached the sliding mode neighborhood in a finite time when the convergence curve of the system is smaller than an appropriate scalar. Therefore, the systems (10), (11) controlled by the controller (40) are stable for a finite time.

SIMULATION

Attitude Stabilization Issue

The superiority of the adaptive finite-time controller (26) proposed in this paper is demonstrated by comparison with a standard sliding-mode controller (46) without inertial matrix uncertainty.

$$\boldsymbol{u} = -k\boldsymbol{s} + \boldsymbol{\omega}^{\times}\boldsymbol{J}\boldsymbol{\omega} - \frac{k_1}{2}\left(q_0\boldsymbol{I}_3 + \boldsymbol{q}_v^{\times}\right)\boldsymbol{\omega} - \bar{d}\boldsymbol{sgn}(\boldsymbol{s})$$
$$\boldsymbol{s} = \boldsymbol{\omega} + k_1\boldsymbol{q}_v \tag{46}$$

Set the simulation parameters as follows

$$\boldsymbol{J} = \boldsymbol{diag}(25, 30, 50)\,\mathrm{kg}\cdot\mathrm{m}^2, k_1 = 0.05, k = 10$$
$$\boldsymbol{\omega}(0) = \begin{bmatrix}-0.03 & -0.04 & 0.05\end{bmatrix}^T \mathrm{rad/s}, \boldsymbol{q}(0) = \begin{bmatrix}0 & \sqrt{6}/6 & \sqrt{3}/3 & \sqrt{2}/2\end{bmatrix}^T \tag{47}$$

Where $\boldsymbol{diag}(\boldsymbol{r})$ is the diagonal matrix transformation of the vector $\boldsymbol{r}$.

Assume the disturbance torque consists of Gauss white noise and sinusoidal signal written as follows

$$d_i = 5\times10^{-4}\,\mathrm{rand}(-1,1) + 5\times10^{-4}\sin t \tag{48}$$

Hence the norm upper bound of disturbance torque satisfies

$$\bar{d} = 10^{-3} \tag{49}$$

The simulation results of the standard sliding mode controller (49) are given as follows.

Based on Figs (**1-4**), it could be found that the system converges to the equilibrium point more than 200s, and the steady accuracy at 200s is about 2×10^{-4}rad/s of angular velocity and $1\,\mathrm{deg}$ of Euler angle. It could be found from the simulation parameters that the total rotate angle is $180\,\mathrm{deg}$ and the maneuver time is longer than 200s, hence the average angular velocity is about $0.9\,\mathrm{deg/s}$. It could be found in Fig. (**1**) that the low convergence rate is caused by the drop in angular velocity.

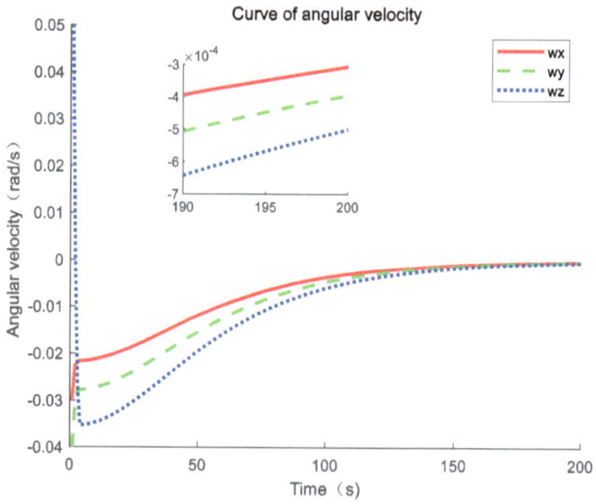

Fig. (1). Curve of angular velocity.

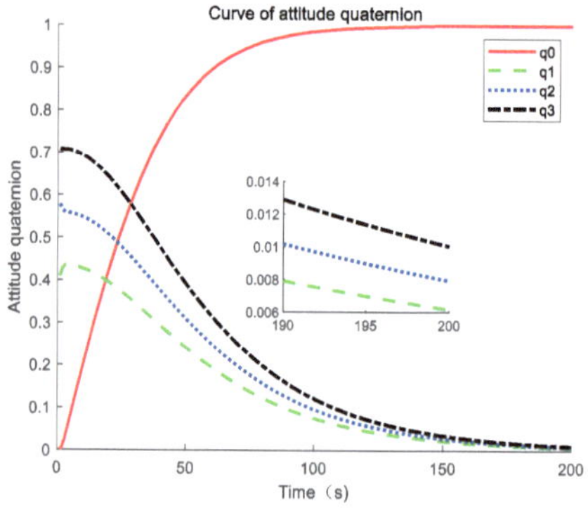

Fig. (2). Curve of attitude quaternion.

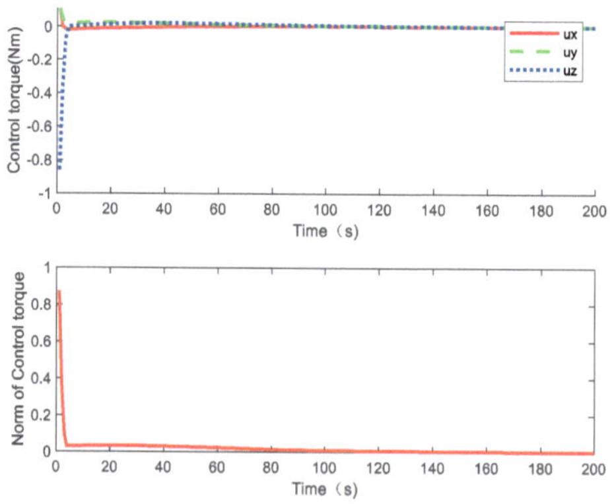

Fig. (3). Curve of control torque and its norm.

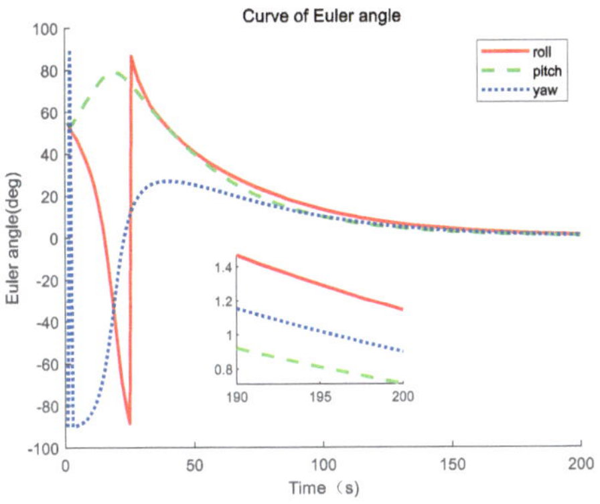

Fig. (4). Curve of Euler angle.

Next the simulation results of the adaptive finite-time controller (26) are given. Set system parameters as follows.

$$\begin{gathered}\hat{J} = \mathrm{diag}(24,31,50)\,\mathrm{kg{\bullet}m^2} \\ \hat{\psi}(0) = [1.5 \quad 1.5 \quad 1.5 \quad 0.5 \quad 0.5 \quad 0.5]^T, \gamma(0) = 3\end{gathered} \tag{50}$$

Generally,larger k_s, k_1 and r_s, r brings a faster convergence rate since the norm of angular velocity on the sliding mode is larger. Moreover, in order to shorten the exponential convergence process, a larger α should be selected. However, if α is selected too far away from the equilibrium point, the reverse of angular velocity and attitude quaternion may not be ensured and the singularity issue may occur. In this paper, α is selected to be 0.2, which means the sliding mode has about 20 deg margin from the equilibrium point, and this is inspired by the work in [1] and [2]. Moreover, considering that the maximum angular velocity of a satellite is about 5der per second, *i.e.* 0.1rad/s, k_1 could be selected as 0.1 to utilize system capability. Considering that the convergence rate of inertia estimation system is not the main focus of this paper, r could be selected as relatively small to reduce the norm of sign function terms in the controller. Above all, the control parameters are selected as follows.

$$k_s = 5, r_s = 1/3, k_1 = 0.1, \alpha = 0.2, r = 1/3, \varepsilon = 0.002, k_\gamma = 2 \tag{51}$$

Assume when $\|\boldsymbol{q}_v\| \leq 5\times10^{-3} = \varepsilon_0$ is satisfied, the system could be treated as converge to its equilibrium point. Based on the discussion in the previous section, system convergence time T_0 could be obtained as follows

$$T_s \leq 2\frac{V_0^{\frac{1}{2}}}{\varepsilon\left(\lambda_M\left(\boldsymbol{J}\right)/2\right)^{1/2}} \approx 14s, T_q \leq 2\ln\frac{1}{\alpha k_2} + 4\frac{1}{\varepsilon_0^{r/2}k_2 r} \approx 114s$$
$$T_0 \leq T_s + T_q = 128s \tag{52}$$

The simulation results of the controller (26) are shown as follows.

It could be found from Figs (**5-8**) that the convergence time is half of the standard sliding mode convergence time, which is about 100s. The convergence time is proved according to equation (52), and the finite time stability of the system is also confirmed. Considering that the total rotate angle is 180 deg, and the average angular velocity is about 1.8 deg/ s, the improvement in convergence time indicates that the key to improving the system convergence speed is to design the angular velocity curve correctly. Furthermore, comparing Figs. (**3** and **7**), it can be found that the control torque norm of the two controllers in the stable phase (close to the equilibrium point) is approximately the same, so by optimizing the efficiency of the control torque, the system performance can be improved, and can be more Make good use of the actuator function. Fig. (**7**) can illustrate that by choosing the correct control parameters, the influence of the sign function term in the controller (26) on the smooth curve of the control torque can become very small. From Figs. (**5-7**), it can also be found that the system is maneuvering along the expected sliding mode, which suggests that the adaptive law (27) can suppress this disturbance. According to Fig. (**7**), the control torque has an upper limit, so in this case no singularity problem occurs. Furthermore, as the system converges, the sign function term tends to zero, indicating that the system can converge smoothly to its equilibrium point. Generally, the controller (26) proposed in this paper could largely improve the performance of the standard sliding mode controller and the system state could converge to its equilibrium point within finite-time, meanwhile it is robust to disturbance torque and inertia matrix uncertainty.

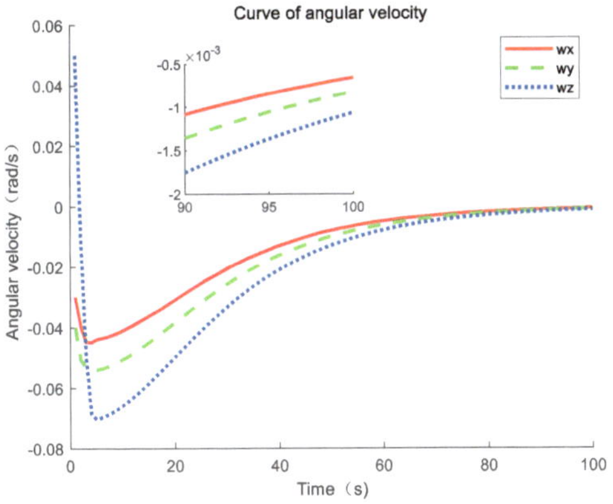

Fig. (5). Curve of angular velocity.

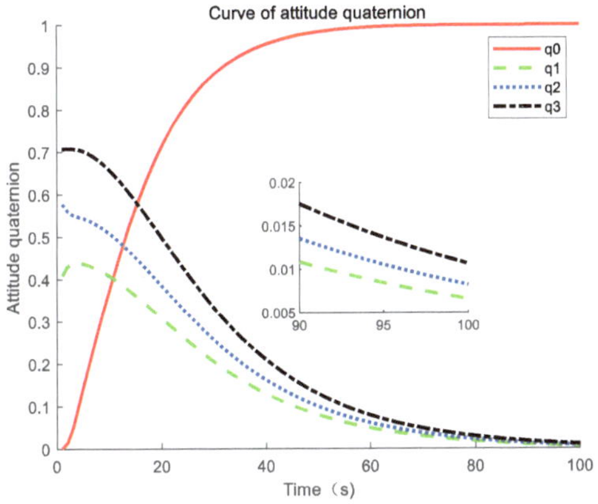

Fig. (6). Curve of attitude quaternion.

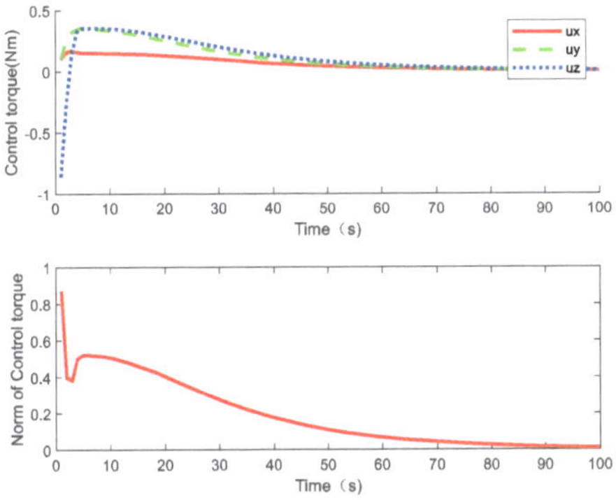

Fig. (7). Curve of control torque and its norm.

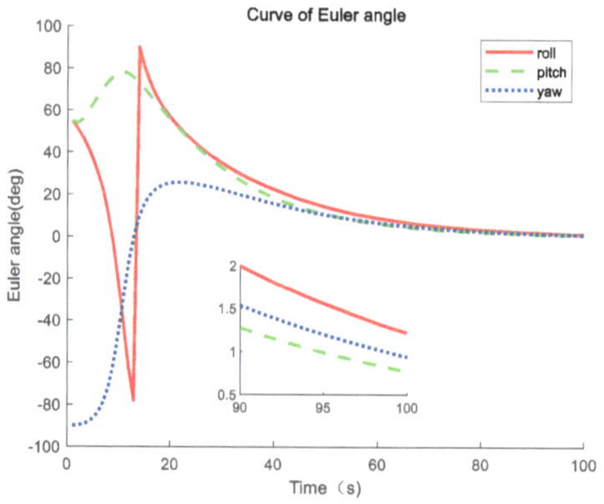

Fig. (8). Curve of Euler angle.

As mentioned in the previous section, sliding mode with such structure of (13) has inherent robustness. In order to demonstrate this robustness, set the norm upper bound of disturbance torque as follows

$$\boldsymbol{d} = 1\times10^{-2} \times \mathrm{rand}(3\times1)\,\mathrm{Nm} \tag{53}$$

Where $\boldsymbol{rand}(3\times1)$ is a 3-dimension vector and its components are all Gauss white noise with the variance equals to 1. Ignore the disturbance torque term and set the controller as follows

$$\boldsymbol{u} = \begin{cases} -k_s \boldsymbol{sig}^{r_s}(\boldsymbol{s}) + \boldsymbol{\omega}^{\times}\hat{\boldsymbol{J}}\boldsymbol{\omega} - \dfrac{1}{2}k_1\hat{\boldsymbol{J}}\boldsymbol{F}\boldsymbol{\omega} - \gamma \boldsymbol{sign}(\boldsymbol{s}), \|\boldsymbol{q}_v\| > \alpha \\ -k_s \boldsymbol{sig}^{r_s}(\boldsymbol{s}) + \boldsymbol{\omega}^{\times}\hat{\boldsymbol{J}}\boldsymbol{\omega} + \dfrac{1}{2}k_2^2 q_0 (1-r)\hat{\boldsymbol{J}}\boldsymbol{sig}^{2r}(\boldsymbol{q}_v) - \gamma \boldsymbol{sign}(\boldsymbol{s}), \|\boldsymbol{q}_v\| \le \alpha \end{cases} \tag{54}$$

Moreover, set the actual control output as follows

$$\boldsymbol{u}' = \mathrm{diag}(0.9, 0.8, 0.7)\boldsymbol{u} \tag{55}$$

Based on (53)-(55), it could be found that the term of suppressing disturbance torque in formula (54) disappears, and the disturbance torque is much larger than the comparison group (46). Considering the output error of the actuator, this analog configuration has a strong disturbance to the control system. Under this condition, the simulation results are as follows:

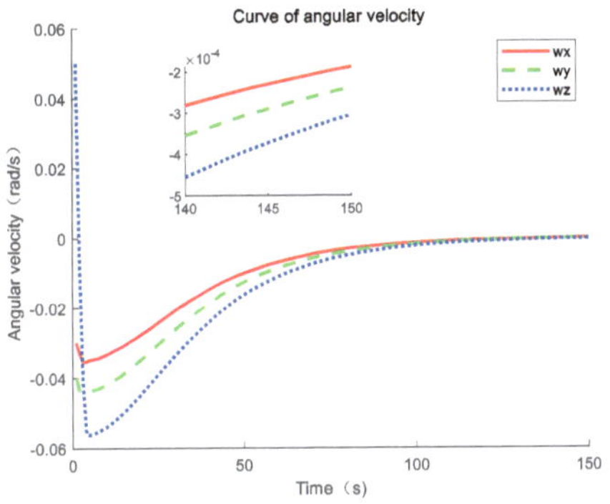

Fig. (9). Curve of angular velocity.

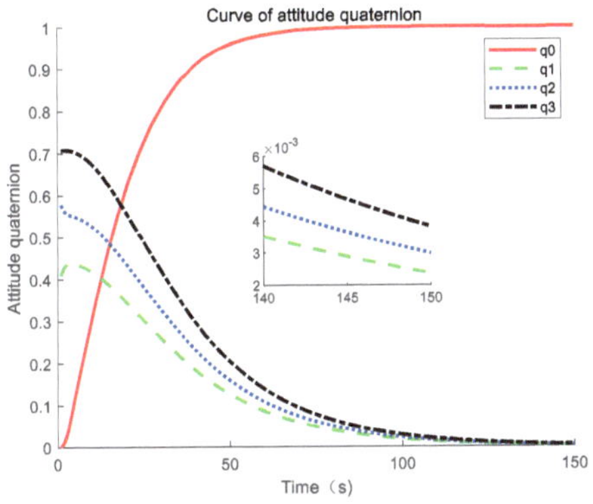

Fig. (10). Curve of attitude quaternion.

It could be found from Figs. (**9** and **10**) that the system still converges smoothly to its equilibrium point, and the convergence speed and steady-state accuracy of the system are also kept at a good level. Furthermore, this set of simulations demonstrates the inherent robustness of standard sliding mode and its corrections, considering the large disturbance torque and output error. The finite-time sliding mode based on equation (12) could maintain its own robustness, so the controller proposed in this paper could be regarded as a controller to improve the robustness of the system.

Attitude Tracking Issue

Similar as the previous section, in order to demonstrate the superiority of adaptive finite-time controller (40) presented in this paper, standard sliding mode controller (53) without inertia matrix uncertainty constructed as follows is compared.

$$\boldsymbol{u} = -k\boldsymbol{s} + \boldsymbol{J}\boldsymbol{R}\dot{\boldsymbol{\omega}}_d - \boldsymbol{J}\boldsymbol{\omega}_e^{\times}\boldsymbol{R}\boldsymbol{\omega}_d + \left(\boldsymbol{\omega}_e + \boldsymbol{R}\boldsymbol{\omega}_d\right)^{\times}\boldsymbol{J}\left(\boldsymbol{\omega}_e + \boldsymbol{R}\boldsymbol{\omega}_d\right) - \frac{1}{2}k_1\boldsymbol{J}\left(q_{e0}\boldsymbol{I}_3 + \boldsymbol{q}_{ev}^{\times}\right)\boldsymbol{\omega}_e \tag{56}$$

$$\boldsymbol{s} = \boldsymbol{\omega}_e + k_1\boldsymbol{q}_{ev}$$

Set system parameters as follows.

$$\begin{aligned} &\boldsymbol{J} = \mathrm{diag}\left(25,30,50\right)\mathrm{kg\cdot m^2}, \bar{d} = 1e-4 \\ &\boldsymbol{\omega}(0) = \begin{bmatrix}0 & 0 & -0.1\end{bmatrix}^T \mathrm{rad/s}, \boldsymbol{q}(0) = \begin{bmatrix}1/2 & -1/2 & \sqrt{3}/3 & \sqrt{6}/6\end{bmatrix}^T \\ &\boldsymbol{\omega}_d = \begin{bmatrix}0.01 & -0.005 & 0\end{bmatrix}^T \mathrm{rad/s}, \dot{\boldsymbol{\omega}}_d = \boldsymbol{0}_{3\times1}, \boldsymbol{q}_d = \begin{bmatrix}1 & 0 & 0 & 0\end{bmatrix}^T \end{aligned} \tag{57}$$

The simulation results of the standard sliding mode controller are shown as follows.

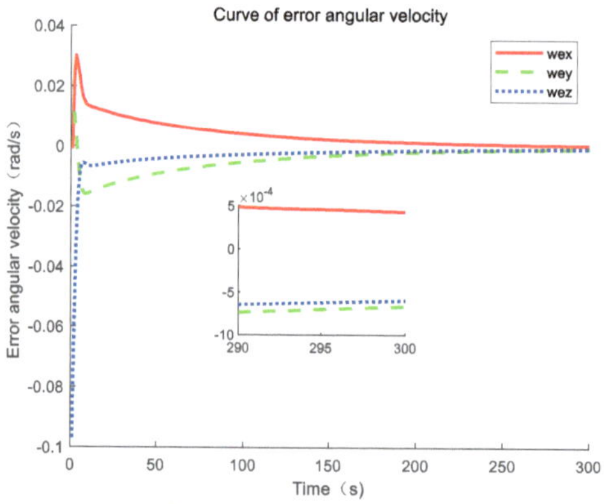

Fig. (11) Curve of error angular velocity.

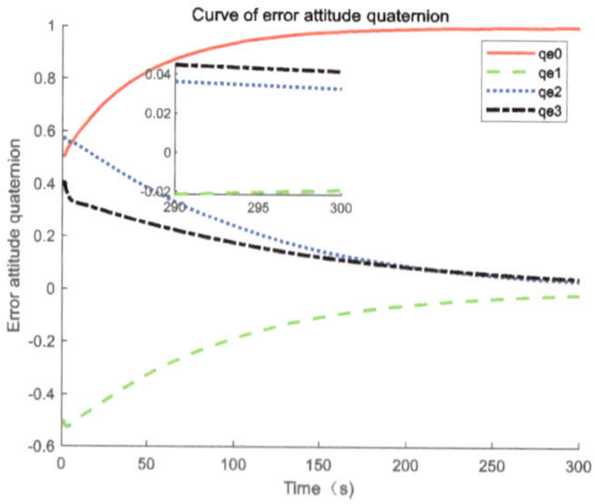

Fig. (12). Curve of error quaternion.

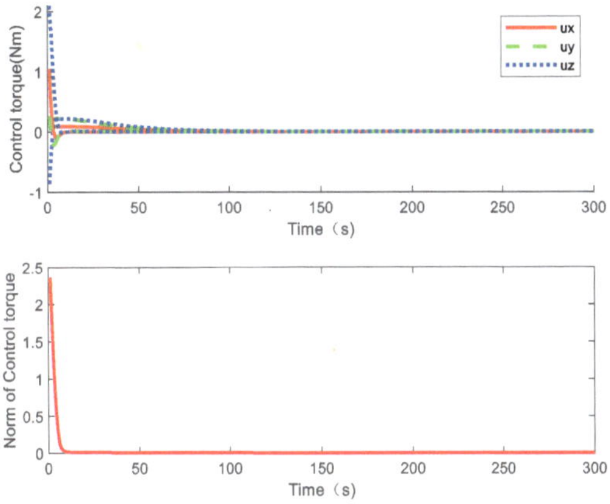

Fig (13). Curve of control torque and its norm.

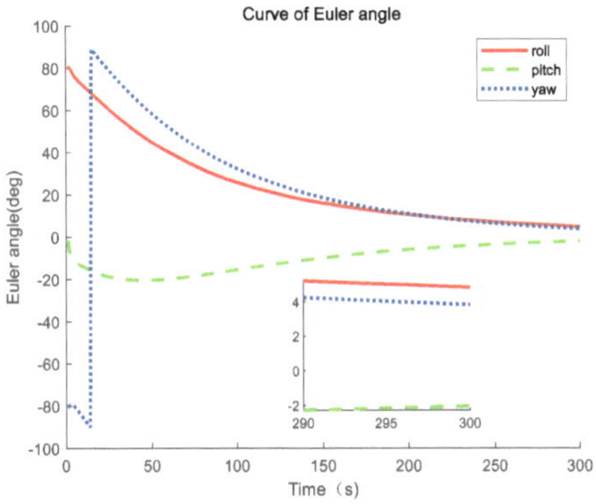

Fig.(14). Curve of error Euler angle.

Based on Figs. (**11-14**), it could be found that the system converges to the equilibrium point more than 300s, and the steady accuracy at 300s is about 5×10^{-2} rad/s of error angular velocity and $5\,\text{deg}$ of error Euler angle. According to the simulation parameters, it could be found that the total rotation angle is $120\,\text{deg}$, and the maneuvering time is greater than 300 s, so the average angular velocity is about $0.6\,\text{deg/s}$. The low convergence velocity is caused by the drop in angular velocity, which could be found in Fig. (**11**).

Next the simulation results of the adaptive finite-time controller (40) are given. Similar as the previous section, set system parameters as follows

$$\hat{J} = \mathrm{diag}(24,31,50)\,\mathrm{kg{\bullet}m^2}$$
$$\hat{\psi}(0) = \begin{bmatrix}1.5 & 1.5 & 1.5 & 0.5 & 0.5 & 0.5\end{bmatrix}^T, \gamma(0) = 3 \tag{58}$$
$$k_s = 5, r_s = 1/3, k_1 = 0.1, \alpha = 0.2, r = 1/3, \varepsilon = 0.002$$

Similar as the previous section, assume when $\|\boldsymbol{q}_{ev}\| \le 5\times10^{-3} = \varepsilon_0$ is satisfied, the system could be treated as converge to its equilibrium point, hence system convergence time could be got as follows

$$T_s \le 2\frac{V_0^{\frac{1}{2}}}{\varepsilon\left(\lambda_M(\boldsymbol{J})/2\right)^{1/2}} \approx 9s, T_q \le 2\ln\frac{1}{\alpha k_2} + 4\frac{1}{\varepsilon_0^{r/2}k_2 r} \approx 92s \tag{59}$$
$$T_0 \le T_s + T_q = 101s$$

The simulation results of the controller (40) are shown as follows.

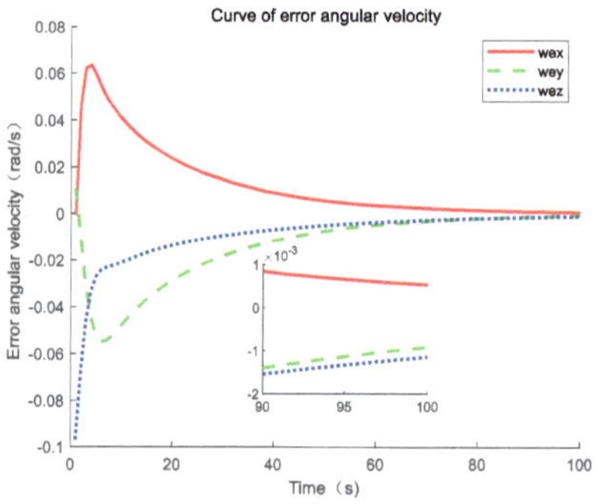

Fig. (15). Curve of error angular velocity.

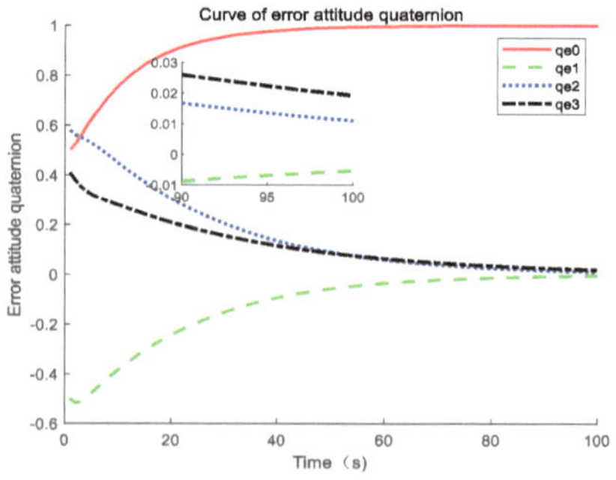

Fig. (16). Curve of error quaternion.

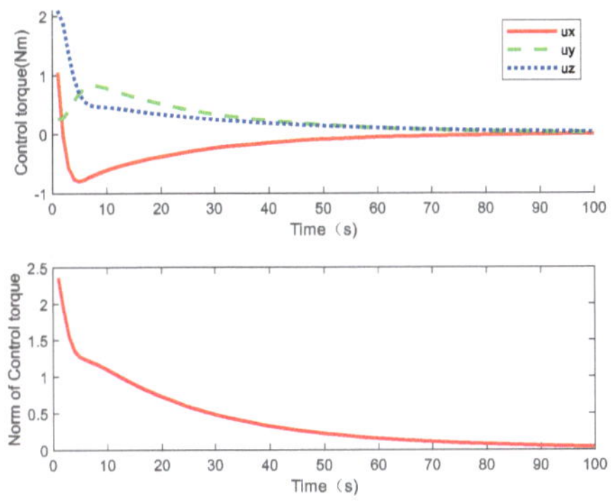

Fig. (17). Curve of control torque and its norm.

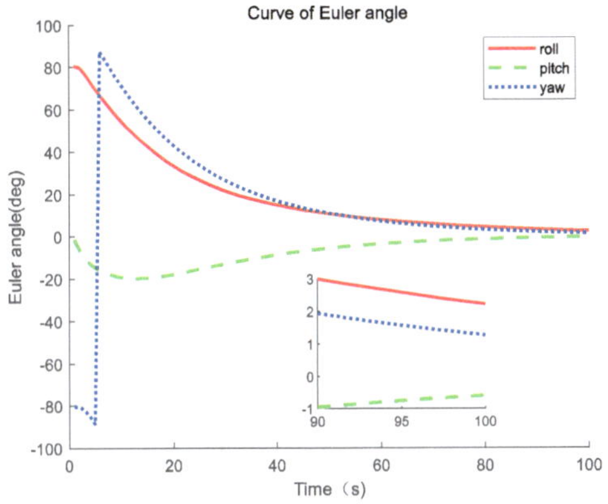

Fig. (18). Curve of error Euler angle.

Based on Figs. (**15-18**) it could be found that the convergence time is about 80s, which is a quarter of the converge time of standard sliding mode. The convergence times of Fig. (**16**) demonstrate the correctness of Equation (59) and the finite-time stability of the system. According to Fig. (**18**), the control torque is the upper norm, in which case the singularity problem does not arise. Based on Figs. (**15 and 18**), the accuracy at steady stage is about 2×10^{-4}rad/s of error angular velocity and 0.02deg of error Euler angle, which is largely improved comparing with standard sliding mode. The improvements in convergence time and steady-state accuracy demonstrate the effectiveness of finite-time sliding modes. Furthermore, the simulation results also demonstrate that the adaptive law could suppress the disturbance caused by the uncertainty of the inertial matrix without causing high-frequency vibrations.

Similar to the previous section, set disturbance torque and control output error as follows

$$\begin{aligned} \boldsymbol{d} &= 5\times10^{-3}\times\text{rand}(3\times1)\,\text{Nm} \\ \boldsymbol{u}' &= \text{diag}(0.9,0.8,0.7)\,\boldsymbol{u} \end{aligned} \tag{60}$$

The simulation results under this condition are shown as follows.

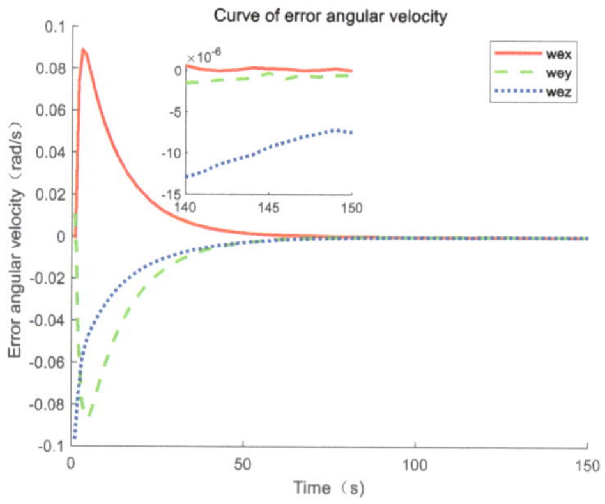

Fig. (19). Curve of error angular velocity.

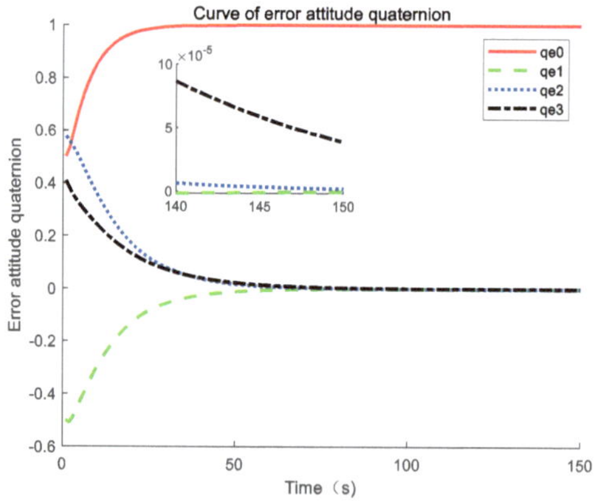

Fig. (20). Curve of error quaternion.

Based on Figs. (**19** and **20**), it could be found that the finite-time sliding mode proposed in this paper not only achieves the purpose of finite-time convergence, but also maintains the robustness of classical sliding-mode control, which could show that it is robust to large disturbance torque and control torque output error.

CONCLUSION

Based on standard sliding mode and adaptive control, an adaptive finite-time controller is proposed for satellite attitude fast maneuver, which greatly improves the convergence speed of standard sliding mode. It maintains the inherent robustness of the system by maintaining the physical meaning and main structure of the standard sliding mode, and designs a finite-time controller, so the controller designed in this paper not only has finite-time stability, but also has strong robustness.

Furthermore, in order to solve more control problems, the fractional power of the attitude quaternion is used to design a finite-time convergence stage, and the singularity problem is solved by using the angular velocity and the property that the orientation of the attitude quaternion does not change when the attitude quaternion is reversed. By properly designing the angular velocity, the convergence speed could be greatly improved and the singularity problem could be avoided. In order to deal with the system inertia matrix uncertainty issue, an adaptive law for inertia matrix estimation is proposed. By rationally designing auxiliary states, the difficulty that there is no direct feedback for error states could be overcome. The finite-time stability is proved considering both sliding mode state and inertia matrix estimation state. The disturbance caused by system uncertainty is largely suppressed by the adaptive law, hence the controller is robustness to model uncertainty. Moreover, the controller based on the adaptive law is continuous, hence the high-frequency vibration is avoided.

The above shows that the advantages of mature methods and newer methods could be effectively combined with appropriate modifications to the standard sliding mode.

CONSENT FOR PUBLICATION

Not applicable.

CONFLICT OF INTEREST

The authors declare no conflict of interest, financial or otherwise.

ACKNOWLEDGEMENT

Declared none.

REFERENCES

[1] Li Y. Ye D. Sun Z.W. Robust finite time control algorithm for satellite attitude control. Aerospace Science and Technology, 2017, 68, 46-57.

[2] Li Y. Ye D. Time efficient sliding mode controller based on Bang-Bang logic for satellite attitude control. Aerospace Science and Technology, 2018, 75, 342-352.

[3] Liu C, Yue X, Shi K, *et al.* Spacecraft Attitude Control: A Linear Matrix Inequality Approach. Elsevier/ Science Press, 2022, ISBN: 978-0-323-99005-9.

[4] Li Y. Sun Z.W. Ye D. Time Efficient Robust PID Plus Controller for Satellite Attitude Stabilization Control Considering Angular Velocity and Control Torque Constraint. Journal of Aerospace Engineering, 2017, 30(5).

[5] Verbin D. Lappas V.J. Rapid rotational maneuvering of rigid satellites with hybrid actuators configuration. Journal of Guidance, Control and Dynamics, 2013, 36(2), 532-547.

[6] Verbin D. Lappas V.J. Joseph Z.B. Time efficient angular steering laws for rigid satellites. Journal of Guidance， Control and Dynamics, 2011, 34(3), 878-892.

[7] Cao X.B. Yue C.F. Liu M. Flexible satellite attitude maneuver via constrained torque distribution and active vibration suppression. Aerospace Science and Technology, 2017, 67, 387-397.

[8] Ye D. Zhang X. Wang X.C. Sun Z.W. Finite time control strategy for satellite attitude maneuver based on hybrid actuator. Transactions of the Institute of Measurement and Control, 2018, 40(9), 2798-2806.

[9] Liu C, Ye D, Shi K, *et al.* Robust high-precision attitude control for flexible spacecraft with improved mixed H_2/H_∞ control strategy under poles assignment constraint. Acta Astronautica, 2017, 136: 166-175.

[10] Wu S.N. Radice G. Gao Y.S. Sun Z.W. Quaternion-based finite time control for spacecraft attitude tracking. Acta Astronautica, 2011, 69(1-2), 48-58.

[11] Wu S.N. Radice G. Gao Y.S. Sun Z.W. Robust finite-time control for flexible spacecraft attitude maneuver. Journal of Aerospace Engineering, 2014, 27(1), 185-190.

[12] Liang H.Z. Sun Z.W. Wang J.Y. Finite-time attitude synchronization controllers design for spacecraft formations via behavior-based approach. Part G，Journal of Aerospace Engineering, 2013, 227(11), 1737-1753.

[13] Wang J.Y. Liang H.Z. Sun Z.W. Dual-quaternion-based finite-time control for spacecraft tracking in six degrees of freedom. Part G， Journal of Aerospace Engineering, 2013, 227(3), 528-545.

[14] Wang J.Y. Liang H.Z. Sun Z.W. Zhang S.J. Liu M. Finite-time control for spacecraft formation with dual-number-based description. Journal of Guidance, Control and Dynamics, 2012, 35(3), 950-962.

[15] Xiao B. Yin S. Wu L.G. A structure simple controller for satellite attitude tracking maneuver. IEEE Transactions on Industrial Electronics, 2017, 64(2), 1436-1446.

[16] Xiao B. Yin S. Gao H.J. Tracking control of robotic manipulators with uncertain kinematics and dynamics. IEEE Transactions on Industrial Electronics, 2016, 63(10), 6439-6449.

[17] Xiao B. Yin S. Velocity-free Fault tolerant and uncertainty attenuation control for a class of nonlinear systems. IEEE Transactions on Industrial Electronics, 2016, 63(7), 4400-4411.

[18] Shi K, Liu C, Biggs J D, *et al.* Observer-based control for spacecraft electromagnetic docking. Aerospace Science and Technology, 2020, 99: 105759.

[19] Hu Q.L. Sliding mode maneuvering control and active vibration damping of three-axis stabilized flexible spacecraft with actuator dynamics. Nonlinear Dynamics, 2008, 52(3), 227-248.

[20] Liu C, Shi K, Yue X, *et al.* Inertia-free saturated output feedback attitude stabilization for uncertain spacecraft. International Journal of Robust and Nonlinear Control, 2020, 30(13): 5101-5121.

[21] Liu C, Sun Z, Shi K, *et al.* Robust dynamic output feedback control for attitude stabilization of spacecraft with nonlinear perturbations. Aerospace Science and Technology, 2017, 64: 102-121.

[22] Shi K, Liu C, Sun Z, *et al.* Disturbance observer-based attitude stabilization for rigid spacecraft with input MRCs. Advances in Space Research, 2020, 66(3): 689-701.

[23] Shi K, Liu C, Sun Z, *et al.* Coupled orbit-attitude dynamics and trajectory tracking control for spacecraft electromagnetic docking. Applied Mathematical Modelling, 2022, 101: 553-572.

[24] Koksal N. Jalalmaab M. Fidan B. Adaptive Linear Quadratic Attitude Tracking Control of a Quadrotor UAV Based on IMU Sensor Data Fusion. Sensors, 2018, 19(1).

[25] Liu C, Yue X, Zhang J, *et al.* Active Disturbance Rejection Control for Delayed Electromagnetic Docking of Spacecraft in Elliptical Orbits. IEEE Transactions on Aerospace and Electronic Systems, 2021.

[26] Sun L. Zhao G.W. Huang H. Zhang N. Attitude Tracking Control of Gravity Gradient Microsatellite in Maneuvering for Space Exploration. Microgravity Science and Technology, 2018, 30(6), 1011-1020.

[27] Jiang K, Zhang H, Karimi H R, *et al.* Simultaneous input and state estimation for integrated motor-transmission systems in a controller area network environment via an adaptive unscented Kalman filter. IEEE Transactions on Systems, Man, and Cybernetics: Systems, 2018, 50(4): 1570-1579.

[28] Jin X, Du W, He W, *et al.* Twisting-based finite-time consensus for Euler–Lagrange systems with an event-triggered strategy. IEEE Transactions on Network Science and Engineering, 2019, 7(3): 1007-1018.

[29] Guo Q, Wang Q, Li X. Finite-time convergent control of electrohydraulic velocity servo system under uncertain parameter and external load. IEEE Transactions on Industrial Electronics, 2018, 66(6): 4513-4523.

[30] Sun Z Y, Dong Y Y, Chen C C. Global fast finite-time partial state feedback stabilization of high-order nonlinear systems with dynamic uncertainties. Information Sciences, 2019, 484: 219-236.

[31] Yang X, Yu J, Wang Q G, *et al.* Adaptive fuzzy finite-time command filtered tracking control for permanent magnet synchronous motors. Neurocomputing, 2019, 337: 110-119.

[32] Luan F, Na J, Huang Y, *et al.* Adaptive neural network control for robotic manipulators with guaranteed finite-time convergence. Neurocomputing, 2019, 337: 153-164.

[33] Dong Y, You L, Bing X, *et al.* Robust finite-time adaptive control algorithm for satellite fast attitude maneuver. Journal of the Franklin Institute, 2020, 357(16): 11558-11583.

[34] Wang Y, Karimi H R, Yan H. An adaptive event-triggered synchronization approach for chaotic lur'e systems subject to aperiodic sampled data. 2019.

[35] Wang Y, Xia Y, Ahn C K, *et al.* Exponential stabilization of Takagi–Sugeno fuzzy systems with aperiodic sampling: An aperiodic adaptive event-triggered method. IEEE Transactions on Systems, Man, and Cybernetics: Systems, 2018, 49(2): 444-454.

CHAPTER 5

Attitude Stabilization of Flexible Spacecraft Using Output Feedback Controller

Ziyu Yang[1,*], **Xiaokui Yue**[1] and **Chuang Liu**[1]

[1]*Northwestern Polytechnical University, Xi'an 710072, China*

Abstract: Spacecraft in space may have some certain non-cooperative characteristics due to the service life limit, fuel exhaustion, component fault, structural fatigue damage, or after performing certain space tasks such as capturing non-cooperative targets. In modeling, these non-cooperative characteristics are often manifested in uncertain and unknown inertia, model parameters uncertainty, actuator faults, etc. In this paper, aiming at the attitude stability control problem of such flexible spacecraft, the attitude dynamics modeling is completed by introducing the nominal inertia to construct the comprehensive disturbance term including external disturbance, inertia uncertainty and actuator failure. Then, a static output feedback (SOF) controller is applied to model the closed-loop attitude control system a stable negative imaginary (NI) system with H_∞ performance constraints according to NI theory. As long as the optimization variables approach zero, the LMI-based iterative algorithm can find such the static output feedback controller to stabilize the flexible spacecraft. It is worth mentioning that an event-trigger mechanism is introduced into the control scheme to reduce communication pressure. Finally, the numerical simulation is carried out in the presence of controller gain perturbations and model parameter uncertainty. The results of the simulation demonstrate the effectiveness, robustness and non-fragility of the control method.

Keywords: Flexible Spacecraft, H_∞ performance, Inertia-free attitude stabilization, Lumped disturbance , Negative imaginary, Static output feedback.

[*]**Corresponding author Ziyu Yang:** Northwestern Polytechnical University, Xi'an 710072, China; Tel: +86 17795832439; E-mails: ziyuyang@mail.nwpu.edu.cn; xkyue@nwpu.edu.cn and cliu@nwpu.edu.cn

Chuang Liu, Honghua Dai, Xiaokui Yue & Yiqing Ma (Eds.)

INTRODUCTION

With the rapid development of human space activities, there are a large number of failure spacecrafts because of the service life limit, fuel exhaustion, component fault, structural fatigue damage and accidental disintegration in the space environment. Malfunctioning spacecraft and space debris cannot provide state information as well as their inertia parameter information actively, which makes them typical non-cooperative targets, and their high-precision on-orbit operation is seriously affected by attitude-orbit coupling [1]. For this kind of spacecraft, the determination of high-precision inertial parameters is costly, moreover, flexible spacecraft usually need to complete pointing accurately, attitude maneuvers rapidly, and stabilization in the case of existing external interference, actuator failure, measurement errors, and input limits which all makes the attitude stabilization more difficult [2]. Therefore, effective, robust methods of control are needed to deal with these problems at the same time.

In recent years, research on attitude control of flexible spacecraft have made a lot of phased progress. For example, as stated in references [3], a fault-tolerant adaptive control scheme developed for attitude tracking of flexible spacecraft with unknown inertia parameters, external disturbance, and actuator faults shows good robustness. The method considering external disturbances and parameter perturbations as uncertainties in Liu's work provides a new idea for dealing with parameter uncertainty [4]. Then, reference [5] proposed a novel finite time control method for spacecraft attitude control and dealt with inertia uncertainty and input constraints. Recently, reference [6] proposed a new fault-tolerant attitude tracking control with specified performance, in which the inertia uncertainty and input constraints were considered. Besides, in order to have better robustness to uncertain dynamics and external disturbances, some other robust sliding mode controllers for satellite attitude stabilization or synchronization are proposed [7, 8].

In addition to the above unfavorable factors, the elastic vibration of spacecraft's flexible appendages may significantly reduce the control performance and even lead to instability of the closed-loop attitude control system, which makes the problem of elastic vibration suppression becomes a research hotspot [9]. Negative imaginary (NI) systems theory is widely used in robust vibration control of flexible structures, such as the control of large space structures and flexible dynamics of aircraft [10]. Especially for the systems with flexible and lightly damped structures using the piezoelectric actuators and sensors for flexible structural control, the system can be modeled as a NI system by choosing system input and output appropriately [11]. However, there is little application example of negative virtual theory in flexible spacecraft in the relevant literature.

Last but not least, it is worth noting that the attitude stability of spacecraft could be handled properly when the full attitude state can be measured. However, in practical applications, due to sensor failure and cost reduction of airborne sensors, state measurement may not be used in controller design [12].

Based on the above discussion, a new static output feedback H_∞ control scheme is proposed in this paper. This scheme does not need the prior knowledge of inertia information, external disturbance and actuator fault to enable the closed loop attitude control system stable negative imaginary and satisfy H_∞ performance constraints simultaneously. Compared to the previous studies, the important contributions in this paper are stated as follows. First of all, introduce the nominal representation of the valid inertia matrix, and the difference between this inertia matrix and the real inertia matrix is treated as one of the sources of the lumped disturbance, so that the controller design will be inertia-free. Secondly, we propose a novel method to decouple controller gain matrix $\boldsymbol{K}$ and Lyapunov matrix by introducing the slack matrix variables, and establishing sufficient conditions for designing a SOF controller. Thirdly, an iterative algorithm is proposed to calculate the SOF controller. Fourth, an event-trigger mechanism is introduced for control input and reduces the communication pressure to some extent.

This article is organized as follows. In the second section, the dynamics model of flexible spacecraft is given, and then it is transformed into state space form, in which actuator failures and input constraints are considered, and the control objectives of this article are formulated. The third section introduces the design method of SOF H_∞ NI controller and stability analysis for the closed-loop control system. In the fourth section, the numerical simulation shows the superior performance of event-triggered controller under the condition of existing controller gain's perturbations and parameter uncertainty. Finally, the fifth section gives the conclusion of this paper.

DYNAMICS MODELING

In this section, the attitude kinematics model and dynamic model of flexible spacecraft will be given.

After "3-2-1" rotation, the kinematics of a flexible spacecraft can be described by [13]

$$\begin{bmatrix} \omega_x \\ \omega_y \\ \omega_z \end{bmatrix} = \begin{bmatrix} 1 & 0 & -\sin\theta \\ 0 & \cos\phi & \sin\phi\cos\theta \\ 0 & -\sin\phi & \cos\phi\cos\theta \end{bmatrix} \begin{bmatrix} \dot{\phi} \\ \dot{\theta} \\ \dot{\psi} \end{bmatrix} - \omega_0 \begin{bmatrix} \cos\theta\sin\psi \\ \sin\phi\sin\theta\sin\psi + \cos\phi\cos\psi \\ \cos\phi\sin\theta\sin\psi - \sin\phi\cos\psi \end{bmatrix} \quad (1)$$

where ω_x, ω_y and ω_z denote the three components of angular velocity; ϕ, θ and ψ denote the three components of attitude angle; and ω_0 is the orbital angular velocity.

If piezoelectric actuators are bonded to the surface of the flexible appendages to provide input voltage $\boldsymbol{u}_p$ that will generate deformation resulting in control torque, then the spacecraft body and flexible dynamics can be described by the following differential equations [14].

$$\begin{cases} \boldsymbol{J}\dot{\boldsymbol{\omega}} + \boldsymbol{\omega}^{\times}\left(\boldsymbol{J}\boldsymbol{\omega} + \tilde{\boldsymbol{\Delta}}\dot{\tilde{\boldsymbol{\eta}}}\right) + \tilde{\boldsymbol{\Delta}}\ddot{\tilde{\boldsymbol{\eta}}} = \boldsymbol{T}_c + \boldsymbol{T}_d \\ \ddot{\tilde{\boldsymbol{\eta}}} + \tilde{\boldsymbol{C}}\dot{\tilde{\boldsymbol{\eta}}} + \tilde{\boldsymbol{K}}\tilde{\boldsymbol{\eta}} + \tilde{\boldsymbol{\Delta}}^{\mathrm{T}}\dot{\boldsymbol{\omega}} = -\tilde{\boldsymbol{\Delta}}_p\boldsymbol{u}_p \end{cases} \quad (2)$$

where, $\boldsymbol{J} \in \mathbb{R}^{3\times3}$ stands for a moment of inertia; $\boldsymbol{\omega} = \begin{bmatrix} \omega_x & \omega_y & \omega_z \end{bmatrix}^{\mathrm{T}}$ stands for the angular velocity; $\tilde{\boldsymbol{\Delta}} \in \mathbb{R}^{3\times m}$ is the coupling coefficient matrix between the rigid and elastic structures; $\tilde{\boldsymbol{\eta}} \in \mathbb{R}^m$ is the modal coordinate vector relative to the main body; $\boldsymbol{T}_c \in \mathbb{R}^3$ and $\boldsymbol{T}_d \in \mathbb{R}^3$ denote the control input torque and external disturbance torque; $\tilde{\boldsymbol{C}} = \mathrm{diag}\left\{\begin{bmatrix} 2\tilde{\xi}_1\Omega_1 & 2\tilde{\xi}_2\Omega_2 & \cdots & 2\tilde{\xi}_m\Omega_m \end{bmatrix}\right\} \in \mathbb{R}^{m\times m}$ is the modal damping matrix, where $\tilde{\xi}_i, i = 1, 2, \cdots, m$ and $\Omega_i, i = 1, 2, \cdots, m$ are the damping ratios and the natural frequencies; $\tilde{\boldsymbol{K}} = \mathrm{diag}\left\{\begin{bmatrix} \Omega_1^2 & \Omega_2^2 & \cdots & \Omega_m^2 \end{bmatrix}\right\} \in \mathbb{R}^{m\times m}$ is the stiffness matrix, and m is the number of elastic modes considered. Here, $\boldsymbol{T}_d$ includes gravity gradient torque, solar radiation pressure torque and aerodynamic torque; $\boldsymbol{u}_p$ is the piezoelectric input and $\tilde{\boldsymbol{\Delta}}_p$ is the corresponding coupling matrix.

Denoting $\boldsymbol{\Theta} = [\phi \quad \theta \quad \psi]^T$. When spacecraft is doing a small angle maneuver in a very high orbit, one has

$$\boldsymbol{\omega} = \dot{\boldsymbol{\Theta}} \tag{3}$$

Using active vibration suppression strategy, the piezoelectric input can be described by

$$\boldsymbol{u}_p = \boldsymbol{F}_a \tilde{\boldsymbol{\eta}} + \boldsymbol{F}_b \boldsymbol{q} \tag{4}$$

Where $\boldsymbol{F}_a$ and $\boldsymbol{F}_b$ are the feedback coefficients.

Defining an auxiliary variable $\boldsymbol{q} = \dot{\tilde{\boldsymbol{\eta}}} + \tilde{\boldsymbol{\Delta}}^{\mathrm{T}} \boldsymbol{\omega}$, one has

$$\dot{\boldsymbol{q}} = \ddot{\tilde{\boldsymbol{\eta}}} + \tilde{\boldsymbol{\Delta}}^{\mathrm{T}} \dot{\boldsymbol{\omega}} = -\tilde{\boldsymbol{C}} \boldsymbol{q} + \tilde{\boldsymbol{C}} \tilde{\boldsymbol{\Delta}}^{\mathrm{T}} \boldsymbol{\omega} - \tilde{\boldsymbol{K}} \tilde{\boldsymbol{\eta}} - \tilde{\boldsymbol{\Delta}}_p \boldsymbol{u}_p \tag{5}$$

Then, the first Equation of (2) becomes

$$\left(\boldsymbol{J} - \tilde{\boldsymbol{\Delta}} \tilde{\boldsymbol{\Delta}}^{\mathrm{T}}\right) \dot{\boldsymbol{\omega}} + \boldsymbol{\omega}^{\times} \left(\left(\boldsymbol{J} - \tilde{\boldsymbol{\Delta}} \tilde{\boldsymbol{\Delta}}^{\mathrm{T}}\right) \boldsymbol{\omega} + \tilde{\boldsymbol{\Delta}} \boldsymbol{q}\right) + \tilde{\boldsymbol{\Delta}} \left(-\tilde{\boldsymbol{C}} \boldsymbol{q} + \tilde{\boldsymbol{C}} \tilde{\boldsymbol{\Delta}}^{\mathrm{T}} \boldsymbol{\omega} - \tilde{\boldsymbol{K}} \tilde{\boldsymbol{\eta}} - \tilde{\boldsymbol{\Delta}}_p u_p\right) = \boldsymbol{T}_c + \boldsymbol{T}_d \tag{6}$$

Denoting $\boldsymbol{J}_0 = \boldsymbol{J} - \tilde{\boldsymbol{\Delta}} \tilde{\boldsymbol{\Delta}}^{\mathrm{T}}$, with $\tilde{\boldsymbol{\Delta}} \tilde{\boldsymbol{\Delta}}^{\mathrm{T}}$ as the contribution of the flexible parts to the total inertia matrix. Therefore, the attitude dynamics model for flexible spacecraft can be written in the state space form [15].

$$\begin{cases} \dot{\boldsymbol{x}}(t) = (\boldsymbol{A} + \boldsymbol{\Delta A} + \boldsymbol{\Delta A}_p) \boldsymbol{x}(t) + (\boldsymbol{B}_1 + \boldsymbol{\Delta B}_1) \boldsymbol{u}(t) + (\boldsymbol{B}_2 + \boldsymbol{\Delta B}_2) \boldsymbol{w}_0(t) \\ \boldsymbol{z}(t) = \boldsymbol{C}_1 \boldsymbol{x}(t) \\ \boldsymbol{y}(t) = \boldsymbol{C}_2 \boldsymbol{x}(t) \end{cases} \tag{7}$$

where $\boldsymbol{x}(t) = \left[\boldsymbol{\Theta}^{\mathrm{T}} \quad \boldsymbol{\omega}^{\mathrm{T}} \quad \boldsymbol{q}^{\mathrm{T}} \quad \tilde{\boldsymbol{\eta}}^{\mathrm{T}}\right]^{\mathrm{T}}$ stands for the state vector, $\boldsymbol{y}(t)$ is chosen as the measured output vector, $\boldsymbol{z}(t)$ is chosen as the controlled output vector, $\boldsymbol{u}(t) = \boldsymbol{T}_c$ stands for the control input torque, $\boldsymbol{w}_0(t) = \boldsymbol{T}_d - \boldsymbol{\omega}^{\times} (\boldsymbol{J}_0 \boldsymbol{\omega} + \tilde{\boldsymbol{\Delta}} \boldsymbol{q})$ is the lumped disturbance. $\boldsymbol{q} = \dot{\tilde{\boldsymbol{\eta}}} + \tilde{\boldsymbol{\Delta}}^{\mathrm{T}} \boldsymbol{\omega}$ is the defined auxiliary variable. The coefficient matrices which are known are denoted by

$$A=\begin{bmatrix} 0 & I & 0 & 0 \\ 0 & -J_{0n}^{-1}\tilde{\Delta}\tilde{C}\tilde{\Delta}^{\mathrm{T}} & J_{0n}^{-1}\tilde{\Delta}\left(\tilde{C}+\tilde{\Delta}_p F_b\right) & J_{0n}^{-1}\tilde{\Delta}\left(\tilde{K}+\tilde{\Delta}_p F_a\right) \\ 0 & \tilde{C}\tilde{\Delta}^{\mathrm{T}} & -\left(\tilde{C}+\tilde{\Delta}_p F_b\right) & -\left(\tilde{K}+\tilde{\Delta}_p F_a\right) \\ 0 & -\tilde{\Delta}^{\mathrm{T}} & I & 0 \end{bmatrix}, B_1=B_2=\begin{bmatrix} 0 \\ J_{0n}^{-1} \\ 0 \\ 0 \end{bmatrix}$$

$$C_1=\begin{bmatrix} I_{3\times 3} & I_{3\times 3} & \begin{bmatrix} I_{1\times m} \\ 0_{1\times m} \\ 0_{1\times m} \end{bmatrix} & \begin{bmatrix} 0_{1\times m} \\ 0_{1\times m} \\ I_{1\times m} \end{bmatrix} \end{bmatrix}, C_2=9\times\left[I_{6\times 6} \quad 0_{2m}\right].$$

and the unknown coefficient matrices are denoted by

$$\begin{aligned}\Delta A &= A(J)-A(J_n)\\ &=\begin{bmatrix} 0 & 0 & 0 & 0 \\ 0 & (J_{0n}^{-1}-J_0^{-1})\tilde{\Delta}\tilde{C}\tilde{\Delta}^{\mathrm{T}} & (J_0^{-1}-J_{0n}^{-1})\tilde{\Delta}\left(\tilde{C}+\tilde{\Delta}_p F_b\right) & (J_0^{-1}-J_{0n}^{-1})\tilde{\Delta}\left(\tilde{K}+\tilde{\Delta}_p F_a\right) \\ 0 & 0 & 0 & 0 \\ 0 & 0 & 0 & 0 \end{bmatrix}\end{aligned}$$

$$\begin{aligned}\Delta B_1 = \Delta B_2 &= B_1(J)-B_1(J_n)\\ &=\begin{bmatrix} 0 \\ J_0^{-1}-J_{0n}^{-1} \\ 0 \\ 0 \end{bmatrix}\end{aligned}$$

where $J_{0n}=J_n-\tilde{\Delta}\tilde{\Delta}^{\mathrm{T}}$ with $J_0=J-\tilde{\Delta}\tilde{\Delta}^{\mathrm{T}}$, and J_n is the nominal inertia matrix.

and ΔA_p represents the model parameter uncertainty in which inertia uncertainty is not included, and satisfies matching condition, *i.e.*

$$\Delta A_p(t)=M_1F_1(t)N_1, \|F_1(t)\|\le 1, \ \forall t \tag{8}$$

where, $\boldsymbol{M}_1$ and $\boldsymbol{N}_1$ are known constant matrices having appropriate dimensions, $\boldsymbol{F}_1(t)$ is a Lebesgue measurable matrix function.

In addition, the following key issues need to be emphasized.

1) Measurement Errors

The output model in Eq. (7) is based upon the assumption that true attitude information of the spacecraft is known prior, but in any real case, only the sensor-measured states of spacecraft are available for feedback purposes. Hence, the sensor measurement error $\boldsymbol{v}(t) \in L_2\left[0,+\infty\right)$ inevitably exists, which must be taken into account for output feedback controller design. It is worth noting that it represents an aggregation of systematic bias in errors, sensor calibration errors, and some stochastic disturbances present in any real sensor measurement.

2) Actuator Saturation

Due to the physical and safe limits, the actuator constraint, if unaccounted for, will often have a negative impact on the stability and performance of an attitude stabilization system designed for the "ideal" case [16, 17]. In order not to violate the requirements necessary for input constraint, the actual control torque $sat(\boldsymbol{u})$ should be described as

$$\mathrm{sat}(u_{i=1,2,3}(t)) = \mathrm{sign}(u_i(t))\min\left\{\left|u_i(t)\right|, u_{mi}\right\} \tag{9}$$

where u_{mi} is the upper bound of the control input caused by the functional limitation of the actuator.

3) Actuator Faults

Because the improper action of the actuator will seriously affect the attitude control system performance, actuator fault is considered to be one of the important problems to be solved. Considering this problem, let the full column rank $\boldsymbol{E}$ similar to $\boldsymbol{B}_1$ the structure represent the distribution matrix of the fault signal in the input. Specifically, if $\boldsymbol{E} \neq \boldsymbol{B}_1$, it indicates process failure; If $\boldsymbol{E} = \boldsymbol{B}_1$, it indicates actuator failure.

Considering the above factors, the actual attitude control system is represented by

$$\begin{cases} \dot{\boldsymbol{x}}(t) = (\boldsymbol{A} + \Delta\boldsymbol{A}_p)\boldsymbol{x}(t) + \boldsymbol{B}_1\boldsymbol{u}(t) + \boldsymbol{B}_2\boldsymbol{w}(t) \\ z(t) = \boldsymbol{C}_1\boldsymbol{x}(t) \\ y(t) = \boldsymbol{C}_2\boldsymbol{x}(t) \end{cases} \tag{10}$$

where, $\boldsymbol{w}(t) = \boldsymbol{w}_0(t) + \boldsymbol{B}_2^*\left(\Delta\boldsymbol{A}\boldsymbol{x}(t) + \Delta\boldsymbol{B}_1\boldsymbol{u}(t) + \Delta\boldsymbol{B}_2\boldsymbol{w}_0(t) + \boldsymbol{E}\boldsymbol{f}(t)\right)$ represents the lumped disturbance constructed by external disturbances, actuator faults and uncertainties matrix, where $\boldsymbol{B}_2^*$ represents the pseudoinverse of matrix $\boldsymbol{B}_2$.

Remark 1. Inertia uncertainty exists widely in actual physical systems because of perturbations of parameter, errors in modeling, and environmental situation changes. If inertia uncertainty is not considered when designing the attitude control system, the obtained controller is probably to fail in practice. Specially, the prior information of the inertia matrix sometimes is not precise, and it is usually costly to determine the high-precision inertia matrix, therefore, the nominal matrix of the effective inertia $\boldsymbol{J}_n$ is used to design the controller, then the inertia matrix will have the uncertainty caused by the difference between $\boldsymbol{J}_n$ and $\boldsymbol{J}$.Hence the combined uncertain term, *i.e.* $\Delta\boldsymbol{A}\boldsymbol{x}(t) + \Delta\boldsymbol{B}_1\boldsymbol{u}(t) + \Delta\boldsymbol{B}_2\boldsymbol{w}_0(t)$, will contribute to the disturbance and the same applies to $\boldsymbol{E}\boldsymbol{f}(t)$. Together they form the lumped disturbance $\boldsymbol{w}(t)$.

Under the premise that the model parameter uncertainty and input saturation are not considered temporarily, the static output feedback controller is designed as follows

$$\boldsymbol{u}(t) = \boldsymbol{K}\boldsymbol{y}(t) \tag{11}$$

Substitute it into (10) and obtain the closed-loop control system

$$\begin{cases} \dot{\boldsymbol{x}}(t) = \boldsymbol{A}_c\boldsymbol{x}(t) + \boldsymbol{B}_1\boldsymbol{w}(t) \\ z(t) = \boldsymbol{C}_1\boldsymbol{x}(t) \end{cases} \tag{12}$$

where $\boldsymbol{A}_c = \boldsymbol{A} + \boldsymbol{B}_2\boldsymbol{K}\boldsymbol{C}_2$, the transfer function of the closed-loop control system is

$$\boldsymbol{G}(s) = \boldsymbol{C}_1(s\boldsymbol{I} - \boldsymbol{A}_c)^{-1}\boldsymbol{B}_1 \tag{13}$$

this paper's goal is to design an SOF controller whose gain matrix is $\boldsymbol{K}$ to enable the closed-loop control system (12) to be asymptotically stable and negative imaginary with H_∞ norm of the transfer function $\boldsymbol{G}$ bounded by the given $\gamma_1 > 0$.

PRELIMINARIES

In order to provide support for further theoretical analysis, the necessary assumptions and lemmas will be given below.

Assumption 1. Suppose that both the external disturbance $\boldsymbol{w}_0(t)$ and the first derivative of the disturbance to time are bounded and piecewise continuous.

Remark 2. The main reasonable support of Assumption 1 is that the external disturbances mainly caused by radiation of solar, Abnormal gravitational potential, magnetic effects, and atmospheric residues are all continuous and bounded [18, 19]. In addition, because the total control authority is limited, all three available control input torques should be continuous and bounded, which is reasonable in practice. It is important to note that the norm of $\boldsymbol{w}(t)$ and its components probably are unknown. Hence, Therefore, it is more versatile than the bang-bang logic-based sliding mode control method in Ref. [20] and fault-tolerant method in Ref. [21], both of which require a preliminary understanding of the upper bound of disturbance.

Lemma 1. (Negative Imaginary Lemma) [10]. Let (***A, B, C, D***) be a minimal state-space realization of a transfer function matrix $\boldsymbol{R}(s) \in \Re^{m\times m}$, where $\boldsymbol{A} \in \mathbb{R}^{n\times n}$, $\boldsymbol{B} \in \mathbb{R}^{n\times m}$, $\boldsymbol{C} \in \mathbb{R}^{m\times n}$, $\boldsymbol{D} \in \mathbb{R}^{m\times m}$. $\boldsymbol{R}(s) \in \Re^{m\times m}$ is Negative imaginary, if and only if $\boldsymbol{A}$ is Hurwitz, $\boldsymbol{D} = \boldsymbol{D}^{\mathrm{T}}$, and there exists a matrix $\boldsymbol{P}_1 \in \mathbb{R}^{n\times n}$, $\boldsymbol{P}_1 = \boldsymbol{P}_1^{\mathrm{T}} > 0$, such that $\boldsymbol{A}\boldsymbol{P}_1 + \boldsymbol{P}_1\boldsymbol{A}^{\mathrm{T}} \le 0$ and $\boldsymbol{B} + \boldsymbol{A}\boldsymbol{P}_1\boldsymbol{C}^{\mathrm{T}} = 0$.

Remark 3. In this paper, poles at the imaginary axis including the origin, are not considered, because the purpose of the controller designed is to stabilize the system, that is to say, as long as the system is stable, its poles will all be located in the open left-half plane.

Lemma 2. (Bounded Real Lemma) [22]. For a given $\gamma_1 > 0$, consider a continuous-time LTI system $\boldsymbol{G}(s) = \boldsymbol{C}(s\boldsymbol{I} - \boldsymbol{A})^{-1}\boldsymbol{B}$, where $\boldsymbol{A} \in \mathbb{R}^{n\times n}$, $\boldsymbol{B} \in \mathbb{R}^{n\times m}$, $\boldsymbol{C} \in \mathbb{R}^{r\times n}$, then

$G(s)$ is asymptotically stable and $\|G(s)\|_{\infty} < \gamma_1$,if and only if there exists a matrix $P_2 \in \mathbb{R}^{n\times n}, P_2 = P_2^{\mathrm{T}} > 0$, such that

$$\begin{bmatrix} AP_2 + P_2A^{\mathrm{T}} + BB^{\mathrm{T}} & P_2C^{\mathrm{T}} \\ CP_2 & -\gamma_1 I \end{bmatrix} < 0 \quad (14)$$

Lemma 3. (see the work of Wu [23]). Let $H \in \mathbb{R}^{n\times n}, H = H^{\mathrm{T}} < 0, N \in \mathbb{R}^{m\times n}$ and $P \in \mathbb{R}^{m\times m}$, then $H + N^{\mathrm{T}}PN < 0$,if and only if there exists a matrix $Z \in \mathbb{R}^{m\times m}$, such that

$$\begin{bmatrix} H & N^{\mathrm{T}}Z \\ Z^{\mathrm{T}}N & -Z - Z^{\mathrm{T}} + P \end{bmatrix} < 0 \quad (15)$$

Lemma 4. (Schur Complement Lemma). Let the partitioned matrix.

$$A = \begin{bmatrix} A_{11} & A_{12} \\ * & A_{22} \end{bmatrix}$$

be symmetric. Then

$$A < 0 \Leftrightarrow A_{11} < 0, A_{22} - A_{12}^{\mathrm{T}}A_{11}^{-1}A_{12} < 0 \Leftrightarrow A_{22} < 0, A_{11} - A_{12}A_{22}^{-1}A_{12}^{\mathrm{T}} < 0$$

CONTROLLER DESIGN

We will introduce constraint formulas based on NI and H_∞ performance in this section. In order to calculate the gain matrix K of the static output feedback controller, a sufficient condition for the decoupling of variables is given.

SOF H_∞ NI Controller Design

Based on **Lemma 1** and **Lemma 2**, it can be obtained by the equivalent transformation that

Lemma 5. For the given scalar $\gamma_1 > 0$, the closed-loop attitude control system (12) is asymptotically stable and negative imaginary with $\|G(s)\|_{\infty} < \gamma_1$, if and only if, there exist $\boldsymbol{P}_1 \in \mathbb{R}^{n\times n}$, $\boldsymbol{P}_1 = \boldsymbol{P}_1^{\mathrm{T}} > 0$, $\boldsymbol{P}_2 \in \mathbb{R}^{n\times n}$, $\boldsymbol{P}_2 = \boldsymbol{P}_2^{\mathrm{T}} > 0$ and $\boldsymbol{K} \in \mathbb{R}^{p\times q}$, for all $\varepsilon > 0$, such that

$$\boldsymbol{T}^{\mathrm{T}}\boldsymbol{U}_1^{\mathrm{T}}\boldsymbol{L}_1\boldsymbol{U}_1\boldsymbol{T} < \varepsilon_1\boldsymbol{I} \tag{16}$$

$$\boldsymbol{T}^{\mathrm{T}}\boldsymbol{U}_2^{\mathrm{T}}\boldsymbol{L}_2\boldsymbol{U}_2\boldsymbol{T} < 0 \tag{17}$$

where

$$\boldsymbol{T} = \begin{bmatrix} \boldsymbol{I} & 0 & \boldsymbol{B}_2\boldsymbol{K} \\ 0 & \boldsymbol{I} & 0 \end{bmatrix}^{\mathrm{T}}, \quad \boldsymbol{U}_1 = \begin{bmatrix} \boldsymbol{I} & \boldsymbol{A} & 0 & 0 \\ 0 & 0 & \boldsymbol{B}_1^{\mathrm{T}} & \boldsymbol{C}_1 \\ 0 & \boldsymbol{C}_2 & 0 & 0 \end{bmatrix}^{\mathrm{T}}, \quad \boldsymbol{U}_2 = \begin{bmatrix} \boldsymbol{A} & \boldsymbol{I} & \boldsymbol{B}_1 & 0 \\ \boldsymbol{C}_1 & 0 & 0 & \boldsymbol{I} \\ \boldsymbol{C}_2 & 0 & 0 & 0 \end{bmatrix}^{\mathrm{T}}$$

$$\boldsymbol{L}_1 = \begin{bmatrix} 0 & \boldsymbol{P}_1 & \boldsymbol{I} & 0 \\ \boldsymbol{P}_1 & 0 & 0 & \boldsymbol{P}_1 \\ \boldsymbol{I} & 0 & 0 & 0 \\ 0 & \boldsymbol{P}_1 & 0 & 0 \end{bmatrix}, \quad \boldsymbol{L}_2 = \begin{bmatrix} 0 & \boldsymbol{P}_2 & 0 & 0 \\ \boldsymbol{P}_2 & 0 & 0 & 0 \\ 0 & 0 & \boldsymbol{I} & 0 \\ 0 & 0 & 0 & -\gamma_1^2\boldsymbol{I} \end{bmatrix}.$$

Proof. Substitute ***U1, L1,*** and ***T*** into (16) and equivalently has that

$$\begin{bmatrix} \boldsymbol{A}_{\mathrm{c}}\boldsymbol{P}_1 + \boldsymbol{P}_1\boldsymbol{A}_{\mathrm{c}}^{\mathrm{T}} & \boldsymbol{B}_1 + \boldsymbol{A}_{\mathrm{c}}\boldsymbol{P}_1\boldsymbol{C}_1^{\mathrm{T}} \\ \boldsymbol{B}_1^{\mathrm{T}} + \boldsymbol{C}_1\boldsymbol{P}_1\boldsymbol{A}_{\mathrm{c}}^{\mathrm{T}} & 0 \end{bmatrix} < \varepsilon_1\boldsymbol{I}, \forall\varepsilon_1 > 0 \tag{18}$$

where $\boldsymbol{A}_{\mathrm{c}}$ is the same as the defined in (12). The inequality (18) is equivalent to

$$\begin{bmatrix} \boldsymbol{A}_{\mathrm{c}}\boldsymbol{P}_1 + \boldsymbol{P}_1\boldsymbol{A}_{\mathrm{c}}^{\mathrm{T}} & \boldsymbol{B}_1 + \boldsymbol{A}_{\mathrm{c}}\boldsymbol{P}_1\boldsymbol{C}_1^{\mathrm{T}} \\ \boldsymbol{B}^{\mathrm{T}} + \boldsymbol{C}_1\boldsymbol{P}_1\boldsymbol{A}_{\mathrm{c}}^{\mathrm{T}} & 0 \end{bmatrix} \le 0 \tag{19}$$

According to **Lemma 4**, the above matrix inequality is equivalent to

$$\begin{cases} \boldsymbol{A}_{\mathrm{c}}\boldsymbol{P}_1 + \boldsymbol{P}_1\boldsymbol{A}_{\mathrm{c}}^{\mathrm{T}} \le 0 \\ \boldsymbol{B}_1 + \boldsymbol{A}_{\mathrm{c}}\boldsymbol{P}_1\boldsymbol{C}_1^{\mathrm{T}} = 0 \end{cases} \tag{20}$$

Substitute ***U2, L2*** and ***T*** into (17) and equivalently has that

$$\begin{bmatrix} \boldsymbol{A}_c\boldsymbol{P}_2 + \boldsymbol{P}_2\boldsymbol{A}_c^T + \boldsymbol{B}_1\boldsymbol{B}_1^T & \boldsymbol{P}_2\boldsymbol{C}_1^T \\ \boldsymbol{C}_1\boldsymbol{P}_2 & -\gamma_1^2\boldsymbol{I} \end{bmatrix} < 0 \quad \textbf{(21)}$$

Therefore, based on Lemmas 1 and 2, conditions of **Lemma 5** are equivalent to the attitude control system (12) is asymptotically stable, negative imaginary and $\|\boldsymbol{G}(s)\|_\infty < \gamma_1$.

Decoupling Method Design

The conditions of **Lemma 5** are Bilinear Matrix Inequalities (BMIs). Inspired by the decoupling method in the work of Li *et al.* [24], we develop similar conditions where the controller gain ***K*** is separated from the Lyapunov matrices $\boldsymbol{P}_1$ and $\boldsymbol{P}_2$.

Lemma 6. (See the work of Ren *et al.* [11]). For all given scalars $\varepsilon_1 > 0$,the inequalities in **Lemma 5** hold, if and only if there exist matrices $\boldsymbol{P}_1 \in \mathbb{R}^{n\times n}$, $\boldsymbol{P}_1 = \boldsymbol{P}_1^T > 0$, $\boldsymbol{P}_2 \in \mathbb{R}^{n\times n}$, $\boldsymbol{P}_2 = \boldsymbol{P}_2^T > 0$, $\boldsymbol{S}_1 \in \mathbb{R}^{q\times n}$, $\boldsymbol{S}_2 \in \mathbb{R}^{q\times q}$ and $\boldsymbol{K} \in \mathbb{R}^{p\times q}$, such that

$$\boldsymbol{\Omega}_1 - \boldsymbol{\Pi} - \boldsymbol{\Pi}^T < 0 \quad \textbf{(22)}$$

$$\boldsymbol{\Omega}_2 - \boldsymbol{\Pi} - \boldsymbol{\Pi}^T < 0 \quad \textbf{(23)}$$

where

$$\boldsymbol{\Omega}_1 = \boldsymbol{U}_1^T\boldsymbol{L}_1\boldsymbol{U}_1 - \varepsilon_1\text{diag}(\boldsymbol{I},\boldsymbol{I},0), \boldsymbol{\Omega}_2 = \boldsymbol{U}_2^T\boldsymbol{L}_2\boldsymbol{U}_2$$

$$\boldsymbol{\Pi} = \begin{bmatrix}\boldsymbol{S}_1 & 0 & -\boldsymbol{S}_2\end{bmatrix}^T \begin{bmatrix}\boldsymbol{K}^T\boldsymbol{B}_2^T & 0 & -\boldsymbol{I}\end{bmatrix}$$

U1, U2, L1 and ***L2*** are the same as the definition in **Lemma 5**.

Proof. (Necessity) Suppose that for any given scalar $\varepsilon_1 > 0$, the matrix inequalities in **Lemma 5** hold for $\boldsymbol{P}_1 \in \mathbb{R}^{n\times n}$, $\boldsymbol{P}_1 = \boldsymbol{P}_1^T > 0$, $\boldsymbol{P}_2 \in \mathbb{R}^{n\times n}$, $\boldsymbol{P}_2 = \boldsymbol{P}_2^T > 0$ and $\boldsymbol{K} \in \mathbb{R}^{p\times q}$ .Then the following inequalities will hold for a large enough scalar $\varepsilon_2 > 0$

$$T^{\mathrm{T}}U_1^{\mathrm{T}}L_1U_1T-\varepsilon_1\mathrm{diag}(I,I)+\frac{1}{\varepsilon_2}\begin{bmatrix} P_1C_2^{\mathrm{T}} \\ C_1P_1C_2^{\mathrm{T}} \end{bmatrix}\begin{bmatrix} C_2P_1 & C_2P_1C_1^{\mathrm{T}} \end{bmatrix}<0 \tag{24}$$

$$T^{\mathrm{T}}U_2^{\mathrm{T}}L_2U_2T+\frac{1}{\varepsilon_2}\begin{bmatrix} P_2C_2^{\mathrm{T}} \\ 0 \end{bmatrix}\begin{bmatrix} C_2P_2 & 0 \end{bmatrix}<0 \tag{25}$$

Note that $\varepsilon_1\mathrm{diag}(I,I)=\varepsilon_1T^{\mathrm{T}}\mathrm{diag}(I,I,0)T$, according to **Lemma 4** (24) and (25) are equivalent to

$$\begin{bmatrix} T^{\mathrm{T}}\Omega_1T & \begin{bmatrix} P_1C_2^{\mathrm{T}} \\ C_1P_1C_2^{\mathrm{T}} \end{bmatrix} \\ \begin{bmatrix} C_2P_1 & C_2P_1C_1^{\mathrm{T}} \end{bmatrix} & -\varepsilon_2I \end{bmatrix}<0 \tag{26}$$

$$\begin{bmatrix} T^{\mathrm{T}}\Omega_2T & \begin{bmatrix} P_2C_2^{\mathrm{T}} \\ 0 \end{bmatrix} \\ \begin{bmatrix} C_2P_2 & 0 \end{bmatrix} & -\varepsilon_2I \end{bmatrix}<0 \tag{27}$$

Define a new matrix

$$H=\begin{bmatrix} I & \begin{bmatrix} 0 \\ 0 \end{bmatrix} \\ \begin{bmatrix} -K^{\mathrm{T}}B_2^{\mathrm{T}} & 0 \end{bmatrix} & I \end{bmatrix}$$

Multiplying the both sides of the inequalities in (26) and (27) by H^{T} and H, respectively, yields

$$\Omega_1-\varepsilon_2\begin{bmatrix} K^{\mathrm{T}}B_2^{\mathrm{T}} & 0 & -I \end{bmatrix}^{\mathrm{T}}\begin{bmatrix} K^{\mathrm{T}}B_2^{\mathrm{T}} & 0 & -I \end{bmatrix}<0 \tag{28}$$

$$\Omega_2-\varepsilon_2\begin{bmatrix} K^{\mathrm{T}}B_2^{\mathrm{T}} & 0 & -I \end{bmatrix}^{\mathrm{T}}\begin{bmatrix} K^{\mathrm{T}}B_2^{\mathrm{T}} & 0 & -I \end{bmatrix}<0 \tag{29}$$

The inequalities in (28) and (29) mean that those in **Lemma 6** by choosing $S_1=\frac{\varepsilon_2}{2}K^{\mathrm{T}}B_2^{\mathrm{T}}$ and $S_2=\frac{\varepsilon_2}{2}I$.

(sufficiency). Suppose that for all given scalar $\varepsilon_1 > 0$,the conditions in **Lemma 6** hold for $\boldsymbol{P}_1 \in \mathbb{R}^{n\times n}$, $\boldsymbol{P}_1 = \boldsymbol{P}_1^{\mathrm{T}} > 0$, $\boldsymbol{P}_2 \in \mathbb{R}^{n\times n}$, $\boldsymbol{P}_2 = \boldsymbol{P}_2^{\mathrm{T}} > 0$, $\boldsymbol{S}_1 \in \mathbb{R}^{q\times n}$, $\boldsymbol{S}_2 \in \mathbb{R}^{q\times q}$ and $\boldsymbol{K} \in \mathbb{R}^{p\times q}$. Note that

$$\boldsymbol{\Pi T} = 0,\ \boldsymbol{T}^{\mathrm{T}}\mathrm{diag}(\boldsymbol{I},\boldsymbol{I},\boldsymbol{0})\boldsymbol{T} = \mathrm{diag}(\boldsymbol{I},\boldsymbol{I}).$$

where $\boldsymbol{T}$ has the same definition as **Lemma 5.** Multiplying both sides of the inequalities (28) and (29) by $\boldsymbol{T}^{\mathrm{T}}$ and $\boldsymbol{T}$ produces (16) and (17), respectively, so the sufficiency is obvious.

Remark 4. As long as $\varepsilon_1 > 0$ it approaches zero, the magnitudes of ε_2 will approach positive infinity. Meanwhile, the values of $\boldsymbol{S}_1 = \frac{\varepsilon_2}{2}\boldsymbol{K}^{\mathrm{T}}\boldsymbol{B}_2^{\mathrm{T}}$ and $\boldsymbol{S}_2 = \frac{\varepsilon_2}{2}\boldsymbol{I}$ will also approach positive infinity. However, as the conclusion in [24], the controller gain matrix $\boldsymbol{K}$ could be found as long as ε_1 is small enough which will be shown in the numerical simulations. Based on **Lemma 6,** the controller gain matrix $\boldsymbol{K}$ can be decoupled from Lyapunov matrices $\boldsymbol{P}_1$ and $\boldsymbol{P}_2$. It is worth pointing out that matrices $\boldsymbol{S}_1$ and $\boldsymbol{S}_2$ are introduced slack matrix variables for decoupling and inequalities **Lemma 6** are called dilated inequalities. In the result after decoupling, comparing to other dilated controller design conditions, there are no additional restrictions on $\boldsymbol{S}_1$ and $\boldsymbol{S}_2$, which means our conditions are less conservative. For instance, in the work of Chang *et al.* [25], the slack variables need to have zero blocks to meet the decoupling conditions. It is worth mentioning that the conditions in **Lemma 6** are necessary and sufficient, but the conditions in [25] are not.

Based on the preceding discussions, the following theorem gives a design condition of the SOF controller to ensure that the close-loop attitude control system (12) is asymptotically stable and negative imaginary with H_∞ performance.

Theorem 1. Consider the system in (12), for all $\varepsilon_1 > 0$ and a given $\gamma_1 > 0$, if there exist, such that $\boldsymbol{P}_1 \in \mathbb{R}^{n\times n}$, $\boldsymbol{P}_1 = \boldsymbol{P}_1^{\mathrm{T}} > 0$, $\boldsymbol{P}_2 \in \mathbb{R}^{n\times n}$, $\boldsymbol{P}_2 = \boldsymbol{P}_2^{\mathrm{T}} > 0$, $\bar{\boldsymbol{K}} \in \mathbb{R}^{n\times q}$, $\boldsymbol{W} \in \mathbb{R}^{p\times q}$ and $\boldsymbol{S}_2 \in \mathbb{R}^{q\times q}$,such that

$$\begin{cases} \boldsymbol{\Omega}_1 - \boldsymbol{Q} - \boldsymbol{Q}^{\mathrm{T}} < 0 \\ \boldsymbol{\Omega}_2 - \boldsymbol{Q} - \boldsymbol{Q}^{\mathrm{T}} < 0 \end{cases} \quad \textbf{(30)}$$

where the $\boldsymbol{\Omega}_1$ and $\boldsymbol{\Omega}_2$ has the same definition as that in **Lemma 6**, and

$$\boldsymbol{Q}=\begin{bmatrix}\bar{\boldsymbol{K}}^{\mathrm{T}} & 0 & -\boldsymbol{I}\end{bmatrix}^{\mathrm{T}}\begin{bmatrix}\boldsymbol{W}^{\mathrm{T}}\boldsymbol{B}_2^{\mathrm{T}} & 0 & -\boldsymbol{S}_2^{\mathrm{T}}\end{bmatrix}$$

then the closed-loop attitude control system in (12) is asymptotically stable and negative imaginary with $\|\boldsymbol{G}(s)\|_\infty < \gamma_1$.At the same time, the controller gain matrix will be given by $\boldsymbol{K}=\boldsymbol{W}\boldsymbol{S}_2^{-1}$.

Proof. It is worth emphasizing that the condition in (30) means $\boldsymbol{S}_2^{\mathrm{T}}+\boldsymbol{S}_2>0$, which will guarantee the matrix $\boldsymbol{S}_2$ invertible. Substitute $\boldsymbol{W}=\boldsymbol{K}\boldsymbol{S}_2$ into $\boldsymbol{Q}$ and yields

$$\boldsymbol{Q}=\begin{bmatrix}\boldsymbol{S}_2\bar{\boldsymbol{K}}^{\mathrm{T}} & 0 & -\boldsymbol{S}_2\end{bmatrix}^{\mathrm{T}}\begin{bmatrix}\boldsymbol{K}^{\mathrm{T}}\boldsymbol{B}_2^{\mathrm{T}} & 0 & -\boldsymbol{I}\end{bmatrix}$$

Then, the conditions in (30) will be reduced to those in **Lemma 6** by making variable $\boldsymbol{S}_1=\boldsymbol{S}_2\bar{\boldsymbol{K}}^{\mathrm{T}}$. From **Lemma 6**, (22) and (23) hold for any $\varepsilon_1>0$. As a result, the $\boldsymbol{K}$ is the desired controller gain from **Lemma 5**.

LMI-based Iterative Algorithm

In this section, a two-stage method is developed to compute the SOF H_∞ NI controller gain matrix. In the first step, the controller gain matrix $\bar{\boldsymbol{K}}$ with the desired form will be get through two optimization processes. In the second step, the SOF controller gain matrix $\boldsymbol{K}$ is obtained by solving the conditions in **Theorem 1**.

Stage 1. Initialization of the controller

Lemma 7. Consider the system in (10) without ΔA_p an initial controller gain $\boldsymbol{K}_I$, the closed-loop attitude control system is NI and its H_∞-norm is less than γ_1, if for all $\delta>0$, there exist scalar $\sigma_1>0$ and matrices $\boldsymbol{P}_1>0$, $\boldsymbol{P}_2>0$, $\boldsymbol{Z}$ and $\bar{\boldsymbol{Z}}$, such that

$$\begin{bmatrix}-\delta\boldsymbol{I} & \bar{\boldsymbol{Z}} \\ \bar{\boldsymbol{Z}}^{\mathrm{T}} & -\boldsymbol{Z}^{\mathrm{T}}-\boldsymbol{Z}+\boldsymbol{U}_1^{\mathrm{T}}\boldsymbol{L}_1\boldsymbol{U}_1\end{bmatrix}<0 \tag{31}$$

$$\begin{bmatrix} -\delta \boldsymbol{I} & \bar{\boldsymbol{Z}} \\ \bar{\boldsymbol{Z}}^{\mathrm{T}} & -\boldsymbol{Z}^{\mathrm{T}} - \boldsymbol{Z} + \boldsymbol{U}_2^{\mathrm{T}} \boldsymbol{L}_2 \boldsymbol{U}_2 + \boldsymbol{\Sigma} \end{bmatrix} < 0 \tag{32}$$

where $\boldsymbol{L}_1$, $\boldsymbol{L}_2$, $\boldsymbol{U}_1$ and $\boldsymbol{U}_2$ are defined in **Lemma 5**, $\boldsymbol{\Sigma} = \sigma \mathrm{diag}(\boldsymbol{I}, \boldsymbol{I}, \boldsymbol{0})$, and

$$\boldsymbol{Z} = \begin{bmatrix} \boldsymbol{Z}_{11} & \boldsymbol{Z}_{12} & 0 \\ \boldsymbol{Z}_{21} & \boldsymbol{Z}_{22} & 0 \\ 0 & 0 & \boldsymbol{Z}_{33} \end{bmatrix}, \bar{\boldsymbol{Z}} = \begin{bmatrix} \boldsymbol{Z}_{11} & \boldsymbol{Z}_{12} & \boldsymbol{Z}_{13} \\ \boldsymbol{Z}_{21} & \boldsymbol{Z}_{22} & 0 \end{bmatrix}.$$

Then, the initial controller gain is obtained by $\boldsymbol{K}_1 = \boldsymbol{Z}_{13} \boldsymbol{Z}_{33}^{-1}$.

Proof. The (3,3) block of the matrix $-\boldsymbol{Z}^{\mathrm{T}} - \boldsymbol{Z} + \boldsymbol{U}_1^{\mathrm{T}} \boldsymbol{L}_1 \boldsymbol{U}_1$ in (31) is $\boldsymbol{Z}_{33} + \boldsymbol{Z}_{33}^{\mathrm{T}}$, one has that $\boldsymbol{Z}_{33} + \boldsymbol{Z}_{33}^{\mathrm{T}} > 0$ which implies that $\boldsymbol{Z}_{33}$ is invertible. Define $\tilde{\boldsymbol{T}} = \begin{bmatrix} \boldsymbol{I} & 0 & \boldsymbol{K}_1 \\ 0 & \boldsymbol{I} & 0 \end{bmatrix}^{\mathrm{T}}$.

Note that $\bar{\boldsymbol{Z}} = \tilde{\boldsymbol{T}}^{\mathrm{T}} \boldsymbol{Z}$. Replacing $\bar{\boldsymbol{Z}}$ in (31) and (32) with $\tilde{\boldsymbol{T}}^{\mathrm{T}} \boldsymbol{Z}$ and based on the results in **Lemma 3**, yields that for all $\delta > 0$,

$$\tilde{\boldsymbol{T}}^{\mathrm{T}} \boldsymbol{U}_1^{\mathrm{T}} \boldsymbol{L}_1 \boldsymbol{U}_1 \tilde{\boldsymbol{T}} < \delta \boldsymbol{I}, \ \tilde{\boldsymbol{T}}^{\mathrm{T}} (\boldsymbol{U}_2^{\mathrm{T}} \boldsymbol{L}_2 \boldsymbol{U}_2 + \boldsymbol{\Sigma}) \tilde{\boldsymbol{T}} < \delta \boldsymbol{I}.$$

Substitute $\boldsymbol{U}_1$, $\boldsymbol{U}_2$, $\boldsymbol{L}_1$, $\boldsymbol{L}_2$, $\boldsymbol{\Sigma}$ and $\tilde{\boldsymbol{T}}$ into above inequalities and produce

$$\begin{bmatrix} \tilde{\boldsymbol{A}}_c \boldsymbol{P}_1 + \boldsymbol{P}_1 \tilde{\boldsymbol{A}}_c^{\mathrm{T}} & \boldsymbol{B}_1 + \tilde{\boldsymbol{A}}_c \boldsymbol{P}_1 \boldsymbol{C}_1^{\mathrm{T}} \\ \boldsymbol{B}_1^{\mathrm{T}} + \boldsymbol{C}_1 \boldsymbol{P}_1 \tilde{\boldsymbol{A}}_c^{\mathrm{T}} & 0 \end{bmatrix} < 0$$

$$\begin{bmatrix} \tilde{\boldsymbol{A}}_c \boldsymbol{P}_2 + \boldsymbol{P}_2 \tilde{\boldsymbol{A}}_c^{\mathrm{T}} + \boldsymbol{B}_1 \boldsymbol{B}_1^{\mathrm{T}} & \boldsymbol{P}_2 \boldsymbol{C}_1^{\mathrm{T}} \\ \boldsymbol{C}_1 \boldsymbol{P}_2 & -\gamma_1^2 \boldsymbol{I} \end{bmatrix} < -\sigma \boldsymbol{I}$$

where $\tilde{\boldsymbol{A}}_c = \boldsymbol{A} + \boldsymbol{K}_1 \boldsymbol{C}_2$, according to the **Lemma 1** and **Lemma 2**, the system (12) is asymptotically stable, satisfies the H_∞ performance and negative imaginary.

Hence, the initial controller gain matrix $\boldsymbol{K}_1$ can be numerically found through the following optimization.

OP 1:

$$
\begin{aligned}
&\min && \delta \\
&s.t. && \begin{cases} \begin{bmatrix} -\delta \boldsymbol{I} & \bar{\boldsymbol{Z}} \\ \bar{\boldsymbol{Z}}^{\mathrm{T}} & -\boldsymbol{Z}^{\mathrm{T}} - \boldsymbol{Z} + \boldsymbol{U}_1^{\mathrm{T}} \boldsymbol{L}_1 \boldsymbol{U}_1 \end{bmatrix} < 0 \\ \begin{bmatrix} -\delta \boldsymbol{I} & \bar{\boldsymbol{Z}} \\ \bar{\boldsymbol{Z}}^{\mathrm{T}} & -\boldsymbol{Z}^{\mathrm{T}} - \boldsymbol{Z} + \boldsymbol{U}_2^{\mathrm{T}} \boldsymbol{L}_2 \boldsymbol{U}_2 + \boldsymbol{\Sigma} \end{bmatrix} < 0 \end{cases}, \boldsymbol{P}_1 > 0, \boldsymbol{P}_2 > 0 \\
&for && \boldsymbol{P}_1, \boldsymbol{P}_2, \boldsymbol{Z}, \delta, \sigma
\end{aligned}
$$

The $\boldsymbol{K}_1$ obtained can not be ensured to have the desired form through **OP 1**, but it still can be used to compute the controller gain $\bar{\boldsymbol{K}}$ with desired form by using the approach in the work of Li *et al* [24].

OP 2:

Fix $\bar{\boldsymbol{K}} = \boldsymbol{K}_1$, and replace $\boldsymbol{B}_2\boldsymbol{W}$ with $\boldsymbol{V}$.

$$
\begin{aligned}
&\min && \lambda_1 \alpha + \lambda_2 \varepsilon_1 \\
&s.t. && \begin{cases} \Omega_1 - \boldsymbol{Q} - \boldsymbol{Q}^{\mathrm{T}} < 0 \\ \Omega_2 - \boldsymbol{Q} - \boldsymbol{Q}^{\mathrm{T}} < 0 \end{cases}, \boldsymbol{P}_1 > 0, \boldsymbol{P}_2 > 0, \begin{bmatrix} -\alpha \boldsymbol{I} & \boldsymbol{\Phi}^{\mathrm{T}} \\ \boldsymbol{\Phi} & -\boldsymbol{I} \end{bmatrix} < 0 \\
&for && \boldsymbol{P}_1, \boldsymbol{P}_2, \boldsymbol{V}, \boldsymbol{S}_2, \alpha, \varepsilon_1
\end{aligned}
$$

where $\boldsymbol{\Phi} = \boldsymbol{V}^{\mathrm{T}} (\boldsymbol{B}_2^{\mathrm{T}})^{\perp}$, λ_1, λ_2 are given scalars named penalty factors [11]. $(\boldsymbol{B}_2^{\mathrm{T}})^{\perp}$ is an orthonormal basis of the null space of $\boldsymbol{B}_2^{\mathrm{T}}$, *i.e.* $(\boldsymbol{B}_2^{\mathrm{T}})(\boldsymbol{B}_2^{\mathrm{T}})^{\perp} = 0$, $((\boldsymbol{B}_2^{\mathrm{T}})^{\perp})^{\mathrm{T}}((\boldsymbol{B}_2^{\mathrm{T}})^{\perp}) = \boldsymbol{I}$. Then, the controller gain $\bar{\boldsymbol{K}} = \boldsymbol{V}\boldsymbol{S}_2^{-1}$. As long as α it approaches to 0, we can get the $\bar{\boldsymbol{K}}$ with the desired form [24].

Stage 2: Iterative Algorithm for SOF controller.

Fix the $\bar{\boldsymbol{K}}$ which is obtained from **stage 1**,

$$\begin{aligned} &\min && \varepsilon_1 \\ &s.t. && \begin{cases} \boldsymbol{\Omega}_1 - \boldsymbol{Q} - \boldsymbol{Q}^{\mathrm{T}} < 0 \\ \boldsymbol{\Omega}_2 - \boldsymbol{Q} - \boldsymbol{Q}^{\mathrm{T}} < 0 \end{cases}, \boldsymbol{P}_1 > 0, \boldsymbol{P}_2 > 0 \\ &for && \boldsymbol{P}_1, \boldsymbol{P}_2, \boldsymbol{W}, \boldsymbol{S}_2, \varepsilon_1 \end{aligned}$$

as long as the ε_1 approaches 0, the SOF controller gain $\boldsymbol{K}$ will be given, and

$$\boldsymbol{K} = \boldsymbol{W}\boldsymbol{S}_2^{-1}.$$

The algorithm flow is as follows:

Algorithm Flow
1. Define $\delta_2 > 0$, $\delta_3 > 0$ as the stopping tolerances. Set $j = 0$ which stand for the number of iterations, and computing the controller gain $\bar{\boldsymbol{K}}^{(0)} = \bar{\boldsymbol{K}}$ through the stage 1.
2. Set $\bar{\boldsymbol{K}} = \bar{\boldsymbol{K}}^{(j)}$ to solve **OP 3**, and get $\varepsilon_1^{(j)} = \varepsilon_1$.
3. If $\varepsilon_1^{(j)} < \delta_2$, return $\boldsymbol{K}$ and skip to step 6, otherwise skip to step 4.
4. If $j \geq 1$ and $\dfrac{\left\lvert \varepsilon_1^{(j)} - \varepsilon_1^{(j-1)} \right\rvert}{\varepsilon_1^{(j)}} < \delta_3$, return $\boldsymbol{K}$ and skip to step 6, otherwise skip to step 5.
5. Set $j = j+1$, $\bar{\boldsymbol{K}}^{(j)} = \boldsymbol{B}_2\boldsymbol{K}$ and skip to step 2.
6. Set $\boldsymbol{K}$ as the one obtained from step 2 and solve $$\begin{cases} \boldsymbol{A}_c\boldsymbol{P}_1 + \boldsymbol{P}_1\boldsymbol{A}_c^{\mathrm{T}} \leq 0 \ , \ \boldsymbol{B}_1 + \boldsymbol{A}_c\boldsymbol{P}_1\boldsymbol{C}_1^{\mathrm{T}} = 0 \\ \begin{bmatrix} \boldsymbol{A}_c\boldsymbol{P}_2 + \boldsymbol{P}_2\boldsymbol{A}_c^{\mathrm{T}} + \boldsymbol{B}_1\boldsymbol{B}_1^{\mathrm{T}} & \boldsymbol{P}_2\boldsymbol{C}_1^{\mathrm{T}} \\ \boldsymbol{C}_1\boldsymbol{P}_2 & -\gamma_1^2\boldsymbol{I} \end{bmatrix} < 0 \end{cases} \quad \textbf{(33)}$$
7. If (33) is solved, then the algorithm success, otherwise the algorithm failure.

Event-trigger Scheme Design

Some kinds of controller errors as well as unknown actuator dynamics, will lead to the actuator uncertainty or nominal controller perturbation, which is called controller's gain perturbation $\boldsymbol{\Delta K}$ with the norm $\lVert \boldsymbol{\Delta K} \rVert \leq \eta_0$. Two kinds of perturbation are taken into consideration:

When $\boldsymbol{\Delta K}$ stands for the additive perturbation, it meets

$$\boldsymbol{\Delta K}_a = \boldsymbol{M}_2\boldsymbol{F}_2(t)\boldsymbol{N}_2, \|\boldsymbol{F}_2(t)\| \le 1$$

When $\boldsymbol{\Delta K}$ stands for the multiplicative perturbation, it meets

$$\boldsymbol{\Delta K}_b = \boldsymbol{M}_3\boldsymbol{F}_3(t)\boldsymbol{N}_3\boldsymbol{K}, \|\boldsymbol{F}_3(t)\| \le 1$$

where $\boldsymbol{M}_2$, $\boldsymbol{N}_2$, $\boldsymbol{M}_3$, $\boldsymbol{N}_3$ are known constant real matrices of appropriate dimensions, $\boldsymbol{F}_2$(t) and $\boldsymbol{F}_3$(t) are Lebesgue measurable matrix function.

Based on the above SOF controller, when both additive and multiplicative perturbations exist simultaneously, the event trigger mechanism is introduced as

$$\boldsymbol{u}(t) = \left\lfloor \boldsymbol{K} + \bar{\rho}\left(\boldsymbol{\Delta K}_a + \boldsymbol{\Delta K}_b\right) \right\rfloor \boldsymbol{y}(t_i), \quad for\ t \in \left[t_i,\ \ t_{i+1}\right),\ \ i\text{=0,1,2,3,...} \tag{34}$$

where $\bar{\rho}$ is a scalar introduced here to indicate whether the controller is perturbed or not, so an increase in $\bar{\rho}$ will result in a decrease in system performance, which is probably the maximum allowable perturbed parameter. In addition, the introduction of $\bar{\rho}$ simulates random gain perturbations.

The control torque is held constant in inter-update time *i.e.*, $\left[t_i,\ \ t_{i+1}\right)$.

Define the measurement error as

$$\boldsymbol{e} \triangleq \boldsymbol{y}(t_i) - \boldsymbol{y}, \quad t \in \left[t_i,\ \ t_{i+1}\right) \tag{35}$$

The measurement error $\boldsymbol{e}$ plays a crucial role in the implementation of the event-triggered control scheme. The triggering instant t_{i+1} is decided by monitoring the evolution of e in an approximately continuous way until it crosses a state-dependent threshold.

The following triggering event is considered

$$t_{i+1} = \min\left\{t \ge t_i : \|\boldsymbol{e}\| \ge \beta\|\boldsymbol{y}\| + \mu\right\}, \quad i\text{=0,1,2,3,...} \tag{36}$$

where $\beta \in (0,1)$ and $\mu > 0$ is a small parameter that is designed to ensure the controller avoids Zeno behavior [26].

NUMERICAL SIMULATIONS

In this part, the robustness and effectiveness of the proposed controller are verified by simulation for the closed-loop control system. Four bending modes (see Table **1**) are considered in all the simulations below, because in flexible spacecraft the lower order modes usually dominate. Main parameters refer to the works of Yang [27] and Di Gennaro [28].

Table 1. Flexible dynamics parameters.

Natural Frequency (rad/s)		Damping
Mode 1	0.7681	0.005607
Mode 2	1.1038	0.008620
Mode 3	1.8733	0.012830
Mode 4	2.5496	0.025160

The nominal inertia matrix of the flexible spacecraft is chosen as the coupling matrices are chosen as,

$$\boldsymbol{J}_n = \begin{bmatrix} 350 & 3 & 4 \\ 3 & 280 & 10 \\ 4 & 10 & 190 \end{bmatrix} \text{kgm}^2$$

$$\tilde{\boldsymbol{\Delta}} = \begin{bmatrix} 6.45637 & -1.25619 & 1.11687 & 1.23637 \\ 1.27814 & 0.91756 & 2.48901 & -0.83674 \\ 2.15629 & -1.67264 & -0.83674 & -1.12503 \end{bmatrix} \sqrt{\text{kg}} \cdot \text{m}$$

$$\tilde{\boldsymbol{\Delta}}_p = \begin{bmatrix} 2.342552 \\ -0.422537 \\ 3.912984 \\ 7.026176 \end{bmatrix} \times 10^{-2} \sqrt{\text{kg}} \cdot \text{m} / \text{s}^2 / \text{V}$$

Initial values of state variables in all simulations are chosen as Table **2**, and then the initial state is chosen as $\boldsymbol{x}(0) = \left[\boldsymbol{\Theta}^{\mathrm{T}}(0) \quad \boldsymbol{\omega}^{\mathrm{T}}(0) \quad \boldsymbol{q}^{\mathrm{T}}(0) \quad \tilde{\boldsymbol{\eta}}^{\mathrm{T}}(0) \right]^{\mathrm{T}}$. And the upper bound of control torque is set be $u_{\mathrm{m1}} = u_{\mathrm{m2}} = u_{\mathrm{m3}} = 15\text{Nm}$. The inertia uncertainty is assumed to be $\Delta \boldsymbol{J} = (0.1 + 0.02\sin(0.11\pi t))\boldsymbol{J}_n$ .According to the work of Liu [29], it is suitable to set the external disturbance $\boldsymbol{T}_d$ as

$$\boldsymbol{T}_d = 5\times10^{-4}\times\begin{bmatrix}\sin(0.11\pi t)\\ \cos(0.11\pi t)\\ \cos(0.11\pi t + \pi/3)\end{bmatrix}\text{Nm}$$

Table 2 Variables initial value.

-	Initial Value
Θ	$[0.18,\ 0.15,\ -0.15]^{\mathrm{T}}(\mathrm{rad})$
ω	$[-0.02,\ -0.02,\ -0.02]^{\mathrm{T}}(\mathrm{rad/s})$
$\tilde{\eta}$	$[0,\ 0,\ 0,\ 0]^{\mathrm{T}}$
$\dot{\tilde{\eta}}$	$[0,\ 0,\ 0,\ 0]^{\mathrm{T}}$

Choose

$$\boldsymbol{F}_{\mathrm{a}}=[3.1533\ -0.5714\ 5.3674\ 9.3389]$$

$$\boldsymbol{F}_{\mathrm{b}}=[1.0976\ 0.1965\ 1.8086\ 3.0873]$$

as the feedback coefficient matrix.

Similar to Ref. [13], the following parameters are also chosen:

$$\boldsymbol{M}_1 = 0.01\times[8\ 11\ 13\ 15\ 16\ -18\ 8\ 11\ 13\ 15\ 16\ 18\ 8\ 11]^{\mathrm{T}},$$

$$\boldsymbol{N}_1=0.01\times[1\ 2\ 3\ 4\ 2\ 10\ 1\ 2\ 3\ 4\ 2\ 10\ 1\ 2],$$

$$\boldsymbol{F}_1(t)=\sin(0.11\pi t).$$

$$\boldsymbol{M}_2 = 0.1\times ones(3,1),\ \boldsymbol{N}_1=0.01\times ones(1,14),\ \boldsymbol{F}_2(t)=\sin(0.11\pi t + \pi/4).$$

$$\boldsymbol{M}_3 = 0.1\times ones(3,1),\ \boldsymbol{N}_3=0.01\times ones(1,3),\ \boldsymbol{F}_3(t)=\cos(0.11\pi t).$$

and the related constants are chosen as

$$\gamma_1 = 1.5, \lambda_1 = 0.2, \lambda_2 = 3.2, \beta = 0.3, \mu = 10^{-3}.$$

The form of fault signal is shown in Table **3**.

Table 3. The form of $\bar{f}(t)$.

t/s	$[0,10]$	$(10,45]$	$(45,60]$	$(60,75]$	$(75,100]$
$\bar{f}(t)$/Nm	0	$\sin(0.1\pi t)$	1	$\cos(0.1\pi t)$	0

And then the faulty signal is

$$f(t) = [0.005 \;\; 0.006 \;\; 0.007]^{\mathrm{T}} \bar{f}(t)$$

The stop tolerances of iterative process are chosen as

$$\delta_2 = 10^{-4}, \delta_3 = 10^{-3}$$

According to **Stage 1**, Substitute relevant parameters in **OP 1** and **OP 2** to initialize the controller gain. The corresponding optimal $\delta^* = 0.0056$, $\varepsilon_1^* = 0.0029$, $\alpha_1^* = 0.0029$ and the initial controller gain matrix and controller gain matrix with the desired form respectively are

$$K_1 = \begin{bmatrix} -0.1255 & -0.0002 & 0.0090 & -0.0213 & -0.0018 & -0.0072 \\ -0.0020 & -0.1066 & -0.0029 & -0.0015 & 0.0003 & 0.0042 \\ 0.0043 & -0.0043 & -0.1025 & -0.0025 & 0.0085 & -0.0274 \\ -0.0034 & -0.0044 & -0.0102 & -0.0647 & 0.0001 & 0.0116 \\ -0.0002 & -0.0081 & 0.0104 & 0.0117 & -0.0910 & -0.0065 \\ -0.0077 & 0.0069 & -0.0610 & 0.0033 & -0.0085 & -0.0106 \\ 0.0240 & -0.0074 & 0.0158 & -0.0248 & 0.0055 & 0.0100 \\ 0.0040 & 0.0026 & -0.0092 & -0.0078 & -0.0035 & -0.0406 \\ 0.0162 & 0.0181 & 0.0148 & 0.0154 & -0.0184 & 0.0003 \\ 0.0171 & -0.0045 & 0.0045 & 0.0064 & 0.0063 & 0.0045 \\ -0.0002 & -0.0009 & 0.0165 & -0.0337 & -0.0083 & 0.0069 \\ -0.0057 & 0.0042 & 0.0060 & 0.0301 & -0.0270 & 0.0511 \\ 0.0014 & 0.0087 & 0.0074 & -0.0064 & -0.0248 & 0.0179 \\ -0.0076 & -0.0119 & -0.0010 & -0.0114 & 0.0147 & 0.0028 \end{bmatrix}$$

$$\bar{K} = \begin{bmatrix} -0.4320 & 0.1691 & 0.0203 & -0.0653 & 0.0940 & 0.0237 \\ 0.0989 & -0.0782 & 0.0002 & 0.0303 & -0.0193 & 0.0042 \\ -0.0202 & 0.0083 & -0.0448 & 0.0163 & 0.0009 & 0.0135 \\ 0.0798 & -0.0277 & 0.0048 & -0.0644 & -0.0157 & 0.0010 \\ 0.1761 & -0.1292 & -0.0366 & 0.0734 & -0.1325 & -0.0168 \\ -0.1200 & 0.0480 & -0.0584 & -0.0005 & 0.0120 & -0.0017 \\ 0.1444 & -0.0559 & -0.0113 & -0.0141 & -0.0314 & -0.0281 \\ 0.2809 & -0.1432 & -0.0033 & 0.0390 & -0.0562 & -0.0377 \\ 0.1658 & -0.0711 & -0.0057 & 0.0099 & -0.0399 & -0.0082 \\ 0.0595 & -0.0192 & -0.0059 & -0.0024 & -0.0316 & 0.0031 \\ -0.0462 & 0.0312 & 0.0226 & -0.0389 & 0.0124 & -0.0072 \\ 0.1954 & -0.1072 & -0.0356 & 0.0701 & -0.0691 & 0.0317 \\ 0.0660 & -0.0288 & -0.0054 & -0.0019 & -0.0216 & -0.0017 \\ 0.1722 & -0.0629 & 0.0010 & -0.0161 & -0.0074 & -0.0224 \end{bmatrix}$$

After the iterative algorithm of **Stage 2**, when the stop tolerances are satisfied, the gain matrix of the SOF controller is

$$K = \begin{bmatrix} -6.9163 & 1.7784 & -2.4064 & -22.4764 & -0.5174 & 3.3606 \\ 16.1543 & -23.2598 & -0.8008 & 6.3395 & -20.3019 & -20.6762 \\ -6.9416 & 1.9569 & -6.8215 & 0.9956 & 0.7812 & -5.3741 \end{bmatrix}$$

and the non-increasing sequence of ε_1 is

$$\left\{\varepsilon_1^n\right\} = \{0.003924\ 0.001995\ 0.001991\ 0.001987\ 0.0019830$$
$$0.001979\ 0.001975\ 0.001972\ 0.001968\ 0.001965\ 0.001962$$
$$0.001960\ 0.001957\ 0.001954\ 0.001952\ 0.001950\ 0.001948\}$$

Applying the obtained controller to the control system (12) in the presence of model parameter uncertainty with event-trigger mechanism gets the time responses of attitude angle and angular velocity which is shown in Fig. (**1**). The attitude angle and angular velocity were observed to converge to a small range within 150 s. After 250 seconds, the accuracy of the attitude angle is approximately 5×10^{-4} rad and that of angular velocity is approximately 5×10^{-4} rad/s in steady-state error. The modal displacement is shown in Fig. (**2**), from which we can know that the modal displacement tends to be stable after 150s, and its accuracy is approximately 5×10^{-3} after 250s. The time responses of control torque is also shown in Fig. (**2**), from which we can see that it is no more than 15 Nm during the whole simulation.

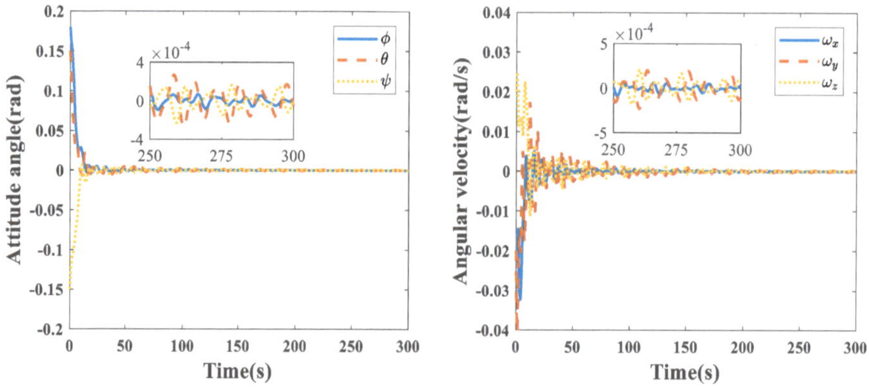

Fig. (1). Time response of attitude angle and angular velocity.

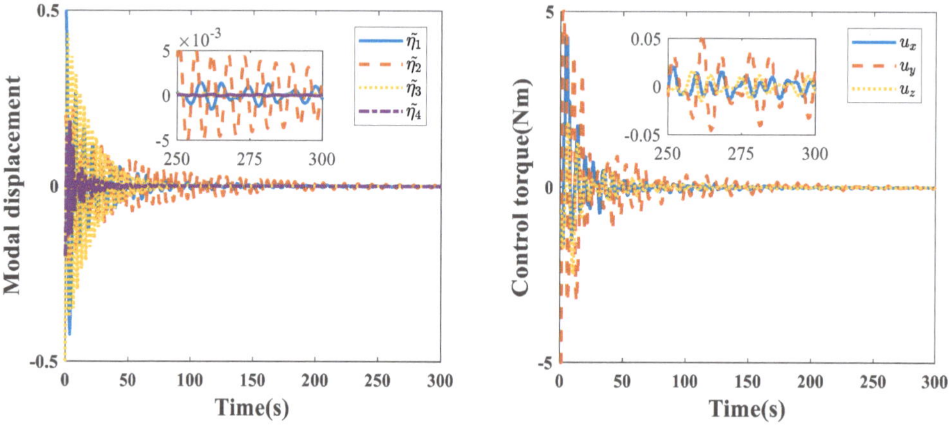

Fig. (2). Time response of modal displacement and control torque.

The energy index function $E_n = \frac{1}{2}\int_0^{\mathrm{T}} \|\boldsymbol{u}\| dt$ is used, where T is simulation time, and T = 300s is set in the simulation, to represent total energy consumption. The total energy consumption index is plotted in Fig. (**3**). In order to further analyze the vibration suppression performance, define the vibration energy index as

$\tilde{E}_n=\int_0^{T}\left(\dot{\tilde{\eta}}^{T}\dot{\tilde{\eta}}+\tilde{\eta}^{T}\tilde{K}\tilde{\eta}\right)dt$, which physically represents the vibration intensity during the entire simulation process, where T represents the simulation time and T=300s is also selected in the simulation. The right picture of (Fig. **3**) shows the total vibration energy.

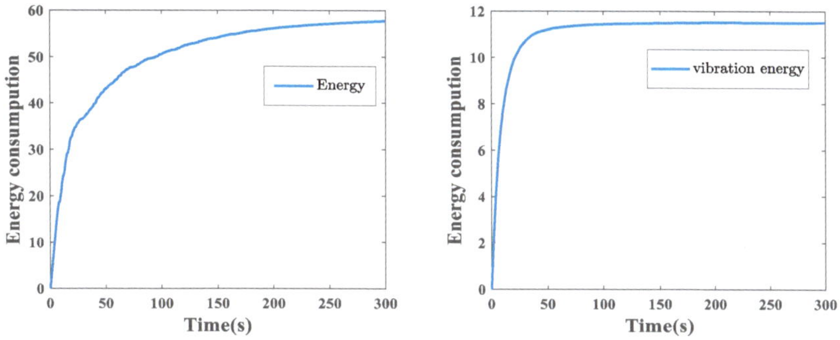

Fig. (3). Total energy consumption and vibration energy.

In order to analyze the effect of the event-trigger mechanism, the triggering instants and intervals are plotted in Figs. (**4** and **5**), and the communication times under the 0.1s-based fixed step sample algorithm and the event trigger mechanism are shown in Fig. (**6**). As can be seen from the figure, the communication times are greatly reduced.

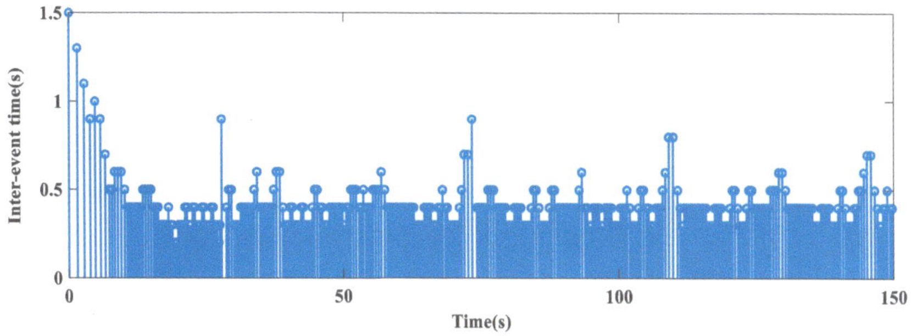

Fig. (4). Triggering instants and intervals in the first 150s.

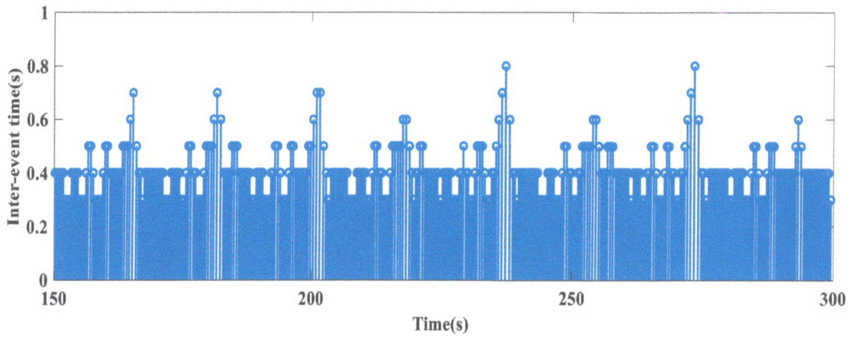

Fig. (5). Triggering instants and intervals in the last 150s.

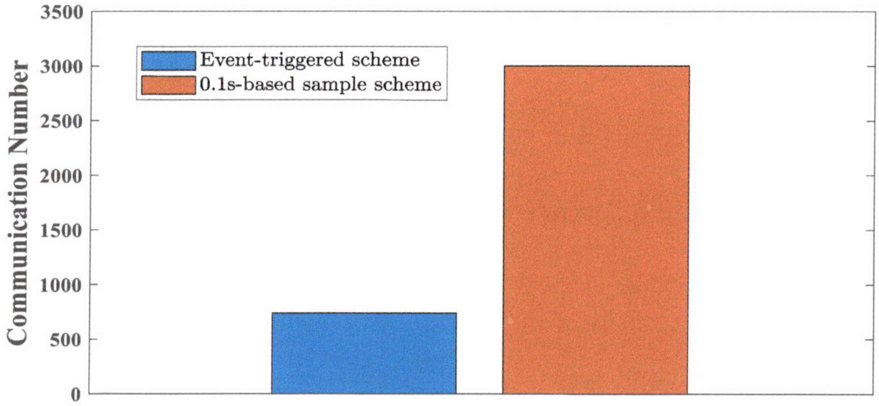

Fig. (6). Communication number of two scheme.

CONCLUSION

This paper proposes a SOF H_∞ NI controller for the uncertain flexible spacecraft. From the results of numerical simulation, the controller can quickly stabilize the closed-loop attitude control system of the spacecraft and tolerate certain model parameter uncertainty which shows the robustness of the controller. In the presence of controller gain perturbation, the controller can still work stably and effectively, which shows that the controller has good non-fragility. The comparison of communication times shows the effectiveness of the event-trigger mechanism introduced for the control input.

CONSENT FOR PUBLICATION

Not applicable.

CONFLICT OF INTEREST

The authors declare no conflict of interest, financial or otherwise.

ACKNOWLEDGEMENT

Declared none.

REFERENCES

[1] Liu C, Yue X, Shi K, et al. Inertia-free attitude stabilization for flexible spacecraft with active vibration suppression. International Journal of Robust and Nonlinear Control, 2019, 29(18): 6311-6336.

[2] Dai H, Jing X, Wang Y, et al. Post-capture vibration suppression of spacecraft via a bio-inspired isolation system. Mechanical Systems and Signal Processing, 2018, 105: 214-240.

[3] Ma Y, Jiang B, Tao G, et al. Uncertainty decomposition-based fault-tolerant adaptive control of flexible spacecraft. IEEE Transactions on Aerospace and Electronic Systems, 2015, 51(2): 1053-1068.

[4] Liu H, Li D, Xi J, et al. Robust attitude controller design for miniature quadrotors. International Journal of Robust and Nonlinear Control, 2016, 26(4): 681-696.

[5] Li Y, Ye D, Sun Z. Robust finite time control algorithm for satellite attitude control. Aerospace science and technology, 2017, 68: 46-57.

[6] Shao X, Hu Q, Shi Y, et al. Fault-tolerant prescribed performance attitude tracking control for spacecraft under input saturation. IEEE Transactions on Control Systems Technology, 2018, 28(2): 574-582.

[7] Sun L. Adaptive fault-tolerant constrained control of cooperative spacecraft rendezvous and docking. IEEE Transactions on Industrial Electronics, 2019, 67(4): 3107-3115.

[8] Hu Q, Shao X, Chen W H. Robust fault-tolerant tracking control for spacecraft proximity operations using time-varying sliding mode. IEEE Transactions on Aerospace and Electronic Systems, 2017, 54(1): 2-17.

[9] He W, Meng T, He X, et al. Unified iterative learning control for flexible structures with input constraints. Automatica, 2018, 96: 326-336.

[10] Petersen I R. Negative imaginary systems theory and applications. Annual Reviews in Control, 2016, 42: 309-318.

[11] Ren D, Xiong J, Ho D W C. Static output feedback negative imaginary controller synthesis with an H_∞ norm bound. Automatica, 2021, 126: 109157.

[12] Liu C, Shi K, Yue X, et al. Inertia-free saturated output feedback attitude stabilization for uncertain spacecraft. International Journal of Robust and Nonlinear Control, 2020, 30(13): 5101-5121.

[13] Liu C, Shi K, Sun Z. Robust H_∞ controller design for attitude stabilization of flexible spacecraft with input constraints. Advances in Space Research, 2019, 63(5): 1498-1522.

[14] Chen T, Shan J. Distributed spacecraft attitude tracking and synchronization under directed graphs. Aerospace Science and Technology, 2021, 109: 106432.

[15] Liu C, Yue X, Yang Z. Are nonfragile controllers always better than fragile controllers in

attitude control performance of post-capture flexible spacecraft?. Aerospace Science and Technology, 2021, 118: 107053.

[16] Ma Z, Huang P. Discrete-time sliding mode control for deployment of tethered space robot with only length and angle measurement. IEEE Transactions on Aerospace and Electronic Systems, 2019, 56(1): 585-596.

[17] Ye D, Shi M, Sun Z. Satellite proximate pursuit-evasion game with different thrust configurations. Aerospace Science and Technology, 2020, 99: 105715.

[18] Liu C, Vukovich G, Sun Z, et al. Observer-based fault-tolerant attitude control for spacecraft with input delay. Journal of Guidance, Control, and Dynamics, 2018, 41(9): 2041-2053.

[19] Feng X, Jia Y, Xu S. Inertia-free minimum-angular-motion detumbling control of spacecraft. Journal of Guidance, Control, and Dynamics, 2019, 42(3): 678-683.

[20] Li Y, Ye D, Sun Z. Time efficient sliding mode controller based on Bang–Bang logic for satellite attitude control. Aerospace science and technology, 2018, 75: 342-352.

[21] Gui H, Vukovich G. Adaptive fault-tolerant spacecraft attitude control using a novel integral terminal sliding mode. International Journal of Robust and Nonlinear Control, 2017, 27(16): 3174-3196.

[22] Gahinet P, Apkarian P. A linear matrix inequality approach to H_∞ control. International journal of robust and nonlinear control, 1994, 4(4): 421-448.

[23] Wu Z, Wang Y, Xiong J, et al. Static output feedback stabilization of networked control systems with a parallel-triggered scheme. ISA transactions, 2019, 85: 60-70.

[24] Li X, Gao H. A heuristic approach to static output-feedback controller synthesis with restricted frequency-domain specifications. IEEE Transactions on Automatic Control, 2013, 59(4): 1008-1014.

[25] Chang X H, Xiong J, Li Z M, et al. Quantized static output feedback control for discrete-time systems. IEEE Transactions on Industrial Informatics, 2017, 14(8): 3426-3435.

[26] Wu B, Shen Q, Cao X. Event-triggered attitude control of spacecraft. Advances in Space Research, 2018, 61(3): 927-934.

[27] Yang C D, Sun Y P. Mixed H2/H_∞ state-feedback design for microsatellite attitude control. Control Engineering Practice, 2002, 10(9): 951-970.

[28] Di Gennaro S. Output stabilization of flexible spacecraft with active vibration suppression. IEEE Transactions on Aerospace and Electronic systems, 2003, 39(3): 747-759.

[29] Liu C, Vukovich G, Shi K, et al. Robust fault tolerant nonfragile H_∞ attitude control for spacecraft via stochastically intermediate observer. Advances in Space Research, 2018, 62(9): 2631-2648.

CHAPTER 6

Vibration Control and Energy Harvesting in Aerospace Engineering Using Nonlinear Energy Sinks

Haiqin Li[1,*], **Ang Li**[2] and **Xianren Kong**[2]

[1]*Department of Mechanics, Tianjin University, Tianjin 300350, China.*

[2]*Research Center of Satellite Technology, Harbin Institute of Technology, Harbin 150001, China*

Abstract: The fact that spacecraft faces a rich and complex dynamic environment makes the development of vibration control and energy harvesting techniques a special concern in space engineering. Nonlinear Energy Sink is such a technique that has developed recently. It generally refers to a lightweight nonlinear device, that is attached to a primary system with essential nonlinear couplings. A special one-way energy transfer called targeted energy transfer (TET) could be observed for passive energy localization into itself. By taking advantage of such essential nonlinearities and the TET phenomenon, NES could be designed as smart and lightweight vibration absorbers or energy harvesters, in a broadband manner, which is especially suitable for the need in aerospace engineering. This chapter is thus devoted to the nonlinear dynamics of vibrational systems with coupled NESs and their applications in the field of passive vibration suppression and vibration energy harvesting.

Keywords: Energy harvesting, Nonlinear energy sink, Targeted energy transfer, Vibration control

INTRODUCTION

Spacecraft faces a complex dynamic environment during the process of launch and on-orbit operation, leading to the vibration of its whole and various components. Such vibrations include the acceleration dynamic load of the engine during launch, aerodynamic noise of high-speed flight, transient vibration during engine ignition shutdown and separation, structural and mechanical vibration during orbit and

*Corresponding author Haiqin Li:** Department of Mechanics, Tianjin University, Tianjin 300350, China: Tel: +86 15245102563; E-mail: lihaiqin1992@yahoo.com

Chuang Liu, Honghua Dai, Xiaokui Yue & Yiqing Ma (Eds.)

attitude maneuver, impact load during solar sail deployment and locking, friction and thermally induced vibration during reentry and return. On the one hand, the complex dynamic environments greatly threaten the normal use of the whole spacecraft and its components, implying the significance of controlling unwanted vibrations in aerospace engineering. On the other hand, those vibrations also provide a rich external energy source, leading to the development of the vibrational energy harvesting technique, which represents another critical direction for developing smart spacecraft systems in future space missions. The Nonlinear Energy Sink (NES) [1], is one of the recently developed smart technologies that are lightweight and broadband, and have a very broad prospect in both the aforementioned fields of vibration control and energy harvesting.

As depicted in Fig.(**1**), an NES generally refers to a lightweight nonlinear device that is attached to a primary system with essential nonlinear couplings (usually a cubic nonlinear stiffness and linear damping) for passive energy localization into itself. Thanks to the essential nonlinearities, a system coupled with an NES can generate rich dynamical regimes that are not possible in the ones with a traditional linear Tuned Mass Damper (TMD) [2, 3]. Moreover, under certain conditions, a one-way irreversible energy transfer phenomenon, termed as targeted energy transfer (TET) is observed [4, 5], providing a highly efficient mechanism for broadband vibration suppression. Analytical, numerical and experimental investigations concerning such TET phenomenon in coupled NES systems for highly effective broadband vibration suppression, were extensively demonstrated in [4-11].

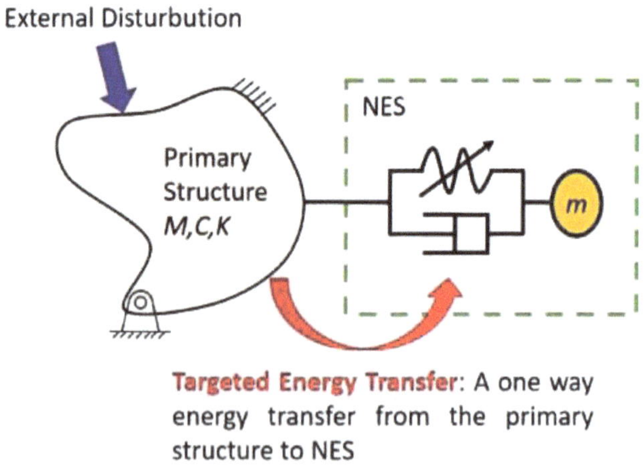

Fig. (1). Schematic of a primary structure coupled to a nonlinear energy sink.

The effectiveness of NES for vibration mitigation could be found in many engineering applications, including beams or plates [12-14], building structures [15], as well as aerospace systems [16, 17]. Especially in the field of aerospace engineering, Yang *et al.* [16] designed conceptually and implemented experimentally a NES to suppress the vibration of whole. Lee *et al.* demonstrated the efficiency of an NES for suppression of aeroelastic instabilities (limit cycle oscillations) in a two-DOF rigid wing model [18, 19]. Recently, NES has also been applied to the flywheel system of a satellite [20]. All of these studies have confirmed the potential of NES for effective vibration suppression in various types of structures in aeronautics and astronautics.

Up to now, various types of NES have been proposed to optimize the suppression performance, the simplest implementation is a lightweight oscillating mass with cubic nonlinear stiffness and linear damping [1], while other kinds of designs could also be found by using nonlinear damping [21, 22], vibro-impact dynamics [23-25], rotating elements, inerter components [26, 27], bistable or piecewise nonlinearities [13, 28-30]. It is proved that for whatever kind of NES and for whatever kind of applications, their key point of design resides in the same principle, *i.e.*, to highlight the importance of TET condition. More precisely, the present studies reveal that the optimization of NESs in the transient regime is to make the TET activated at the initial time. While in the permanent regime, the objective is reflected in tuning the parameters to generate a strongly modulated response (SMR) [31, 32], which is an equivalent representation of TET in the forced responses.

Besides the fruitful developments in vibration suppression, the potential of NES exploring broadband energy harvesting systems is also investigated in recent studies. One of the representative investigations is by Deniel *et al.* [33, 34], where an electromagnetic energy harvester using NES to achieve efficient TET is designed. Their numerical and experimental results have confirmed that, once TET is activated, the NES can achieve highly effective energy harvesting performance over a wide frequency range and hence outperforms the traditional linear harvesters. Other NES based energy harvesting systems by using piezoelectric transduction methods were also proposed and investigated [35-39]. These studies clearly demonstrated the advantage of NES in developing broadband energy harvesting systems. Despite the abovementioned NES related energy harvesting applications that focus only on smooth essential nonlinearities, it is only recently that the potential of the VINES for energy harvesting was addressed by Li *et al.* [40]. In their study, an electromagnetic VINES consisting of a magnet moving inside the clearance of a coil fixed primary structure was proposed. It is thus demonstrated that such electromagnetic VINES can act as an effective harvester with higher instantaneous energy density than the linear device and cubic NES. Moreover, the vibro-impact induced TET phenomenon is responsible for realizing the optimal performance.

This chapter is thus to perform the dynamics of nonlinear energy sinks and their applications in vibration suppression and energy harvesting. For this purpose, two main examples will be presented, the first one presents a nonlinear energy sink with smooth nonlinearity, and its performance in vibration suppression will be evaluated. A second example is devoted to a non-smooth vibro-impact nonlinear energy sink, and its dynamical behaviors and energy harvesting potential will be performed.

VIBRATION SUPPRESSION USING AN NES WITH NONLINEAR DAMPING

Dynamical Modeling of a System Coupled to an NES

We consider a 2-dof nonlinear system consisting of a primary structure coupled to an NES, as depicted in Fig. (**2**). The primary structure is a linear oscillator (LO) characterized by mass m_1, stiffness k_1 and damping c_1, and subjected to a harmonic excitation $F = F_0 \cos(\omega t)$. The NES is of mass m_2, and associated with a cubic stiffness k_2 and a geometric nonlinear damping c_2.

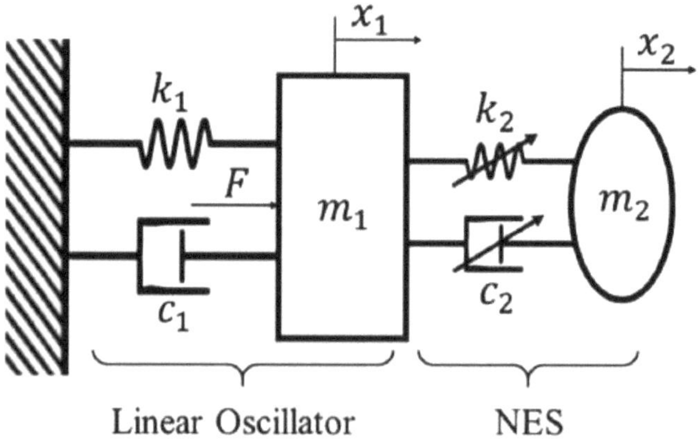

Fig. (2). Schematic of a linear oscillator coupled an NES.

Assuming the damping of the linear oscillator be $c_1 = 0$, and let x_1 and x_2 be the displacements of the LO and the NES, the governing equations of the system then writes

$$\begin{aligned} m_1\ddot{x}_1 + k_1x_1 + c_2(x_1 - x_2)^2(\dot{x}_1 - \dot{x}_2) + k_2(x_1 - x_2)^3 &= F\cos\omega t \\ m_2\ddot{x}_2 + c_2(x_2 - x_1)^2(\dot{x}_2 - \dot{x}_1) + k_2(x_2 - x_1)^3 &= 0 \end{aligned} \quad (1)$$

where the nonlinear damping with the specific form $c_2(x_1 - x_2)^2(\dot{x}_1 - \dot{x}_2)$ shows a geometrically nonlinearity, which can usually be observed when the vibration amplitude of a structure is large, and can be mathematically realized by two linear dampers, see for details in [22]. System (1) can be simplified into a dimensionless form by introducing the following variables:

$$\varepsilon = \frac{m_2}{m_1}, \tau = \sqrt{\frac{k_1}{m_1}}t, \frac{k_2}{k_1} = \varepsilon k, \varepsilon\lambda = c_2\sqrt{\frac{1}{k_1m_1}}, \frac{F_0}{m_1} = \varepsilon A \quad (2)$$

and considering the following two assumptions: Firstly, the mass ratio ε satisfies $0 < \varepsilon \ll 1$, meaning that the NES is lightweight compared to the linear oscillator. Secondly, as a matter of 1:1 resonance condition, the frequency of the external force is assumed to be at the near neighborhood of the eigenfrequency of the linear oscillator, thus one has $\omega = 1 + \varepsilon\delta$ with δ a detuning parameter. One can finally have

$$\begin{aligned} \ddot{u} + \frac{u+\varepsilon v}{1+\varepsilon} &= \varepsilon A\cos(1 + \varepsilon\delta)\,\tau \\ \ddot{v} + (1+\varepsilon)\lambda v^2\dot{v} + \frac{u+\varepsilon v}{1+\varepsilon} + (1+\varepsilon)kv^3 &= \varepsilon A\cos(1 + \varepsilon\delta)\,\tau \end{aligned} \quad (3)$$

with $u = x_1 + \varepsilon x_2$ and $v = x_1 - x_2$.

A complex averaging method [21, 31, 32] is applied to analytically approximate the system dynamics, Introducing

$$\psi_1 e^{i(1+i\varepsilon\delta)\tau} = \dot{u} + iu, \psi_2 e^{i(1+i\varepsilon\delta)\tau} = \dot{v} + iv \quad (4)$$

into (4) and keeping only the slow terms in the resulting equations leads to,

$$\dot{\psi}_1 + i\varepsilon\delta\psi_1 + \frac{i\varepsilon}{2(1+\varepsilon)}(\psi_1 - \psi_2) = \frac{\varepsilon A}{2}$$
$$\dot{\psi}_2 + i\varepsilon\delta\psi_2 + \frac{i}{2(1+\varepsilon)}(\psi_2 - \psi_1) + \frac{(1+\varepsilon)(\lambda - 3ik)}{8}|\psi_2|^2\psi_2 = \frac{\varepsilon A}{2} \tag{5}$$

Equation (5) is thus called the slow dynamics flow of the system, which is a reduced system, allowing one to analytically approximate the dynamics of the previous system.

Bifurcation Analysis

By setting the time derivatives of (5) to zero, the fixed points of the system, ψ_{10} and ψ_{20} can be solved as

$$\psi_{10} = \frac{\psi_{20} - i(1+\varepsilon)A}{2\varepsilon\delta + 2\delta + 1}$$
$$\alpha_3|\psi_{20}|^6 + \alpha_2|\psi_{20}|^4 + \alpha_1|\psi_{20}|^2 = A^2 \tag{6}$$

with

$$\alpha_1 = 4\delta^2, \alpha_2 = -\frac{3k\delta(2\varepsilon\delta+2\delta+1)}{2\varepsilon\delta+1}, \alpha_3 = \frac{(9k^2+\lambda^2)(2\varepsilon\delta+2\delta+1)^2}{16(2\varepsilon\delta+1)^2} \tag{7}$$

Apparently enough, the number of fix points of the system depends on the number of roots in the second equation of (6), and the condition when the number of fix points of the system changes is called a saddle-node bifurcation in nonlinear dynamics. To derive the saddle-node bifurcation condition, we introduce $Z = |\psi_{20}|^2$, then the second equation of (6) can be transformed to a cubic algebraic equation for Z, *i.e.*, $\alpha_3 Z^3 + \alpha_2 Z^2 + \alpha_1 Z = A^2$, whose roots could be determined by the following Cardano discriminant,

$$\Delta = \left(\frac{\alpha_2^3}{27\alpha_3^3} - \frac{\alpha_1\alpha_2}{6\alpha_3^2} - \frac{A^2}{\alpha_3}\right)^2 + \left(\frac{\alpha_1}{3\alpha_3} - \frac{\alpha_2^2}{9\alpha_3^2}\right)^3 \tag{8}$$

If $\Delta < 0$, one has three unequal real roots, if $\Delta > 0$, there is only one real root, when $\Delta = 0$, there are three real roots and at least two of them are equal. As such, the saddle-node bifurcation is characterized by solving $\Delta = 0$, which can finally be written in form of $A = f(\lambda, \delta)$.

On the other hand, once the fixed points are obtained, their stability can consequently be determined. The change of stability of a given fix point, of course, affects significantly the dynamics of the considered system. And this, defines another important bifurcation in nonlinear dynamics: Hopf bifurcation. Consider the perturbation motion near the fixed points and let

$$\psi_1 = \psi_{10} + \Delta_1, \ \psi_2 = \psi_{20} + \Delta_2 \tag{9}$$

inserting these variables into (6), one can obtain

$$\begin{aligned}
\dot{\Delta}_1 &= -i\varepsilon\delta\Delta_1 - \frac{i\varepsilon}{2(1+\varepsilon)}(\Delta_1 - \Delta_2) \\
\dot{\Delta}_1^* &= i\varepsilon\delta\Delta_1^* + \frac{i\varepsilon}{2(1+\varepsilon)}(\Delta_1^* - \Delta_2^*) \\
\dot{\Delta}_2 &= -i\varepsilon\delta\Delta_2 + \frac{i}{2(1+\varepsilon)}(\Delta_1 - \Delta_2) - \frac{(1+\varepsilon)(\lambda-3ik)}{4}|\psi_{20}|^2\Delta_2 \\
&\quad - \frac{(1+\varepsilon)(\lambda-3ik)}{8}\psi_{20}^2\Delta_2^* \\
\dot{\Delta}_2^* &= i\varepsilon\delta\Delta_2^* - \frac{i}{2(1+\varepsilon)}(\Delta_1^* - \Delta_2^*) - \frac{(1+\varepsilon)(\lambda+3ik)}{4}|\psi_{20}|^2\Delta_2^* \\
&\quad - \frac{(1+\varepsilon)(\lambda+3ik)}{8}\psi_{20}^{*2}\Delta_2
\end{aligned} \tag{10}$$

whose characteristic polynomial reads

$$\mu^4 + \gamma_1\mu^3 + \gamma_2\mu^2 + \gamma_3\mu + \gamma_4 = 0 \tag{11}$$

where μ is the unknown eigenvalue, and $\gamma_1, \gamma_2, \gamma_3, \gamma_4$ could be computed using MATLAB. In this regard, the occurrence of Hopf bifurcation implies $\mu = \pm i\Omega$, which by substituting into (11), one can obtain the the Hopf bifurcation condition in the form of $A = f(\lambda)$.

Fig. (**3**) illustrates the bifurcation diagram for system parameters aligned as $\varepsilon = 0.1, k = 4/3, \delta = 0.5$. The blue solid curve defines the Hopf bifurcation on the $[\lambda, A]$ plane, and the region bounded by this blue curve is the unstable. On the other hand, the red dash-dotted lines, depict the saddle-node bifurcation, that appears to be a 'triangle' shape inside which the system processes three different fixed points. A critical value of damping $\lambda = \sqrt{3}k$ could be noted from Fig. (**3**), beyond which the saddle-node and Hopf bifurcations vanish simultaneously, and only single stable fixed point exists. We called this value as truncation damping.

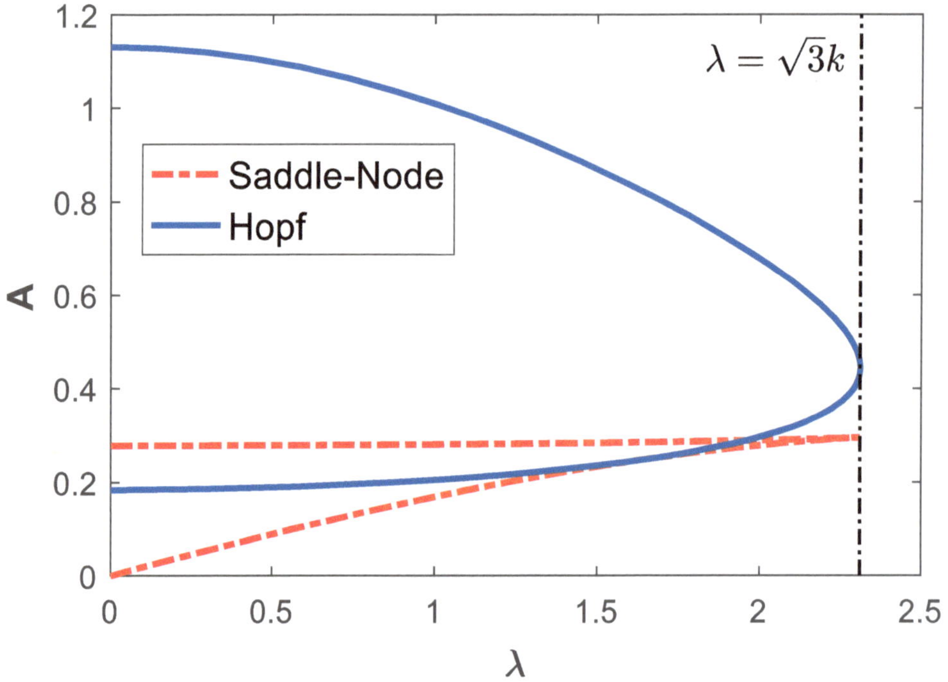

Fig. (3). Bifurcation diagram of the system for $\varepsilon = 0.1$, $k = 4/3$, $= 0.5$.

Owing to the rich bifurcations, the system could thus perform rich response regimes, this fact can be easily observed from the frequency response shown in Fig. (**4**) which shows the frequency-response of the system. One can easily note that near the resonance frequency, the system generally shows multiple solutions with coexistence of both stable and unstable states. As a result, some of the solutions with large amplitudes become unstable, leading to a considerable reduction in the vibrational response of the system. However, for some values of δ, the system has only one stable periodic solution, undesired periodic solutions then form up a sharp peak on the frequency response, leading a performance failure for the purpose of vibration suppression, thus should be avoided as much as possible. To emphasize this phenomenon, we may call the interval δ as the failure frequency, in Fig. (**4**) for example, the failure frequency is observed as $\delta \in [-0.75, -0.45]$.

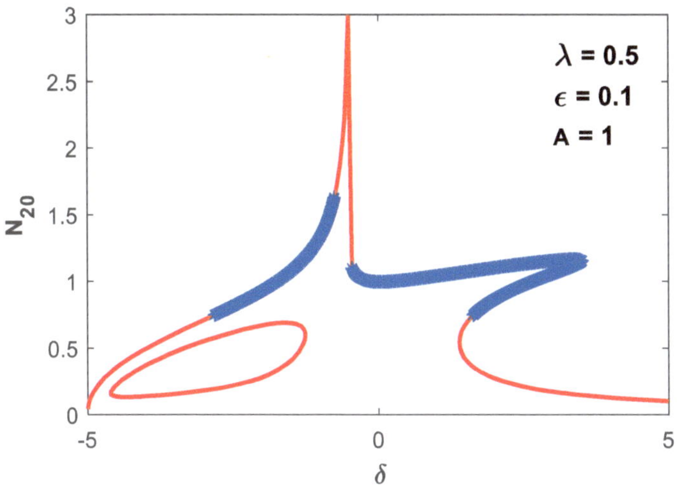

Fig. (4). Frequency response of the system for $\varepsilon = 0.1, k = 4/3, \lambda = 0.5, A = 1$.

Strongly Modulated Response

Among the numerous nonlinear phenomena that could be brought by the rich bifurcations in such a system with coupled NES, one representative regime is called the strongly modulated response (SMR), which is firstly observed by Starosvetsky and Gendelman and has now been extensively studied and proved to be one of the most significant response regimes for effective vibration suppression using NESs. The purpose of the following section is to determine the frequency range for occurrence of SMR.

We first rewrite (6) into a second-order ODE in terms of the single variable ψ_2. And for simplicity, one can drop the subscript of ψ_2 and the resulting equation writes

$$\ddot{\psi} + \frac{d}{dt}\left[\frac{i(1+4\varepsilon\delta)}{2}\psi + \frac{(\lambda - 3ik)(1+\varepsilon)}{8}|\psi|^2\psi\right]$$

$$-\frac{\varepsilon}{2}\left[\delta(1+2\varepsilon\delta) - \frac{(3k+i\lambda)(1+2\delta+2\varepsilon\delta)}{8}|\psi|^2\right]\psi = \frac{i\varepsilon(1+2\varepsilon\delta)A}{4} \quad \textbf{(12)}$$

Taking into account the assumption that $0 < \varepsilon \ll 1$, (12) can thus be expanded to different scales by means of multiple scales approach with respect to the order this small parameter ε, giving respectively

$$\varepsilon^0: D_0^2\psi + D_0\left(\frac{i}{2}\psi + \frac{\lambda - 3ik}{8}|\psi|^2\psi\right) = 0 \tag{13}$$

and

$$\begin{aligned}\varepsilon^1: 2D_0D_1\psi + D_0\left(2i\delta\psi + \frac{\lambda - 3ik}{8}|\psi|^2\psi\right) + D_1\left(\frac{i}{2}\psi + \frac{\lambda - 3ik}{8}|\psi|^2\psi\right)\\ -\frac{\delta}{2}\psi + \frac{(i\lambda + 3k)(1+2\delta)}{16}|\psi|^2\psi = \frac{iA}{4}\end{aligned} \tag{14}$$

Considering first the ε^0 order equation in (13), the fixed points are given as

$$\frac{9k^2+\lambda^2}{16}Z^3 - \frac{3k}{2}Z^2 + Z = 4C^2 \tag{15}$$

where $Z = N^2 = |\psi|^2$. In the literature, (15) defines the famous Slow Invariant Manifold (SIM) of the system. As shown in Fig. (**5**), one can see that the SIM generally shows a 'S' shape with two folding points N_1 and N_2, they are calculated as

$$Z_{1,2} = N_{1,2}^2 = \frac{1}{9k^2+\lambda^2}\left(8k \mp 4\sqrt{k^2 - \frac{\lambda^2}{3}}\right) \tag{16}$$

Due to the invariant property of the SIM, the system state can move only on the SIM. Following Fig. (**5**), the whole SIM is divided by three parts, among which the left and right branches (solid) are stable, while the middle part (dashed) is unstable. As such, when a state moves up along the left stable branch to N_1, since the instablity of the middle branch, the state may jump to the point N_u on the right branch with the same value of C. Similarly, a state that approaches to N_2 may also jump to N_d. The jump on the SIM between its stable branches gives a rise for the amplitude modulation on the system response, and hence provides a fundamental mechanism of SMR.

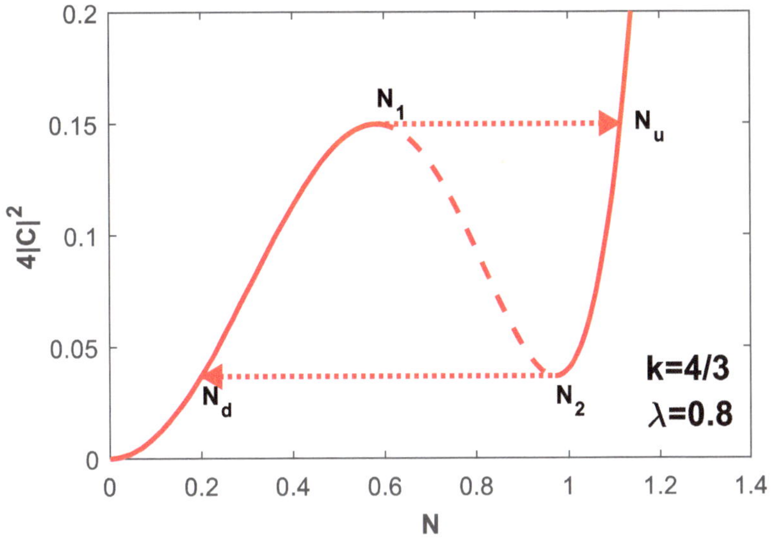

Fig. (5). SIM projection, solid: stable branch, dashed: unstable branch, arrow: jump between the stable branches.

In order to determine the condition of these jumps and hence to predict the occurrence of SMR, one must study the ε^1 order dynamics in (14), assuming $\tau_0 \rightarrow +\infty$, the ε^1 equation can be simplified to derive

$$\frac{\partial\psi}{\partial\tau_1} = \frac{2\Gamma\left(\frac{\lambda+3ik}{2}|\psi|^2-i\right)-\frac{\lambda-3ik}{2}\psi^2\Gamma^*}{\frac{3}{16}(9k^2+\lambda^2)|\psi|^4-3k|\psi|^2+1} \tag{17}$$

with

$$\Gamma = \frac{\delta}{2}\psi + \frac{iA}{4} - \frac{(i\lambda+3k)(1+2\delta)}{16}|\psi|^2\psi \tag{18}$$

Considering $\psi = Ne^{i\theta}$, one can then have

$$2g(N)\frac{\partial N}{\partial\tau_1} = -\frac{\lambda}{4}N^3 - \frac{3k}{4}N^2A\cos\theta + \frac{\lambda}{4}N^2A\sin\theta + A\cos\theta$$

$$2g(N)\frac{\partial\theta}{\partial\tau_1} = -\frac{27k^2+3\lambda^2}{16}(1+2\delta)N^4 + \frac{3k}{4}(1+8\delta)N^2 + \frac{9kNA\sin\theta}{4} \tag{19}$$

$$+\frac{3\lambda NA\cos\theta}{4}-2\delta-\frac{A\sin\theta}{N},$$

where

$$g(N)=\frac{3}{16}(9k^2+\lambda^2)N^4-3kN^2+1 \tag{20}$$

Let $f_1(N,\theta)$ and $f_2(N,\theta)$ be the right hands of the first and second equation for (19). As the fold lines occur when $g(N)=0$, the equations are therefore can be rescaled time by the function $g(N)$ to the following flow without singularities

$$N'=f_1(N,\theta),\ \ \theta'=f_2(N,\theta) \tag{21}$$

The phase portraits for system with different parameters are shown in Fig. (**6**). In Fig. (**6a**), there is an interval of θ bounded by the folded singularities for which all phase trajectories can arrive to and jump from N_1. See also in Fig. (**6b**) where an stable attractor exists in the lower or the upper branch, not every trajectory that starts from the lower fold of the SIM could reach the initial interval $[\Theta_1,\Theta_2]$, this inereval is called the jump interval J. For those points mapped into J, The phase trajectory jumps from a point of J to the upper branch of the SIM, then it moves along the line of the super-slow flow to the upper fold line, then jumps back to the lower branch and moves to the lower fold line, commencing on one of the points of the interval J in order to enable the next jump. This fact suggests the possible existence of SMR, to simplify this procedure, a 1-D map from J to itself can be conducted to determine the parametric zones of existence of the SMR.

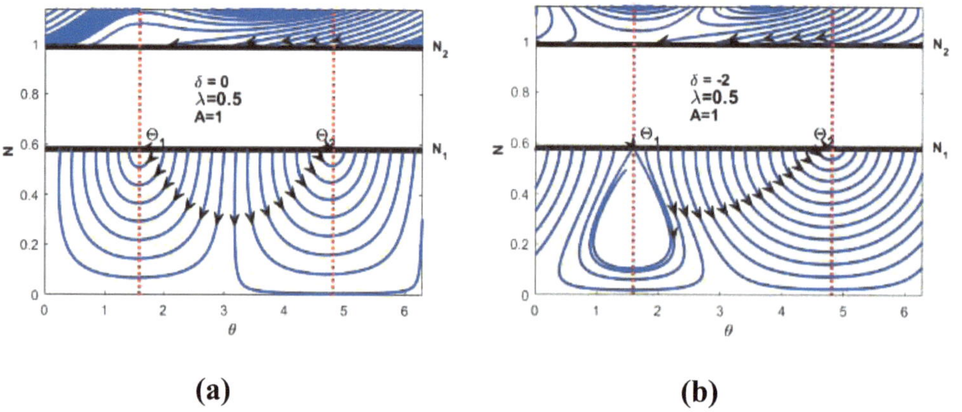

Fig. (6). Phase portraits for $\varepsilon=0.1, k=4/3, \lambda=0.5, A=1$, and for (a): $\delta=0$, (b): $\delta=-2$.

Note that C is constant and by applying (15) and (16), one can obtain the values of N_u and N_d as

$$N_u = \sqrt{\frac{8k+8\sqrt{k^2-\frac{\lambda^2}{3}}}{9k^2+\lambda^2}}, \quad N_d = \sqrt{\frac{8k-8\sqrt{k^2-\frac{\lambda^2}{3}}}{9k^2+\lambda^2}} \tag{22}$$

In addition, according to [22], the phase angle of the fixed point can be calculated

$$\theta + \tan^{-1}\left[\frac{4-3kZ}{\lambda Z}\right] = \arg C \tag{23}$$

then, manipulations between (22) and (23) yield the phase angle at N_u on the upper stable branch from the jump at N_1, together with N_d on the lower stable branch from the jump at N_2, let

$$P_i = \frac{4-3kZ_i}{\lambda Z_i}, i = 1,2,u,d \tag{24}$$

the relations then write

$$\theta_{u,d} = \theta_{1,2} + \tan^{-1}\left[\frac{P_{1,2}-P_{u,d}}{1+P_{1,2}P_{u,d}}\right] \tag{25}$$

The 1-D map is shown in Fig. (7). From varying δ and observing when trajectories from the Θ_1 and Θ_2 interval no longer returned, it can be concluded that the SMR may exist in $\delta \in [-5.85,2.68]$. The significance of SMR for effective vibration suppression will be discussed in next section.

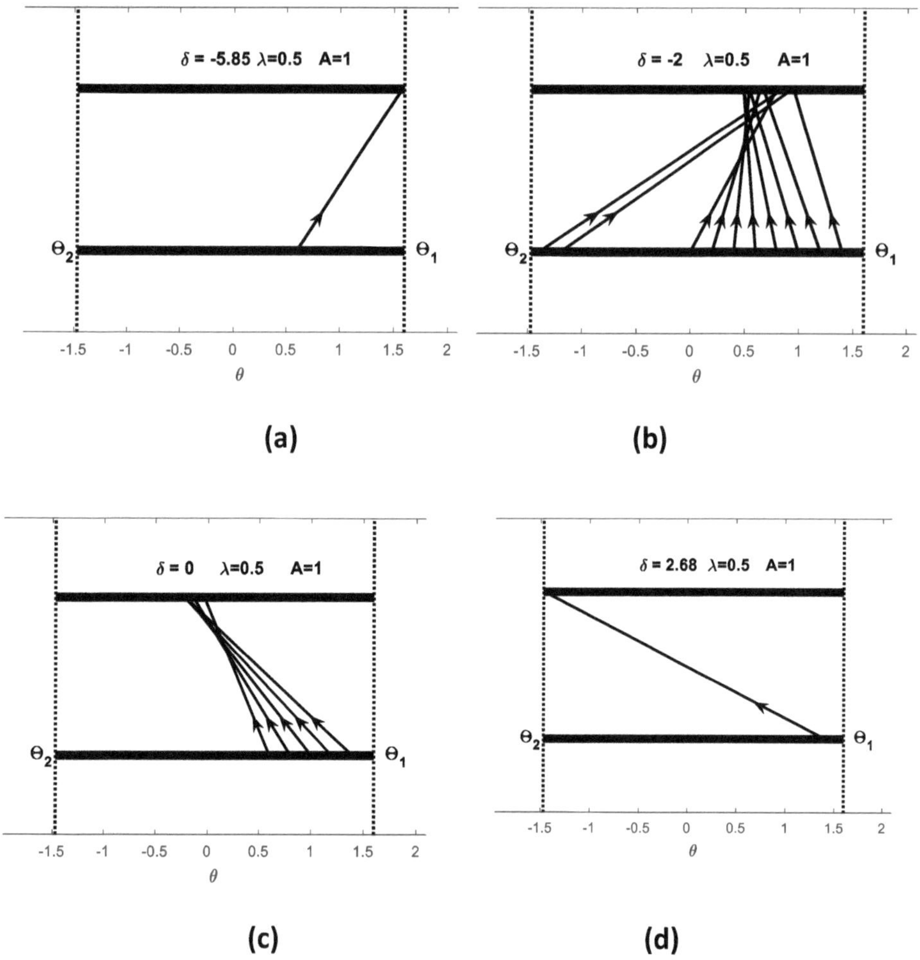

Fig. (7). 1-D mapping with varying δ, (a): $\delta = -5.85$, (b): $\delta = -2$, (c): $\delta = 2.5$, (d): $\delta = 2.68$.

Vibration Suppression Performance

Due to the rich non-periodic response regimes that can exist in such a nonlinear system, the vibration amplitude is also time dependent. We will use the energy spectrum instead of the amplitude to assess the vibration damping efficiency of the NES. The energy spectrum is obtained averaging the energy of the primary structure over a sufficiently long period of time in the permanent response, and is expressed as,

$$E = \left\langle \frac{1}{2}\dot{x}_1^2 + \frac{1}{2}x_1^2 \right\rangle_t \tag{26}$$

In this regard, the main objective for the performance optimization of the NES is to reduce the energy spectrum expressed in (26). For illustrating purpose, let us fix $\varepsilon = 0.1, A = 0.3$, focusing on the effects of varying damping and stiffness, and the value of E is then calculated over the time interval $t \in [2000,3000]$.

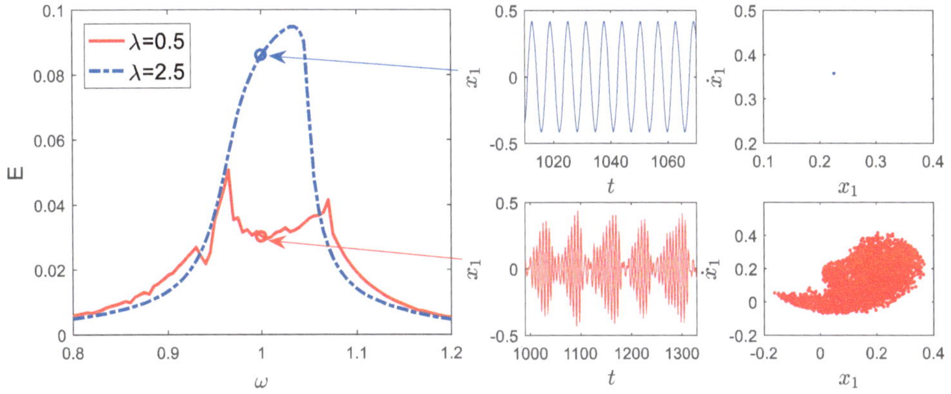

Fig. (8). Energy spectrum and time response of the main structure through λ variation, with $k = 4/3$.

Considering at first the effect of damping, Fig. (**8**) compares the energy spectrum at two different cases with$\lambda = 2.5$ and $\lambda = 0.5$, respectively. As the truncation damping can be calculated as $\lambda = 2.3$, hence for $\lambda = 2.5$, there is no bifurcation and the system possesses only periodic response in the vicinity of the resonance frequency, whereas for $\lambda = 0.5$, a SMR is presented. One can see that the vibrational energy is largely reduced when the system generates SMR, as verified by the drastic reduction on the energy spectrum for $\lambda = 0.5$.

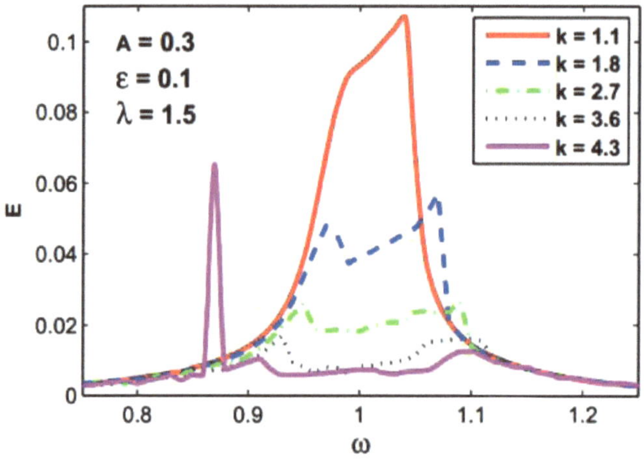

Fig. (9). Energy spectrum for the main structure through k variation.

The influence of stiffness is then reported in Fig. (9), the obtained results shows that, by varying k in an interval of modulated response, the energy of the primary mass can be monotonically reduced (in average). However, when the value of k is larger than $k = 4.3$, an undesired periodic response occurred in the left side of the resonance frequency. This is an effect of the failure frequency that was previously observed, and it should be avoided when design the NES for vibration control purpose.

ENERGY HARVESTING BY MEANS OF A VINES

Formulation of an Energy Harvesting System with a VINES

We consider the energy harvesting system proposed in [40], which is formulated based on the general configuration of a vibro-impact nonlinear energy sink (VINES) [23], associated also with an electromagnetic transduction technique. The resulting implementation is depicted in Fig. (10), which consists of a primary mass m_1 coupled to a magnet attachment m_2. The primary mass is characterized as a linear oscillator with linear stiffness k_1 and damping c_1. The magnet attachment undergoes two-sided inelastic impacts when it reaches the limits of the clearance 2Δ, acting as a VINES. The VINES magnet creates a magnetic field, and a coil is fixed to the primary mass so that to extract electric energy.

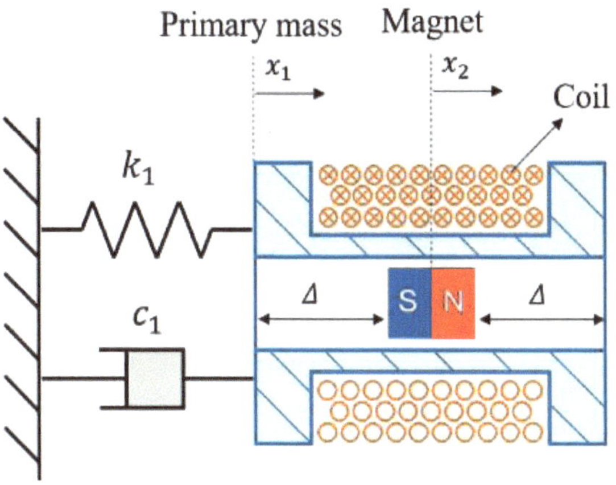

Fig. (10). Layout of an electromagnetic VINES energy harvester.

For such a coupled nonlinear system, its governing equations could be easily written as

$$\begin{gathered} m_1\ddot{x}_1 + c_1\dot{x}_1 + k_1x_1 = F + F_e + F_c \\ m_2\ddot{x}_2 + F_e + F_c = 0, \end{gathered} \tag{27}$$

in which x_1 and x_2 are respectively the motion of the primary mass and the VINES magnet, F_e is the electromagnetic force between magnet and the coil, and F_c represents is the contact force. Finally, F is the external excitation induced to the primary mass.

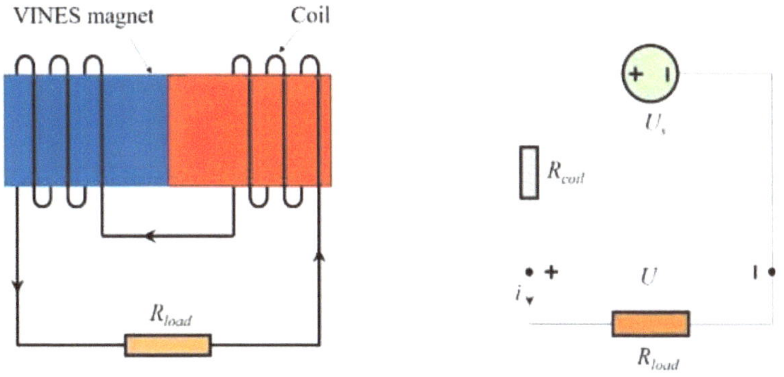

Fig. (11). Schematic of the harvesting circuit.

Electric power is harvested by connecting the coil with a load resistance R_{load}, resulting current i to flow in, see Fig. (**11**). This current i gives rise to the electromagnetic force F_e shown in (27). Applying Lenz's law, F_e is computed *via*:

$$F_e = k_t i \quad \mathbf{(28)}$$

where k_t is called the transduction factor or flux linkage gradient, and is assumed as constant for sake of simplicity. Furthermore, the current i reads

$$i = \frac{U_s}{R_{load}+R_{coil}} \quad \mathbf{(29)}$$

where U_s is the generated voltage, which as stated in the principle of Faraday's law, can be expressed as:

$$U_s = k_t(\dot{x}_2 - \dot{x}_1) \quad \mathbf{(30)}$$

Combining these relationships leads to $F_e = k_t^2(\dot{x}_2 - \dot{x}_1)/(R_{load} + R_{coil})$, meaning that the electromagnetic force acts as a linear damping force with an equivalent damping coefficient $c_e = k_t^2/(R_{load} + R_{coil})$.

By introducing $u_1 = \frac{x_1+\varepsilon x_2}{\Delta}$ and $u_2 = \frac{x_2-x_1}{\Delta}$ as the normalized mass center motion and relative motion, and letting

$$\varepsilon = \frac{m_2}{m_1}, \tau = \sqrt{\frac{k_1}{m_1}}t, \lambda = \frac{c_1}{m_2}\sqrt{\frac{m_1}{k_1}}, \kappa = \frac{k_t}{\sqrt{\varepsilon k_1 R_{coil}}}, \eta = \frac{R_{load}}{R_{coil}}, f = \frac{F}{\varepsilon k_1 \Delta}, f_c = \frac{F_c}{\varepsilon k_1 \Delta}, \quad \mathbf{(31)}$$

the original system can then be transformed into the following dimensionless model

$$\begin{aligned} \ddot{u}_1 + \varepsilon\lambda\frac{\dot{u}_1-\varepsilon\dot{u}_2}{1+\varepsilon} + \frac{u_1-\varepsilon u_2}{1+\varepsilon} &= \varepsilon f \\ \ddot{u}_2 - \varepsilon\lambda\frac{\dot{u}_1-\varepsilon\dot{u}_2}{1+\varepsilon} - \frac{u_1-\varepsilon u_2}{1+\varepsilon} + \frac{1+\varepsilon}{1+\eta}\kappa^2\dot{u}_2 &= -\varepsilon f - (1+\varepsilon)f_c \end{aligned} \quad \mathbf{(32)}$$

to generalize the analysis and reduce the parameter space.

Non-Smooth Contact Force Modeling

Contact occurs when the VINES approaches either side of the clearance limits, given by $|u_2| = 1$. The Signorini contact law can then be introduced to mathematically describe such relationship as

$$0 \leq 1 - |u_2| \perp |f_c| \geq 0 \tag{33}$$

representing that either $|u_2| = 1$ and $|f_c| \geq 0$, or $|u_2| \leq 1$ and $|f_c| = 0$.

The contact dynamics involves velocity jumps, which in terms of the relative velocity $v_2 = \dot{u}_2$, is then formulated by the Newton impact law as

$$v_2^+(t) = -\rho v_2^-(t), \quad if \ |u_2| = 1 \tag{34}$$

where $v_2^+(t)$ and $v_2^-(t)$ represent the right and left limit for the velocity v_2 at the contact time t, and $\rho \in [0, 1]$ is the restitution coefficient defining the contact loss. The complete contact dynamics is characterized by combining (33) and (34), it reads

$$\begin{array}{ll} 0 \leq {v_2}^+ + \rho {v_2}^- \perp f_c \geq 0 & if \ u_2 \geq 1 \\ 0 \leq -{v_2}^+ - \rho {v_2}^- \perp -f_c \geq 0 & if \ u_2 \leq -1 \\ f_c = 0 & otherwise \end{array} \tag{35}$$

Owing to the discontinuities on the relative velocity v_2, the acceleration does not have a classical form as $\dot{v}_2$. A differential measure dv_2, is thus introduced for a more rigorous definition of the acceleration:

$$dv_2 = \ddot{u}_2 dt + ({v_2}^+ - {v_2}^-) d\sigma \tag{36}$$

where dt is the Lebesgue measure for t, and $d\sigma = \sum_i a_i \delta_{t_i}$ is a discrete measure with given sequences a_i and δ_{t_i}. δ_s stands for the Dirac measure supported at t. In this expression, $\dot{v}_2(t) = \ddot{u}_2(t)$ almost everywhere, while at contact instants t_* one has $d\dot{v}_2 = ({v_2}^+ - {v_2}^-) d\delta_{t_*}$. In this way, the contact force can also be expressed in terms of $d\sigma$ as,

$$dI = f_c dt + \Gamma d\sigma \tag{37}$$

where f_c stands for the continuous contact force and Γ is the impulse corresponding to velocity jumps. (32) finally reads:

$$\begin{aligned} dv_1 + \varepsilon\lambda \frac{\dot{u}_1 - \varepsilon \dot{u}_2}{1+\varepsilon} dt + \frac{u_1 - \varepsilon u_2}{1+\varepsilon} dt = \varepsilon f dt \\ dv_2 - \varepsilon\lambda \frac{\dot{u}_1 - \varepsilon \dot{u}_2}{1+\varepsilon} dt - \frac{u_1 - \varepsilon u_2}{1+\varepsilon} dt + \frac{1+\varepsilon}{1+\eta} \kappa^2 \dot{u}_2 dt = -\varepsilon f dt - (1+\varepsilon) dI \end{aligned} \tag{38}$$

Injecting (36) and (37) into (38), it is remarked that (32) holds dt −almost everywhere. While at the contact instants, one has

$$-\frac{1}{1+\varepsilon}(v^+ - v^-) = \Gamma \tag{39}$$

and

$$\begin{array}{ll} 0 \le {v_2}^+ + \rho {v_2}^- \perp dI \ge 0 & if \;\; u_2 \ge 1 \\ 0 \le -{v_2}^+ - \rho {v_2}^- \perp -dI \ge 0 & if \;\; u_2 \le -1 \\ dI = 0 & otherwise \end{array} \tag{40}$$

Altogether, the system dynamics can be described by:

$$\begin{array}{c} u = [u_1, u_2]^T, \;\; v = [v_1, v_2]^T, \;\; \dot{u} = v \\ dv + \mathbf{A} u dt + \mathbf{B} v dt = \mathbf{F} dt - \mathbf{Q} dI \\ 0 \le {v_2}^+ + \rho {v_2}^- \perp dI \ge 0 \quad if \;\; u_2 \ge 1 \\ 0 \le -{v_2}^+ - \rho {v_2}^- \perp -dI \ge 0 \quad if \;\; u_2 \le -1 \end{array} \tag{41}$$

which is known as a measure differential complementarity problem, where the associated matrices read

$$\mathbf{A} = \frac{1}{1+\varepsilon}\begin{bmatrix} 1 & -\varepsilon \\ -1 & \varepsilon \end{bmatrix}, \mathbf{B} = \begin{bmatrix} 0 & 0 \\ 0 & \frac{1}{1+\eta}\kappa^2 \end{bmatrix} + \varepsilon\lambda A, \mathbf{F} = \begin{bmatrix} \varepsilon f \\ -\varepsilon f \end{bmatrix}, \mathbf{Q} = \begin{bmatrix} 0 \\ 1+\varepsilon \end{bmatrix} \tag{42}$$

Transient Response Evaluation

Consider the transient response of the system under the situation that the primary has an initial velocity v_0, and all the other initial conditions are zero. The energy convention ability of the VINES, can be first studied by evaluating the percentage of energy dissipated by the equivalent electric damping in the VINES magnet, which is computed via

$$E_{\text{diss}} = \frac{\int_0^{T_f} \varepsilon\kappa^2 v_2^2 dt}{E_{\text{int}}(1+\eta)} \quad \mathbf{(43)}$$

where E_{diss} is defined as the total percentage of energy dissipated by the VINES damping within the time length T_f, and $E_{int} = v_0^2/2$ accounts for the total energy that is initially induced into the system.

The variation of E_{diss} as a function of the initial velocity v_0 is illustrated in Fig. (**12**), three distinct energy levels could be observed: at the low energy level $v_0 \leq 0.95$, the VINES is not excited and E_{diss} is identical to the non-contact case depicted by the dashed line; when v_0 reaches the critical value $v_0 = 0.95$ to activate vibro-impacts between the primary mass and the VINES magnet, E_{diss} exhibits a sudden dramatical increase from less than 50% to nearly 80%, and then remains at a relatively high level over a wide range of energy input before crossing again the dashed line at $v_0 = 2.8$. In general, the vibro-impacts at this mid-energy level bring positive effects to make the VINES behave more efficiently as compared the non-contact situation; Finally, when the system enters into the third energy level described by $v_0 \geq 2.8$, the vibro-impacts cause negative effects and the VINES loses its efficiency again. Hence, in order to obtain optimum energy transduction, the VINES should be designed to work at its mid-energy level.

To better understand the energy-dependent behavior of the VINES, the response regimes in each energy level exampled at Points A, B, and C, are then presented and compared in Fig. (**13**). At the low energy level A, where the VINES magnet is not engaged and the system behaves linearly, the vibration is highly localized in the primary mass (very close to 100%), almost no energy exchange can be observed between the two oscillators. In general, without vibro-impact nonlinearity, the VINES is not activated and the instantaneous harvesting efficiency remains very low.

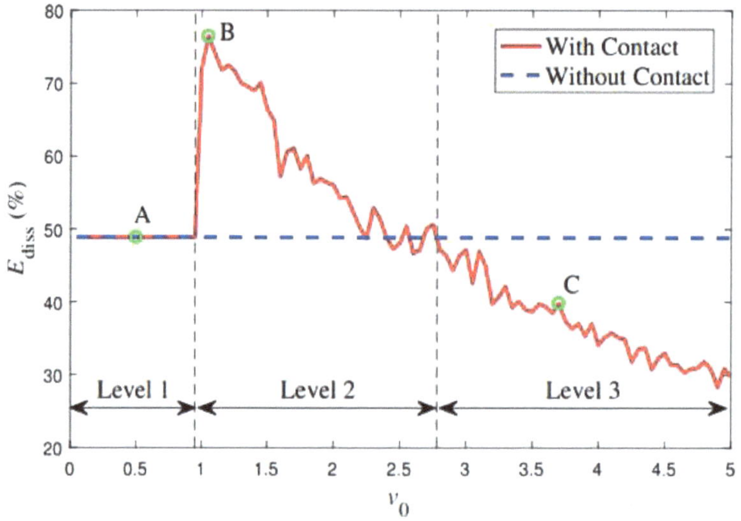

Fig. (12). E_{diss} as a function of v_0, obtained for T_f=200, $\epsilon = 0.1, r = 1, \eta = 4$, and $\kappa = 0.7$.

However, the performance shows a dramatic change as the system enters into the mid-energy level B, strong nonlinear behavior is activated thanks to the vibro-impacts, making the VINES dissipate rapidly the vibrational energy. At beginning, the system undergoes a coexisting state of 1:2 and 1:1 resonance with more than 2 vibro-impacts per cycle, with each contact, a certain amount of energy is transferred from the primary mass to the VINES magnet and finally harvested to the electric energy. As energy decreases, the system dynamics is then captured into a stage of 1:1 resonance after 38 time units with 2 vibro-impacts per period, representing the most effective mechanism. TET is hence realized to effectively transfer the energy into the VINES through the vibro-impacts, leading to a much higher instantaneous harvesting efficiency. This effective stage keeps working until the energy is fully damped, and finally the system enters into the linear regime same as what was observed for point A. At a further high energy level, point C, the system shows a more complicated scenario transfer from coexistence of 1:3 and 1:1 resonance, to coexistence of 1:2 and 1:1 resonance, to efficient 1:1 resonance with TET, to finally the non-impact linear regime. A qualitatively similar conclusion can be drawn that the system behaves most efficiently in the 1:1 resonance vibro-impact regime with TET, but not in the others.

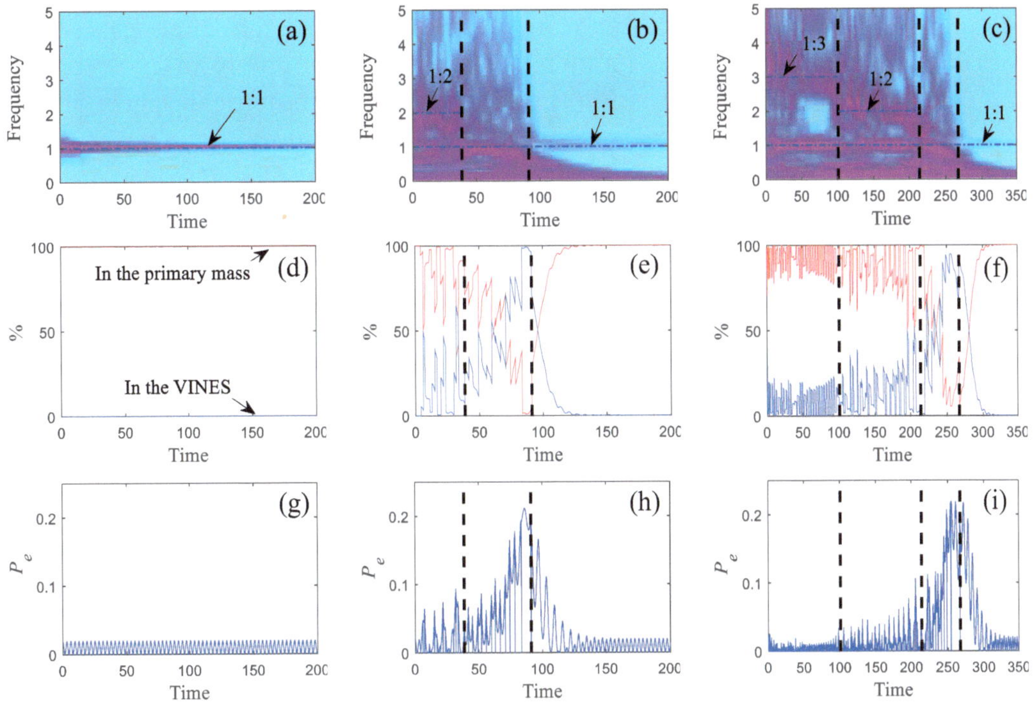

Fig. (13). Response of the system at point A (left, low energy level), point B (center, mid-energy level), and point C (right, high energy level), (a-c): time frequency response of the relative motion, (d-f): the percent of instantaneous energy in the VINES and the primary mass, (g-i): instantaneous power efficiency $P_e(t) = P_{nes}(t)/H(t)$, defined as the ratio between power of the VINES $P_{nes}(t)$ and the total energy $H(t)$.

To sum up, it is clear that the optimal harvesting efficiency is obtained when the VINES is working under the 1:1 resonance condition with 2 vibro-impacts per cycle. In this process, the vibro-impact plays an important role in realizing highly effective TET to convert the mechanical vibrations to electrical energy.

The electric energy is shared by the coil resistance R_{coil} and the load resistance R_{load}, and one may want to maximize the harvesting power. The percentage of energy harvested from R_{load}, denoted by E_{load},

$$E_{load} = \frac{\eta}{1+\eta} E_{diss} = \frac{\int_0^{T_f} \varepsilon \eta \kappa^2 v_2^2 dt}{E_{int}(1+\eta)^2}, \tag{44}$$

is thus then applied as a performance indicator for evaluating the energy harvesting efficiency of the VINES. Fig. (**14**) shows the dependence of E_{load} on the initial velocity v_0 and the resistance ratio η. As η increases, the performance improves at first and then undergoes a deterioration, while the optimum performance could be could be observed for η at around 5 to 10. Moreover, the optimal values of η is almost independent of the initial vibration level v_0. For almost all the η, the vibro-impact is activated at a nearly same critical initial energy level around $v_0 = 1$, beyond which the harvesting efficiency shows a dramatical increase. With a proper design, the value of E_{load} can reach up to 70%, meaning a very high energy harvesting efficiency.

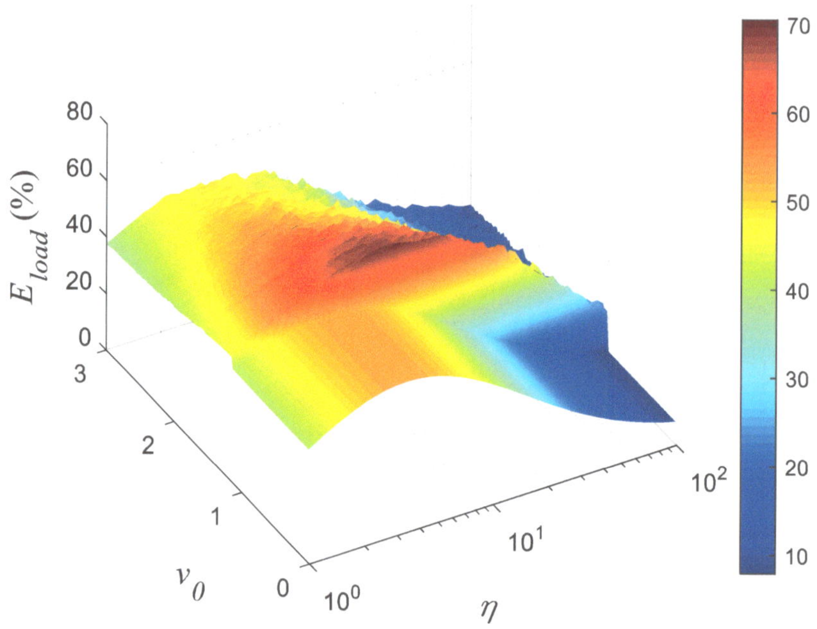

Fig. (14). E_{load} as a function of v_0 and η, for $\epsilon = 0.1$, $r = 1$, and $\kappa = 0.7$.

Forced Response Evaluation

Let us now evaluate the harvesting performance of the proposed system.Fig. (**15**) shows the possible steady-state regimes of the system under harmonic excitation $f = Acos(\omega t)$ when ω and A vary. The variations of A are summarized in Fig. (**15a-d**). For a small amplitude $A = 0.1$, a linear periodic response with no vibro-impact is observed. With forcing amplitude increase up to $A = 0.5$, a chaotic regime with strong amplitude modulation is observed. Characterized by the strong modulation on the vibration amplitudes in both oscillators, this special response regime is referred as the strongly modulated response (SMR), whose significance has been widely pointed out by many researchers as a representation of efficient

TET in the steady-state response with NES systems. Special attention will then be paid to this SMR regime, one will see in coming discussion that the SMR improves a lot the harvesting efficiency. As $A = 1$, a nonlinear periodic response with 2 vibro-impacts per cycle is activated. When the forcing amplitude increases up to $A = 2$, another chaotic regime with weak amplitude modulation is shown, this regime is generally called the weakly modulated response (WMR) in the literature. As frequency changes, in Fig. (**15e-f**), the response regime also varies, at $\omega = 0.95$, the system performs a chaotic WMR, while at $\omega = 0.91$, a periodic response with 4 impacts per cycle is presented.

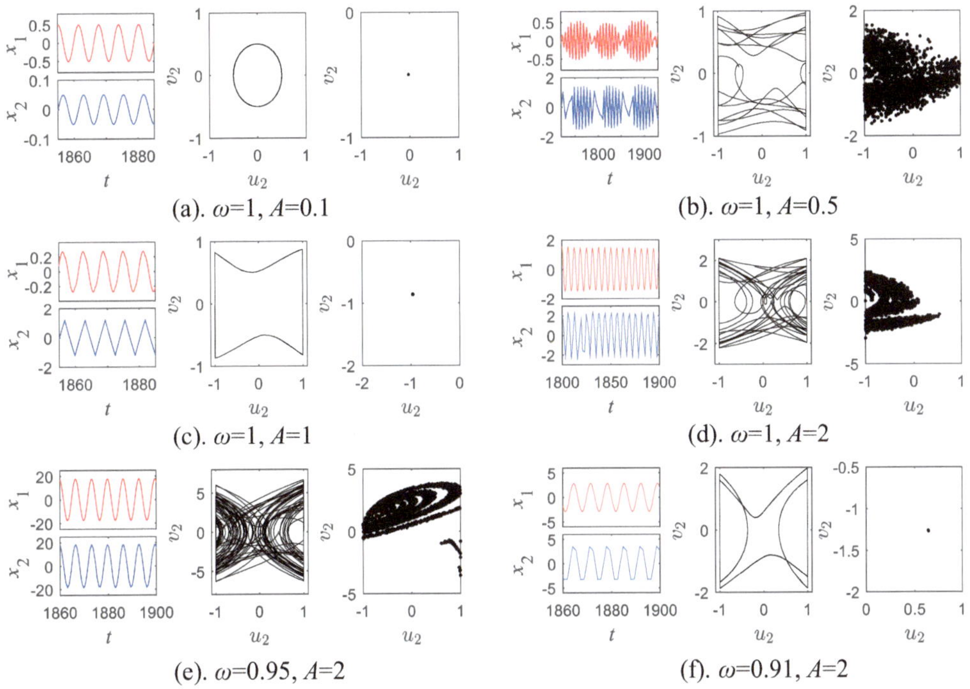

Fig. (15). Responses of the system under different conditions, for $\varepsilon = 0.1$, $\kappa = 0.7$, $\eta = 4$, and $r = 0.95$.

In order to build a correspondence between these different response regimes and the harvesting performance, four different indicators are then defined and presented in Fig. (**16**) over the full $\omega - A$ parameter space.

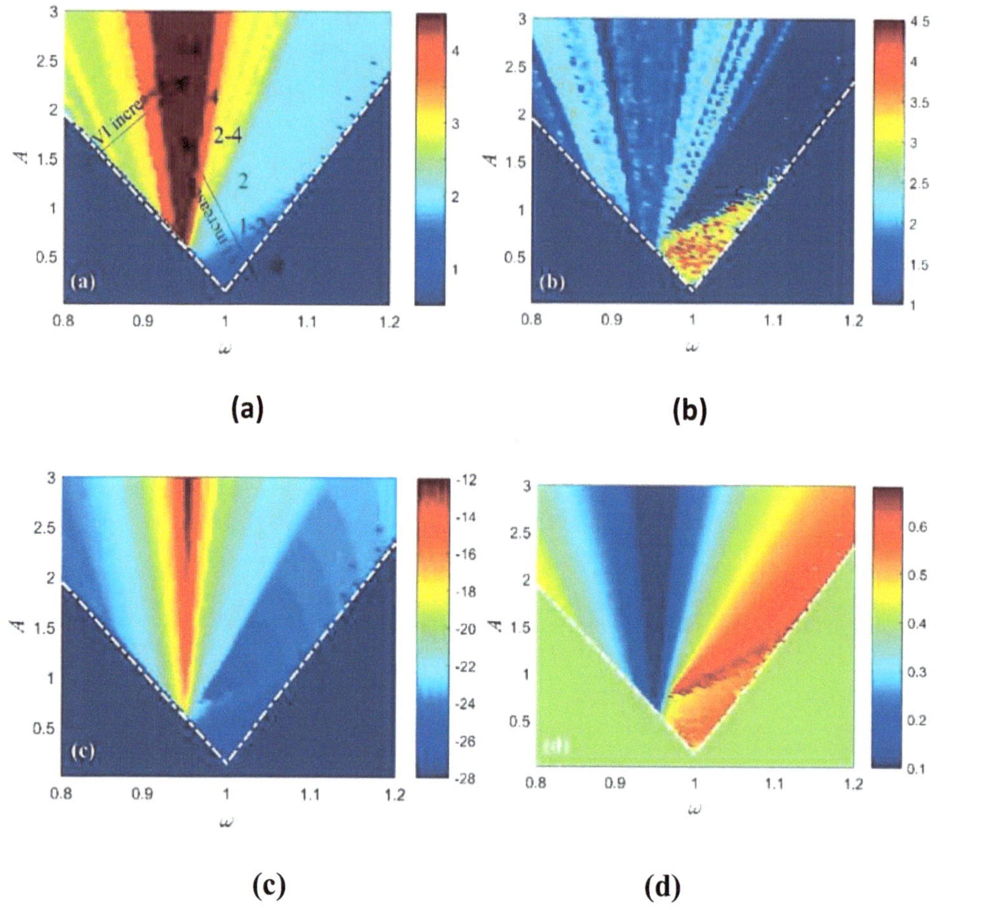

Fig. (16). Response regimes and harvesting performance of the VINES at different forcing condition, (a): average VIs per forcing period, an indicator used to distinguish different parametric regions according to the contact condition, (b): the ratio between the max amplitude and the average amplitude at the permanent regime, as an illustration for the response regimes, (c): average harvesting powerP_{load}, (d): average harvesting efficiency P_{load}/P_{in}. The parameters of the system are $\epsilon = 0.1$, $\kappa = 4$, $\eta = 4$, and $r = 0.95$.

Here, Fig. (**16a**) shows the average vibro-impacts (VIs) per forcing period, and Fig. (**16b**) gives the ratio between the maximum and mean vibration amplitudes of the primary mass. These two indicators thus allow one to easily distinguish the system response regimes of the system under different VI condition. One can see that as A increases or as ω approaches to the resonance frequency, *i.e.*, along the two arrows in Fig. (**16a**), the average impacts increases constantly. As a result, the whole parametric space could be divided into several different regions according to the VI condition. Below the white dashed line is the linear region VI_0, there is no

VI and the system produces only period response. The first nonlinear region VI_{1-2}, with 1-2 VIs per forcing period, and has the maximum amplitude ratio, distinguishes the SMR. Following VI_{1-2} is the region VI_2, in which the amplitude ratio is equal to 1 and there exists 2 VIs per forcing period, thus the period response with 2 symmetrical VIs per period is at hand. With further increasing VIs, the non-periodic response region VI_{2-4}, the periodic region VI_4 with four VIs per forcing period, and the non-periodic region $VI_{>4}$, can be respectively observed.

Once the different response regimes space are identified on the $\omega - A$, the harvesting power and harvesting efficiency are then studied in Figs. (**16c** and **d**), to compare their efficiency. Clearly enough, their performances show strong relevance to the response regions characterized in Figs. (**16a** and **b**). In the VI_0 region, the VINES is not activated, both the power and efficiency are at low level. For the different contact regions, the system also behaves very differently on the harvesting performance. First, in VI_{1-2} region where SMR is at hand, the harvesting power and efficiency shows a strong increase, and then within VI_2 region, periodic orbits with 2 VIs per forcing period are activated, representing another effective regime with the highest harvesting efficiency. For the other three regions with more frequent VIs (VI_{2-4}, VI_4, and $VI_{>4}$), the harvesting power and efficiency undergoes opposite trends when increasing the VIs: the former shows a constant increase, while the later decreases. To conclude, the performance of the VINES improves significantly when vibro-impact is activated. Moreover, the optimal harvesting efficiency shall be achieved in the regions of VI_{1-2} (SMR regime) and VI_2 (VI-2 period response). It is interesting to note that such conclusion consistent well with what is observed in the field of passive vibration control [23-25], where the SMR and the VI-2 period response accounts also for the most effective regimes. This is to say the fundamental role of TET in the effectiveness of VINES is the same, regardless of energy harvesting or vibration suppression.

CONCLUDING REMARKS

Nonlinear energy sink, as being a smart nonlinear technique, provides a new idea for vibration control and energy harvesting in aerospace engineering. It has been proved by many studies that by taking advantage of its essential nonlinearities, an NES could be designed as highly effective vibration absorbers and energy harvesters, in a broadband manner. The main focus when considering an NES for such applications should be paid on the effect of the rich nonlinear phenomena and responses. In this chapter, by considering two examples of NESs with respectively smooth and non-smooth nonlinearities, the main dynamical characteristics and their analysis methods are performed. Meanwhile, the corresponding effect of those nonlinear regimes on the vibration suppression or energy harvesting performance is also briefly evaluated. The presented results perform the main dynamical behaviors of an NES and the related analyzing methods. We sincerely hope these could help open the door to other researches who want to apply NES in the future design of aerospace systems.

CONSENT FOR PUBLICATION

Not applicable.

CONFLICT OF INTEREST

The authors declare no conflict of interest, financial or otherwise.

ACKNOWLEDGEMENT

Declared none.

REFERENCES

[1] Lee Y S, Vakakis A F, Bergman L A, *et al.* Passive non-linear targeted energy transfer and its applications to vibration absorption: a review. Proceedings of the Institution of Mechanical Engineers, Part K: Journal of Multi-body Dynamics, 2008, 222(2): 77-134. Frahm, H.: Device for damping vibrations of bodies., April 18 1911. US Patent 989,958

[2] Den Hartog J P. Mechanical vibrations. Courier Corporation, 1985.

[3] Krenk S, Frequency analysis of the tuned mass damper. Journal of Applied Mechanics, 2005, 72(6), 936–942.

[4] Aubry S, Kopidakis G, Morgante A M, *et al.* Analytic conditions for targeted energy transfer between nonlinear oscillators or discrete breathers. Physica B: Condensed Matter, 2001, 296(1-3): 222-236.

[5] Liu C, Sun Z, Shi K, *et al.* Robust dynamic output feedback control for attitude stabilization of spacecraft with nonlinear perturbations. Aerospace Science and Technology, 2017, 64: 102-121.

[6] Vakakis A F, Gendelman O V. Energy pumping in nonlinear mechanical oscillators: part II—resonance capture. J. Appl. Mech., 2001, 68(1): 42-48.

[7] Vakakis A F. Inducing passive nonlinear energy sinks in vibrating systems. J. Vib. Acoust., 2001, 123(3): 324-332.

[8] Lin D C, Oguamanam D C D. Targeted energy transfer efficiency in a low-dimensional mechanical system with an essentially nonlinear attachment. Nonlinear Dynamics, 2015, 82(1): 971-986.

[9] Lu X, Liu Z, Lu Z. Optimization design and experimental verification of track nonlinear energy sink for vibration control under seismic excitation. Structural Control and Health Monitoring, 2017, 24(12): e2033.

[10] AL-Shudeifat M A, Wierschem N E, Bergman L A, *et al.* Numerical and experimental investigations of a rotating nonlinear energy sink. Meccanica, 2017, 52(4): 763-779.

[11] Hsu Y S, Ferguson N S, Brennan M J. The experimental performance of a nonlinear dynamic vibration absorber//Topics in Nonlinear Dynamics, Volume 1. Springer, New York, NY, 2013: 247-257.

[12] Georgiades F, Vakakis A F. Dynamics of a linear beam with an attached local nonlinear energy sink. Communications in Nonlinear Science and Numerical Simulation, 2007, 12(5): 643-651.

[13] Li H, Li A, Kong X. Design criteria of bistable nonlinear energy sink in steady-state dynamics of beams and plates. Nonlinear Dynamics, 2021, 103(2): 1475-1497.

[14] Taleshi M, Dardel M, Pashaie M H. Passive targeted energy transfer in the steady state dynamics of a nonlinear plate with nonlinear absorber. Chaos, Solitons & Fractals, 2016, 92: 56-72.

[15] Feudo S L, Touzé C, Boisson J, *et al.* Nonlinear magnetic vibration absorber for passive control of a multi–storey structure. Journal of Sound and Vibration, 2019, 438: 33-53.

[16] Yang K, Zhang Y W, Ding H, *et al.* Nonlinear energy sink for whole-spacecraft vibration reduction. Journal of Vibration and Acoustics, 2017, 139(2).

[17] Bichiou Y, Hajj M R, Nayfeh A H. Effectiveness of a nonlinear energy sink in the control of an aeroelastic system. Nonlinear Dynamics, 2016, 86(4): 2161-2177.

[18] Lee Y S, Vakakis A F, Bergman L A, *et al.* Suppression aeroelastic instability using broadband passive targeted energy transfers, part 1: Theory. AIAA journal, 2007, 45(3): 693-711.

[19] Lee, Y.S., Kerschen, G., McFarland, D.M., Hill, W.J., Nichkawde, C., Strganac, T.W., Bergman, L.A., Vakakis, A.F.: Suppressing aeroelastic instability using broadband passive targeted energy transfers, part 2: experiments. AIAA J. 45, 2391–2400 (2007)

[20] Sun Y H, Zhang Y W, Ding H, *et al.* Nonlinear energy sink for a flywheel system vibration reduction. Journal of Sound and Vibration, 2018, 429: 305-324.

[21] Starosvetsky Y, Gendelman O V. Vibration absorption in systems with a nonlinear energy sink: nonlinear damping. Journal of Sound and Vibration, 2009, 324(3-5): 916-939.

[22] Kong X, Li H, Wu C. Dynamics of 1-dof and 2-dof energy sink with geometrically nonlinear damping: application to vibration suppression. Nonlinear Dynamics, 2018, 91(1): 733-754.

[23] Gendelman O V. Analytic treatment of a system with a vibro-impact nonlinear energy sink. Journal of Sound and Vibration, 2012, 331(21): 4599-4608.

[24] Li H, Li A, Zhang Y. Importance of gravity and friction on the targeted energy transfer of vibro-impact nonlinear energy sink. International Journal of Impact Engineering, 2021, 157: 104001.

[25] Li T, Seguy S, Berlioz A. On the dynamics around targeted energy transfer for vibro-impact nonlinear energy sink. Nonlinear Dynamics, 2017, 87(3): 1453-1466.

[26] Zhang Y W, Lu Y N, Zhang W, *et al.* Nonlinear energy sink with inerter. Mechanical Systems and Signal Processing, 2019, 125: 52-64.

[27] Javidialesaadi A, Wierschem N E. An inerter-enhanced nonlinear energy sink. Mechanical Systems and Signal Processing, 2019, 129: 449-454.

[28] Manevitch L I, Sigalov G, Romeo F, *et al.* Dynamics of a linear oscillator coupled to a bistable light attachment: analytical study. Journal of Applied Mechanics, 2014, 81(4).

[29] Romeo F, Sigalov G, Bergman L A, *et al.* Dynamics of a linear oscillator coupled to a bistable light attachment: numerical study. Journal of Computational and Nonlinear Dynamics, 2015, 10(1).

[30] Habib G, Romeo F. The tuned bistable nonlinear energy sink. Nonlinear Dynamics, 2017, 89(1): 179-196.

[31] Liu C, Shi K, Yue X, *et al.* Inertia-free saturated output feedback attitude stabilization for uncertain spacecraft. International Journal of Robust and Nonlinear Control, 2020, 30(13): 5101-5121..

[32] Gendeman, O.V., Starosvetsky, Y., Feldman, M.: Attractors of harmonically forced linear oscillator with attached nonlinear energy sink II: optimization of a nonlinear vibration absorber. Nonlinear Dynamics, 2008, 51(1–2): 47–57.

[33] Kremer, D., Liu, K. A nonlinear energy sink with an energy harvester: transient responses. Journal of Sound and Vibration, 2014, 333 (20): 4859–4880.

[34] Kremer, D., Liu, K. A nonlinear energy sink with an energy harvester: harmonically forced responses. Journal of Sound and Vibration, 2017,410: 287–302.

[35] Li, X., Zhang, Ding, Y. H., Chen, L. Integration of a nonlinear energy sink and a piezoelectric energy harvester. Applied Mathematics and Mechanics, 2017, 38 (7): 1019–1030.

[36] Zhang, Y., Tang, L., Liu, K. Piezoelectric energy harvesting with a nonlinear energy sink. Journal of Intelligent Material Systems and Structures, 2017, 28 (3): 307–322.

[37] Darabi, A., Leamy, M.J. Clearance-type nonlinear energy sinks for enhancing performance in electroacoustic wave energy harvesting. Nonlinear Dynamics, 2017, 87 (4): 2127–2146.

[38] Xiong, L., Tang, L., Liu, K., Mace, B.R. Broadband piezoelectric vibration energy harvesting using a nonlinear energy sink. Journal of Physics D: Applied Physics, 2018, 51 (18): 185502.

[39] Pennisi, G., Mann, B.P., Naclerio, N., Stephan, C., Michon, G. Design and experimental study of a nonlinear energy sink coupled to an electromagnetic energy harvester. Journal of Sound and Vibration, 2018, 437: 340–357.

[40] Li, H. and Li, A. Potential of a vibro-impact nonlinear energy sink for energy harvesting. Mechanical Systems and Signal Processing. 2021, 159:107827.

CHAPTER 7

Configuration Keeping Technology of Partial Space Elevators

Gefei Shi[1,*] and **Zheng H. Zhu**[2,*]

[1]*School of Aeronautics and Astronautics, Sun Yat-sen University, Guangzhou, 510275, P.R. China*
[2]*Department of Mechanical Engineering, York University, 4700 Keele Street, Toronto, Ontario, M3J 1P3, Canada*

Abstract: The partial space elevator (PSE) is a space transportation system that consists of one main satellite and one end body connected to a piece of tether. A climber can move along the tether conveying the cargo. This chapter studies a new configure-keeping technology for the stable cargo transportation of the PSE. The new technology contains two control modules. Module I predicts the optimal climber speed as a reference and suppresses the libration motions using the actuators on the climber. The control law of Module I is designed based on an analytical climber speed function, PPC control is used to compensate for system error. Module II further stabilizes the system by eliminating the possible disturbances in real-time acting on the end body. Two control modes are used given to further ensure the system configuration keeping. To test the validity of the proposed technology, two cases are simulated. The numerical results show that the proposed configuration keeping technology is very effective in dealing with the configuration keeping problems for the partial space elevators and other complex nonlinear dynamic systems in the aerospace engineering area.

Keywords: Aerospace engineering, Configuration keeping, Hybrid control scheme, Nonlinear dynamic system, Partial space elevator.

INTRODUCTION

The partial space elevator (PSE) is a candidate cargo transportation technology for the space station and future extra-large space structure [1, 2]. The general structure of a PSE is shown in Fig. (**1**) [3]. In the transportation period, the climber movement leads to libration motions of a PSE like a double pendulum, due to the action of Coriolis force. The libration may lead to instability of the PSE which is adverse to the transportation mission. Thus, it is desired to keep the configuration in a straight

*__Corresponding authors Gefei Shi and Zheng H. Zhu:__ School of Aeronautics and Astronautics, Sun Yat-sen University, Guangzhou, 510275, P.R. China; Department of Mechanical Engineering, York University, 4700 Keele Street, Toronto, Ontario, M3J 1P3, Canada; Tel: (86) 020-84112828; (416) 7362100 x 77729; E-mails: shigf@mail.sysu.edu.cn and gzhu@yorku.ca

Chuang Liu, Honghua Dai, Xiaokui Yue & Yiqing Ma (Eds.)

line with marginal or no libration. However, it is hard to suppress the libration and keep the configuration simultaneously since PSE is underactuated. Furthermore, this estimation of the unknown tether tensions is also an issue in the control process.

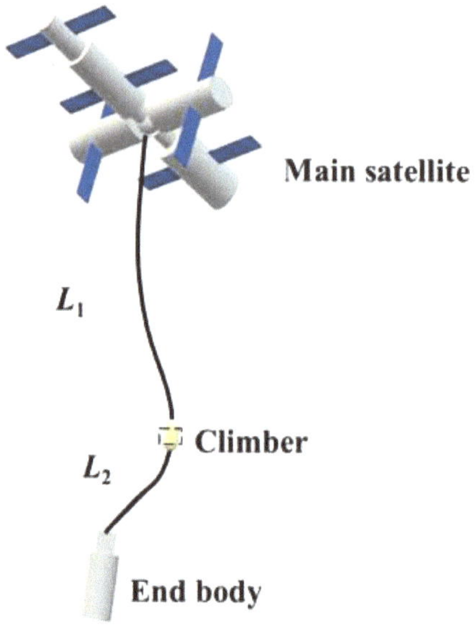

Fig. (1). General scheme of a PSE.

Up to now, many efforts have been devoted to stabilizing the PSE [4, 5]. The concept of PSE was first proposed by Woo and Misra in 2009 [6], the dynamic model matches the earlier studies of Misra [7] and Lorenzini [8]. The dynamic characteristics of a PSE can be analyzed using the classical two-piece dumbbell model (TPDM), which is simple and accurate [9]. In TPDM, the tethers are assumed to light straight rods, ignoring the tether's flexible motion. The flexible motions of the tether can be modeled by multiple nodes and rods [10]. Tang *et al.* [11], Li and Zhu *et al.* [12] proposed high-fidelity PSE models adopting the nodal position finite element method (NPFEM). Such a model is only computationally expensive for concept study. Yamagiwa *et al.* [13] concluded that the tether bend was not significant to a PSE due to the action of tether tension, in the engineering aspect. Using TPDM, Jung [14], Shi and Zhu *et al.* [9] found that the fast-transporting climber leads to significant libration of the PSE, and the libration motion is difficult to be eliminated, even though the libration angles are suppressed. Yu *et al.* [15] analyzed the dynamics of a classical two-body tethered system and found an analytical speed function to ensure a fixed configuration. Shi *et al.* [3] expanded this work and revealed the configuration of PSE is controllable to a fixed angle, relative to the local vertical, by controlling the climber speed and the end body's thrust in the transportation period. The aforementioned investigations also reveal the estimation of uncertain tether tensions is critical to both dynamics and control. One desired way is estimating the unknown tension to enhance the accuracy of the TPDM. The other way is to enhance the robustness of the control method in real-time.

The configuration stabilization of the PSE is also important. Tension control [16-19] is widely used to suppress the libration of a PSE. Yet, limited work has been done to keep the system at a desired stable state during the mission, which is critical to the safety of the on-orbit cargo transfer. Yu *et al.* [15] proposed an analytical deployment control law to hold the stabilization of a two-body tethered satellite. Yet for an underactuated three-body PSE, the method in [15] needs further compensation control. To control an underactuated PSE, the optimal control is a reasonable way [20], due to its ability in handling multiple control objectives and constraints [20-24]. Shi and Zhu proposed a parallel optimization method [17] with good computationally efficiency to control the underactuated PSE. These aforementioned works mainly aim to ensure the final state stability without considering the stability held during the mission period. To this goal, it is desired to add thrust controllers on the climber and the end body. Huang *et al.* [25] studied the stable control of the dexterous tethered space robot (DTSR), systematically, with mission design and observer-based controller design. The ground experiment has been used to validate the effectiveness of the proposed control method.

To release the heavy calculation burden of these optimal control schemes, Shi and Zhu [26] proposed a closed-loop velocity control strategy, yet, it cannot keep the configuration stable. Therefore, improvements are needed. To overcome the disadvantages of the optimal control [3] and its alternative [26], a novel prescribed performance control (PPC) is proposed due to its smooth performance subject to boundary constraints [27, 28]. The aforementioned studies give a way to stabilize the PSE free from the optimal control method. Moreover, they also reveal that the configuration can be kept by controlling tether tensions only. This is an advantage to saving fuel for practical missions.

To keep the configuration of a PSE in cargo transportation, the configuration keeping technology for the stable cargo transportation of the PSE is studied. First, a novel tension observing the law is proposed to estimate the tensions in tether so that the system uncertainties can be partly eliminated from the dynamic aspect. Then a two-module control technology is proposed to keep the configuration of the PSE in cargo transportation. The analytical control law in Module I is compensated by a prescribed performance control law. In Module II, two control schemes are proposed to fit two control modes. For the control mode without deploying/retrieving the end body, the thrusters on the end body with a sliding mode control law are used. To overcome the chattering, which is serious in the engineering application, the super-twisting sliding mode surface is used due to its ability to eliminate the chattering. Such that, the thrusters are driven by a super-twisting sliding model control law (STSM). In the mode, under which the end body can roll in/out through the tether, the tension and thrust control are coordinated at the end body to suppress the end body's libration. An event-triggered loaning control is proposed to redistribute the control inputs.

DYNAMICS OF THE PARTIAL SPACE ELEVATOR

Dynamic Model

A TPDM is used, as shown in Fig. (**2**), and the PSE is subject to a central gravitational field, neglecting all other external perturbations. The main satellite and the end body are simplified as lumped masses (M and m_2). The libration motions are described in an inertial frame OXY with the origin at the center of Earth, which is aligned with the orbital plane. The position of M can be denoted by a vector $\boldsymbol{r}$ measuring from O. The main satellite M, climber m_1, and the end body m_2 are connected by tethers with the lengths of L_1 and L_2, respectively. The libration angles of the climber and the end-body (θ_1 and θ_1) are measured from the unit vector $\boldsymbol{e}_r$ of $\boldsymbol{r}$ as shown in Fig. (**2**). The main satellite is assumed moving in a circular Kepler orbit. Based upon these assumptions, the dynamic equations of the PSE can be written as [9]

$$\ddot{\theta}_1 = -\frac{3\mu \sin\theta_1 \cos\theta_1}{r^3} - \frac{2\omega(\dot{L}_1 + \dot{\theta}_1)}{L_1} - \frac{F_1}{L_1 m_1} - \frac{T_2 \sin(\theta_1 - \theta_2)}{L_1 m_1} \quad \textbf{(1)}$$

$$\begin{aligned}\ddot{\theta}_2 = &-\frac{3\mu \sin(2\theta_2)}{2r^3} - \frac{3\mu \cos\theta_2 \sin(\theta_1 - \theta_2) L_1}{r^4} - \frac{2\dot{L}_2(\omega + \dot{\theta}_2)}{L_2} \\ &+ \frac{F_1 \cos(\theta_1 - \theta_2)}{L_2 m_1} - \frac{F_2}{L_2 m_2} + \frac{T_1 \sin(\theta_1 - \theta_2)}{L_2 m_1}\end{aligned} \quad \textbf{(2)}$$

$$\ddot{L}_1 = \frac{3\mu L_1 \cos^2\theta_1}{r^3} + \frac{3\mu L_1^2 \cos^2\theta_1}{r^4} - \frac{T_1}{m_1} + \frac{T_2\cos(\theta_1-\theta_2)}{m_1} + 2\omega L_1\dot{\theta}_1 + L_1\dot{\theta}_1^2 \quad (3)$$

$$\begin{aligned}\ddot{L}_2 =& \frac{3\mu L_2 \cos^2\theta_2}{r^3} - \frac{9\mu L_1 L_2\cos\theta_1}{2r^4} - \frac{3\mu L_1 L_2\cos(\theta_1-2\theta_2)}{2r^4} - \frac{3\mu L_2^2\cos\theta_2}{r^4}\\ &+ \frac{\cos(\theta_1-\theta_2)T_1}{m_1} - \frac{T_2}{m_1} - \frac{T_2}{m_2} + 2\omega L_2\dot{\theta}_2 + L_2\dot{\theta}_2^2 + \frac{\sin(\theta_1-\theta_2)F_1}{m_1}\end{aligned} \quad (4)$$

where $\mu = 3.9872\times10^{14}$, F_1 and F_2 are the thruster forces acting on the climber and the end body, respectively.

In defining $L_1 + L_2 = L_0 + L_c$, L_0 is the initial total length of two tethers, and L_c is the length increment, or variation, relative to L_0 implemented by deploying or retrieving the end body along the tether. Then the dynamics in Eqs. (**4** and **5**) can be rewritten as

$$\begin{aligned}\ddot{\theta}_2 =& -\frac{3\mu\sin(2\theta_2)}{2r^3} - \frac{3\mu\cos\theta_2\sin(\theta_1-\theta_2)L_1}{r^4} + \frac{T_1\sin(\theta_1-\theta_2)}{(L_0-L_1+L_c)m_1}\\ &-\frac{2(\omega+\dot{\theta}_2)(\dot{L}_c-\dot{L}_1)}{L_0-L_1+L_c} + \frac{F_2}{(L_0-L_1+L_c)m_2}\end{aligned} \quad (5)$$

$$\ddot{L}_1 = \frac{3\mu L_1\cos^2\theta_1}{r^3} + \frac{3\mu L_1^2\cos^2\theta_1}{r^4} + (2\omega+\dot{\theta}_1)L_1\dot{\theta}_1 - \frac{T_1}{m_1} + \frac{\cos(\theta_1-\theta_2)T_2}{m_1} \quad (6)$$

$$\begin{aligned}\ddot{L}_c =& \frac{3\mu(L_0-L_1+L_c)\cos\theta_2}{r^3}\left(\cos\theta_2 - \frac{L_0-L_1+L_c}{r}\right) + (2\omega+\dot{\theta}_2)(L_0-L_1+L_c)\dot{\theta}_2\\ &-\frac{3\mu L_1(L_0-L_1+L_c)}{2r^4}\left[3\cos\theta_1+\cos(\theta_1-2\theta_2)\right] + (2\omega+\dot{\theta}_1)L_1\dot{\theta}\\ &+\frac{3\mu L_1\cos^2\theta_1}{r^3}\left(1+\frac{L_1}{r}\right) + \frac{\left[\cos(\theta_1-\theta_2)-1\right]T_1}{m_1} - \frac{\left[m_1 - m_2\cos(\theta_1-\theta_2)+m_2\right]T_2}{m_1 m_2}\end{aligned} \quad (7)$$

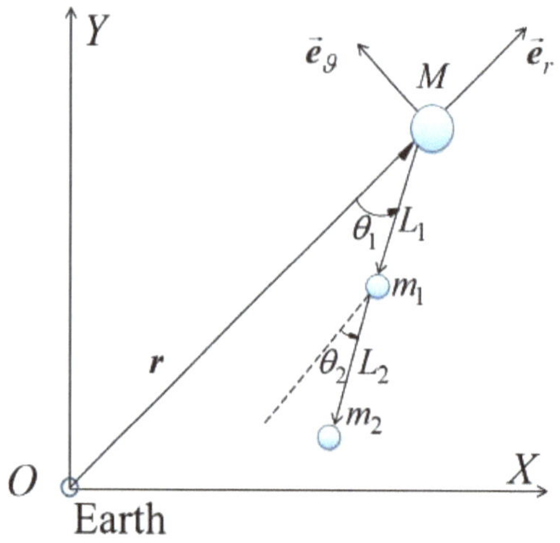

Fig. (2). The simple structure of a PSE.

Speed Function

To make the transportation stable, the climber and the end body should be kept at a constant pitch angle, such that $\theta_1 = \theta_2 = \theta_e$. The corresponding speed function can be derived from Eq. (1) by setting all first and second-order derivatives to zero, such that, $\dot{\theta}_1 = \dot{\theta}_2 = 0$ and $\ddot{\theta}_1 = \ddot{\theta}_2 = 0$. The speed function can be obtained from the dynamic equations when the thrusts are zero, such that [15].

$$\dot{L}_1 = -\frac{3}{4}\omega L_1 \sin 2\theta_e \tag{8}$$

Since a PSE has an additional subsystem, as shown in Eq. (**2**). Under the expected stable condition, $\theta_2 = \theta_e$ and $\dot{\theta}_1 = \dot{\theta}_2 = 0$ cannot be kept if no thrust acts. The thrust control will be discussed in the next section. The desired uncompensated acceleration function is the differential of Eq. (**8**),

$$\ddot{L}_1 = -\frac{3}{4}\omega \dot{L}_1 \sin 2\theta_e \tag{9}$$

The libration effects of the end body need to be compensated by planning the climber's speed. Thus, Eq. (**9**), the speed function, can be written as

$$\ddot{L}_1 = -\frac{3}{4}\omega \dot{L}_1 \sin 2\theta_e + \Delta\ddot{L}_1 \qquad \textbf{(10)}$$

The configuration keeping technology is based on the climber speed function.

CONFIGURATION KEEPING CONTROL

Control Scheme

To keep the desired configuration fixed, the control objectives are,

$$(1).\,\theta_1 \to \theta_e \quad and \quad (2).\,\theta_2 \to \theta_1 \qquad \textbf{(11)}$$

Accordingly, a two-module control scheme is proposed to deal with the configuration keeping issue of a PSE. The scheme is shown in Fig. (**3**). in which $\boldsymbol{x} = [\theta_1, \theta_2]^T$.

The control objective (1), $\theta_1 = \theta_e$, can be achieved by controlling the climber movement by regulating $\ddot{L}_1$ only. The control objective (2) can be achieved by adjusting F_2 only or coordinating controlling thrust F_2, as shown in Eq. (**5**). To avoid the thrust saturation, the deployment or retrieval of the end body is controlled by a coordinated control law for the control inputs $\ddot{L}_c$ and F or using F only to make $\theta_1 - \theta_2 \to 0,\ \dot{\theta}_1 - \dot{\theta}_2 \to 0$.

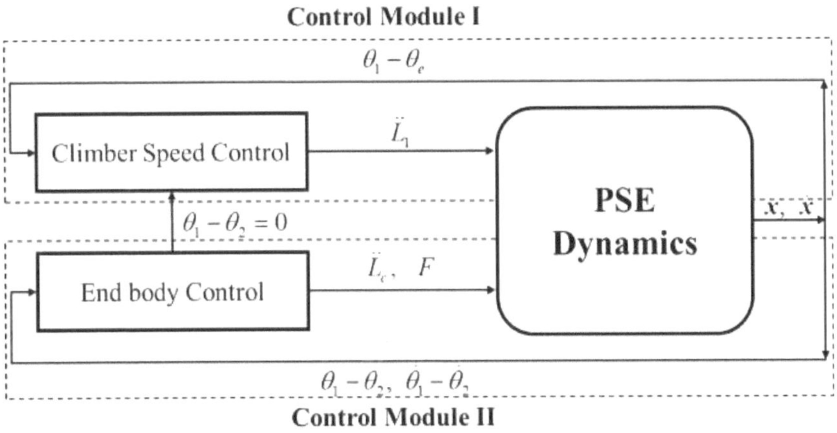

Fig. (3). Scheme of the two-module control.

Disturbance Observer

It is hard to derive the tether tensions by the corresponding terms in Eqs. (**1**) and (**2**). Since in the desired condition, the tension terms are zero, such that

$$-\frac{T_2\sin(\theta_1-\theta_2)}{L_1 m_1},\ \frac{T_1\sin(\theta_1-\theta_2)}{\left(L_0-L_1+L_c\right)m_1}\ \ \textit{and}\ \ \sin(\theta_1-\theta_2)=0$$

Thus, the tension terms can be regarded as unknown terms, and the effects of the tensions on the libration motions can be estimated. Tensions should be bounded due to the limited capability of the rolling device on the climber. Thus, the unknown tension terms can be assumed as bounded disturbances, the dynamics can be rewritten

$$\ddot{\theta}_1=-\frac{3\mu\sin\theta_1\cos\theta_1}{r^3}-\frac{2\omega(\dot{L}_1+\dot{\theta}_1)}{L_1}-\frac{F_1}{L_1 m_1}+d_1 \quad \textbf{(12)}$$

$$\begin{aligned}\ddot{\theta}_2=&-\frac{3\mu\sin(2\theta_2)}{2r^3}-\frac{3\mu\cos\theta_2\sin(\theta_1-\theta_2)L_1}{r^4}-\frac{2\dot{L}_2(\omega+\dot{\theta}_2)}{L_0-L_1+L_c}\\&+\frac{F_1\cos(\theta_1-\theta_2)}{\left(L_0-L_1+L_c\right)m_1}-\frac{F_2}{\left(L_0-L_1+L_c\right)m_2}+d_2\end{aligned} \quad \textbf{(13)}$$

Where

$$\begin{aligned}d_1&=-\frac{T_2\sin(\theta_1-\theta_2)}{L_1 m_1}\\d_2&=\frac{T_1\sin(\theta_1-\theta_2)}{\left(L_0-L_1+L_c\right)m_1}\end{aligned} \quad \textbf{(14)}$$

To estimate d_1 and d_2, a nonlinear disturbance observer is designed as follows [29]

$$\begin{aligned}\hat{d}_i&=z_i+p_i(\boldsymbol{\theta},\dot{\boldsymbol{\theta}})\\\dot{z}_i&=-\Phi_i(\boldsymbol{\theta},\dot{\boldsymbol{\theta}})z_i+\Phi_i(\boldsymbol{\theta},\dot{\boldsymbol{\theta}})[-p_i(\boldsymbol{\theta},\dot{\boldsymbol{\theta}})-f_i-\varpi_i]\end{aligned} \quad \textbf{(15)}$$

where $i=1$ and 2, z_i is the auxiliary variable, $\boldsymbol{\theta}=[\theta_1,\ \theta_2]$, f_i is the known nonlinear section in libration dynamic equitation, ϖ_i is the generalized control input, $p_i(\boldsymbol{\theta},\dot{\boldsymbol{\theta}})$

is a nonlinear function that is to be designed, $\Phi_i(\boldsymbol{\theta},\dot{\boldsymbol{\theta}})$ is the gain of the nonlinear observer, and obeys the relationship of

$$\Phi_i(\boldsymbol{\theta},\dot{\boldsymbol{\theta}})\ddot{\boldsymbol{\theta}} = \Phi_i(\boldsymbol{\theta},\dot{\boldsymbol{\theta}})\left\{\begin{bmatrix} f_1 \\ f_2 \end{bmatrix} + \begin{bmatrix} \omega_1 \\ \omega_2 \end{bmatrix} + \begin{bmatrix} d_1 \\ d_2 \end{bmatrix}\right\} = \frac{dp_i(\boldsymbol{\theta},\dot{\boldsymbol{\theta}})}{dt} \tag{16}$$

Then the error of the nonlinear disturbance observer is

$$\tilde{d}_i = d_i - \hat{d}_i \tag{17}$$

Based on the dynamic characteristics of the PSE near the stable state, the tension sections change slowly, such that

$$\dot{d}_i = 0 \tag{18}$$

The first-order observer error can be written as

$$\dot{\tilde{d}}_i = -\dot{\hat{d}}_i = -\dot{z}_i - \dot{p}_i = -\Phi_i \cdot \tilde{d}_i \tag{19}$$

Designing $\Phi_i(\boldsymbol{\theta},\dot{\boldsymbol{\theta}}) = a_i = const > 0$, and $p_i(\boldsymbol{\theta},\dot{\boldsymbol{\theta}}) = a_i\dot{\theta}_i$, the solution of $\tilde{d}_i$ has the form of

$$\tilde{d}_i(t) = \exp(-a_i t)\cdot\tilde{d}_i(0) \tag{20}$$

When $a_i > 0$, $\tilde{d}_i$ will exponentially converge to zero $t \to \infty$, such that, the observer is asymptotically stable [30]. Then the system dynamics can be written as

$$\begin{bmatrix} \ddot{\theta}_1 \\ \ddot{\theta}_2 \end{bmatrix} = \begin{bmatrix} f_1 \\ f_2 \end{bmatrix} + \boldsymbol{B}\left(\begin{bmatrix} F_1 \\ F_2 \end{bmatrix} + \boldsymbol{B}^{-1}\begin{bmatrix} d_1 \\ d_2 \end{bmatrix}\right) \tag{21}$$

where

$$\boldsymbol{B} = \begin{bmatrix} -\dfrac{1}{L_1 m_1} & 0 \\ \dfrac{\cos(\theta_1 - \theta_2)}{L_2 m_1} & -\dfrac{1}{L_2 m_2} \end{bmatrix} \quad \textbf{(22)}$$

A case study is considered with the initial states and parameters shown in Table **1**, to validate the proposed observer law.

Table 1. System parameters and initial states of PSE.

Parameters	**Values**
Orbital radius of the main satellite, r (m)	7.1×10^5
Initial true anomaly, $\vartheta(0)$	0
Mass of climber, m_1 (kg)	500
Mass of end body, m_2 (kg)	10^3
Fixed total tether length, L_0 (m)	2×10^4
Initial distance between climber and end body, $L_2(0)$, (m)	500
Initial libration angle and angular velocity, ($\theta_i(0)$, $\dot{\theta}_i(0)$)	0.3, 0.3
Initial climber speed, $\dot{L}_1$ (m/s)	-9.0837
Initial auxiliary variable z_i	0

The climber moves upward following the speed function in Eq. (**9**) with $\theta_e = 0.3$, and the total tether length is constant. The initial speed can be obtained through Eq.

(**9**). The transfer period stops by $L_1 \le 500\,m$. All simulations were carried out by MATLAB with the RK-4 integrator. The time step is $0.001s$. The results are shown in Figs. (**4** and **5**). Fig. (**4a**) shows that the climber's movement follows the designed speed function, as shown in Fig. (**4b**), the PSE liberates with the magnitude of 0.3 *rad* in the transfer period. When the climber is moving along the tether, Coriolis forces act on the climber and the end body continuously, as shown in Eqs. (**12**) and (**13**). Thus, θ_1 and θ_2 will not be damped if no active control works. Thus, compensating control is needed. Fig. (**4c**) shows that d_1 increases due to the "near singular". When L_1 is small and decreasing, this enlarges the magnitude of d_1 sharply even though T_2 and $\sin(\theta_1 - \theta_2)$ change slightly, as shown in Eq. (**14**). However, when the climber stops moving along the tether, the magnitudes of d_1 will not enlarge. This is because, no additional energy is injected into the system. On the contrary, by the end of the transfer period, the magnitude of d_2 decreases, as shown in Eq. (**14**). Fig. (**4c**) shows that the real tensions are greater than zero, such that, the dynamic model is effective. Comparing the results in Fig. (**4b**), the changing tendencies of d_1 and d_2 do not match the tendency of the tether tensions. This is because, these two terms are affected by the tensions, libration angles, and tether lengths, simultaneously, as shown in Eq. (**14**). In summary, the estimated tension terms match the real tension terms well, which denotes that the designed disturbance observer is effective, as shown in Fig. (**4b**). Moreover, the estimation of the tension terms also makes the system dynamics which is a benefit to controller design. The proposed observing technology can also be used to handle complex tethered space systems with long flexible tethers.

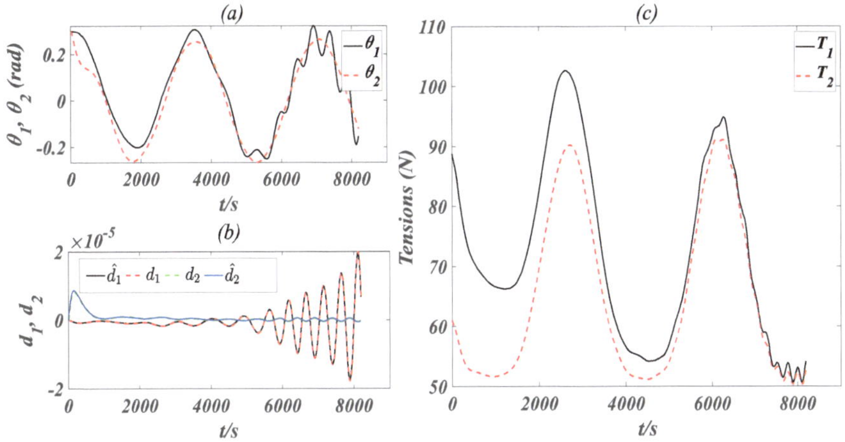

Fig. (4) (a) Libration angles. **(b)** Estimated and real disturbances. **(c)** Real tensions in tethers.

Module I: Moving Speed using PPC

The PPC is proposed to stabilize the configuration of a PSE in the cargo transfer period in a fast manner so that the climber's libration angle θ_1 converges to the desired angle with limited control input and adoptable overshoot.

The control error *z*(*t*) between θ_1 and θ_e are expressed as

$$\begin{aligned} -\delta\rho(t) \le z(t) = \theta_1 - \theta_e + k\dot{\theta}_1 \le \rho(t) \quad \text{in case} \quad z(0) \ge 0 \\ -\rho(t) \le z(t) = \theta_1 - \theta_e + k\dot{\theta}_1 \le \delta\rho(t) \quad \text{in case} \quad z(0) \le 0 \end{aligned} \tag{23}$$

where $\rho(t)$ is the prescribed performance function, $k > 0 = const$ and δ is a constant to limit the overshoot, such that

$$\begin{aligned} 0 \le \delta \le 1, \quad z(0) \ne 0 \\ 0 < \delta \le 1, \quad z(0) = 0 \end{aligned} \tag{24}$$

The smooth prescribed performance function $\rho(t): R^+ \to R^+$ is designed as

$$\begin{aligned} &\rho(t) = (\rho_0 - \rho_\infty)e^{-l \cdot t} + \rho_\infty \\ &\lim_{t\to\infty} \rho(t) = \rho_\infty > 0 \end{aligned} \tag{25}$$

where the constant *l* denotes the convergence speed, ρ_0 and ρ_∞ are the maximum allowable sizes of the tracking error *z* at the steady-state. To structure a convergence domain, the parameters should obey the following condition

$$\rho_0 > |z(0)| > \rho_\infty > 0 \tag{26}$$

Since it is hard to derive a control law from Eq. (**23**), the general error is transformed into

$$\varepsilon(t)=\begin{cases}\ln\left(\dfrac{s(t)+\delta}{1-s(t)}\right), & s(0)\geq 0\\ \ln\left(\dfrac{s(t)+1}{\delta-s(t)}\right), & s(0)<0\end{cases} \tag{27}$$

where $s(t)=z(t)/\rho(t)$. The derivation of ε with respect to time is

$$\begin{aligned}&\dot{\varepsilon}=H\beta\\ &H=\frac{\rho\dot{z}-z\dot{\rho}}{\rho^2}\\ &\beta=\begin{cases}\dfrac{1+\delta}{(1-s)(\delta+s)}, & s(0)\geq 0\\ \dfrac{1+\delta}{(1+s)(\delta-s)}, & s(0)<0\end{cases}\end{aligned} \tag{28}$$

Let

$$H=-\lambda\varepsilon\beta/\rho^2 \tag{29}$$

where $\lambda=const>0$, $\dot{z}$ can be derived as

$$\dot{z}=\frac{z\dot{\rho}-\lambda\varepsilon\beta}{\rho}=\dot{\theta}_1+k\ddot{\theta}_1 \tag{30}$$

When $\theta_1-\theta_2$ is kept near zero, $-\dfrac{T_2\sin(\theta_1-\theta_2)}{L_1m_1}$ is a small scalar that can be ignored [3, 26]. Thus, Eq. (**1**) can be simplified as

$$\ddot{\theta}_1=-\frac{3\omega^2\sin 2\theta_1}{2}-\frac{2\left(\omega+\dot{\theta}_1\right)\dot{L}_1}{L_1} \tag{31}$$

Substituting Eq. (**31**) into (**30**) and then taking time derivative yields

$$\ddot{L}_1 = \frac{\dot{L}_1^2}{L_1} - \frac{3\omega^2 L_1 \dot{\theta}_1 \cos 2\theta_1}{2\left(\omega + \dot{\theta}_1\right)} - \frac{L_1 - 2\dot{L}_1}{2\left(\omega + \dot{\theta}_1\right)k}\left(\frac{3\omega^2 \sin 2\theta_1}{2} + \frac{2\left(\omega + \dot{\theta}_1\right)\dot{L}_1}{L_1}\right) + \frac{L_1\left(\lambda\varepsilon\dot{\beta} + \lambda\dot{\varepsilon}\beta - z\ddot{\rho}\right)}{2\left(\omega + \dot{\theta}_1\right)k\rho} - \frac{L_1\dot{\rho}\left(\lambda\varepsilon\beta - z\dot{\rho} + \dot{z}\rho\right)}{2\left(\omega + \dot{\theta}_1\right)k\rho^2} \quad \textbf{(32)}$$

Equation (32) is the prescribed performance control input in Control Module I.

Design a Lyapunov candidate function

$$V = \varepsilon^2 / 2 \quad \textbf{(33)}$$

Taking the derivative of V with respect to time yields

$$\dot{V} = \dot{\varepsilon} \cdot \varepsilon = H\beta\varepsilon \quad \textbf{(34)}$$

Substituting Eq. (**29**) into Eq. (**34**) yields

$$\dot{V} = -\lambda\varepsilon^2\beta^2 / \rho^2 \le 0 \quad \textbf{(35)}$$

Thus, θ_1 is stable and will converge to the desired steady-state under the proposed PPC law.

Module II: Control for the End Body

The analytical climber moving speed can partly suppress the libration of the system. The further stabilization is handled by a real-time closed-loop control implemented by thrusters and tether tension on the end body.

Using Thrusters Only

Under the analytical climber moving speed, the libration of the system will be suppressed partly. To further stabilize the system, a real-time closed-loop control implemented by thrusters on the climber and the end body is needed. An STSM control is applied. The sliding surface can be presented as

$$\boldsymbol{s} = \boldsymbol{C}\left(\boldsymbol{\theta} - \boldsymbol{\theta}_d\right) + \boldsymbol{K}\left(\dot{\boldsymbol{\theta}} - \dot{\boldsymbol{\theta}}_d\right) \quad \textbf{(36)}$$

where $\boldsymbol{\theta}_d$ and $\dot{\boldsymbol{\theta}}_d$ are desired libration states,

$$\boldsymbol{C} = \begin{bmatrix} c_1 & 0 \\ 0 & c_2 \end{bmatrix}, \ \boldsymbol{K} = \begin{bmatrix} k_1 & 0 \\ 0 & k_2 \end{bmatrix} \tag{37}$$

$\boldsymbol{C}$ and $\boldsymbol{K}$ are the positive constant diagonal matrices. Therefore, the input-output dynamics with unknown disturbances can be presented as:

$$\dot{\boldsymbol{s}}=\boldsymbol{C}\left(\dot{\boldsymbol{\theta}}-\dot{\boldsymbol{\theta}}_d\right)+\boldsymbol{K}\left(\ddot{\boldsymbol{\theta}}-\ddot{\boldsymbol{\theta}}_d\right)=\boldsymbol{C}\left(\dot{\boldsymbol{\theta}}-\dot{\boldsymbol{\theta}}_d\right)+\boldsymbol{K}\left(\boldsymbol{f}+\hat{\boldsymbol{d}}+\boldsymbol{B}\boldsymbol{u}-\ddot{\boldsymbol{\theta}}_d\right) \tag{38}$$

where $\boldsymbol{f}=[f_1, f_2]^T$, $\hat{\boldsymbol{d}}=[\hat{d}_1, \hat{d}_2]^T$ that can be obtained via Eq. (**15**) and $\boldsymbol{u}=[F_1, F_2]^T$. Let

$$\begin{aligned} \dot{\boldsymbol{s}} &= -\boldsymbol{\alpha}\sqrt{|s|}sign(\boldsymbol{s})+\boldsymbol{v} \\ \dot{\boldsymbol{v}} &= -\frac{\boldsymbol{\beta}}{2}sign(\boldsymbol{s}) \end{aligned} \tag{39}$$

where $\boldsymbol{v}=[v_1, v_2]^T$ is an auxiliary vector, and

$$\boldsymbol{\alpha} = \begin{bmatrix} \alpha_1 & 0 \\ 0 & \alpha_2 \end{bmatrix}, \ \boldsymbol{\beta} = \begin{bmatrix} \beta_1 & 0 \\ 0 & \beta_2 \end{bmatrix} \tag{40}$$

are the positive constant diagonal matrices. Therefore, the generalized control input can be written as

$$\varpi = \boldsymbol{K}^{-1}\left[-\boldsymbol{\alpha}\sqrt{|s|}sign(\boldsymbol{s})+\boldsymbol{v}-\boldsymbol{C}\left(\dot{\boldsymbol{\theta}}-\dot{\boldsymbol{\theta}}_d\right)\right]+\ddot{\boldsymbol{\theta}}_d-\boldsymbol{f}-\hat{\boldsymbol{d}} \tag{41}$$

The control input vector is

$$\boldsymbol{u} = \boldsymbol{B}^{-1}\left\{\boldsymbol{K}^{-1}\left[-\boldsymbol{\alpha}\sqrt{|s|}sign(\boldsymbol{s})+\boldsymbol{v}-\boldsymbol{C}\left(\dot{\boldsymbol{\theta}}-\dot{\boldsymbol{\theta}}_d\right)\right]+\ddot{\boldsymbol{\theta}}_d-\boldsymbol{f}-\hat{\boldsymbol{d}}\right\} \tag{42}$$

Stability of the control law

Introducing a new state vector for the *i-th* sliding surface

$$\boldsymbol{\eta}^i = \left[\eta_1^i,\ \eta_2^i\right]^T = \left[\sqrt{|s_i|}sign(s_i),\ v_i\right]^T \tag{43}$$

where $i=1, 2$. Then the sliding surface system (39) can be rewritten as

$$\begin{bmatrix} \dot{\eta}_1^i \\ \dot{\eta}_2^i \end{bmatrix} = \underbrace{\frac{1}{2|\eta_1^i|}\begin{bmatrix} -\alpha_i & 1 \\ -\beta_i & 0 \end{bmatrix}}_{\boldsymbol{A}}\begin{bmatrix} \eta_1^i \\ \eta_2^i \end{bmatrix} \tag{44}$$

It can be seen that when $\boldsymbol{\eta}^i \to 0$ infinite time, then $s_i,\ \dot{s}_i \to 0$ infinite time. The following Lyapunov function is introduced [31]

$$V_i = \left(\boldsymbol{\eta}^i\right)^T \underbrace{\begin{bmatrix} \lambda + 4\varepsilon^2 & -2\varepsilon \\ -2\varepsilon & 1 \end{bmatrix}}_{\boldsymbol{P}} \boldsymbol{\eta}^i \tag{45}$$

where the real number $\lambda > 0$, such that the matrix $\boldsymbol{P}$ is positive definite. The derivative of Eq. (45) is

$$\begin{aligned} \dot{V}_i &= \left(\boldsymbol{\eta}^i\right)^T\left(\boldsymbol{A}^T\boldsymbol{P} + \boldsymbol{P}\boldsymbol{A}\right)\boldsymbol{\eta}^i \\ &\le \frac{-1}{2|\eta_1^i|}\left(\boldsymbol{\eta}^i\right)^T \underbrace{\begin{bmatrix} 2\lambda\alpha_i + 4\varepsilon(2\varepsilon\alpha_i - \beta_i) & \beta_i - 2\varepsilon\alpha_i - \lambda - 4\varepsilon^2 \\ \beta_i - 2\varepsilon\alpha_i - \lambda - 4\varepsilon^2 & 4\varepsilon \end{bmatrix}}_{\boldsymbol{Q}} \boldsymbol{\eta}^i \end{aligned} \tag{46}$$

Let

$$\beta_i = 2\varepsilon\alpha_i \tag{47}$$

then $\boldsymbol{Q}$ will be positive definite if

$$\alpha_i > \frac{\lambda}{8\varepsilon} - \varepsilon + \frac{2\varepsilon^3}{\lambda} \quad \textbf{(48)}$$

Thus, the system is stable, the proposed STSM control will drive the system to the sliding mode surface such that $\boldsymbol{s} = 0$ and $\dot{\boldsymbol{s}} = 0$.

Coordinated Control

The $\theta_1 - \theta_2$ can be kept at zero by two PD controllers. However, the final L_c will not be zero. In some cases, the end body is being deployed or retrieved over 5,000 m. This is undesired to neither the configuration nor mechanical structure of a PSE. To achieve the objective of the Control Module II with limited thrust and a slight change of L_c, a new event-triggered loaning control law is proposed to switch the control mode to reduce the variation of L_c with a limited thrust by the following IF-THEN rule [32]:

Switch Algorithm: Event-triggered loaning control
If[1] $\lvert F \rvert \le F_{\max}$ **Then[1]** run the *Repaying Mode* and the control laws: $$F^{+} = f + 2\dot{L}_c^{+}\left(\omega + \dot{\theta}_2\right)m_2 + u_g^F \quad \textbf{(49)}$$ $$\ddot{L}_c^{+} = -k_{p3}L_c - k_{d3}\dot{L}_c \quad \textbf{(50)}$$ $$u_g^F = -b_1^{-1}\left[-\left(k_{p1} + k_{p2}\right)\Delta\theta - \left(k_{d1} + k_{d2}\right)\Delta\dot{\theta}\right] \quad \textbf{(51)}$$ To deal with this issue, a deeper switching law is given as: **If[2]** $\lvert F^{+} \rvert \le F_{\max}$ **Then[2]** the control continues. **Else[2]** $$F = F_{\max} sign\left(F^{+}\right) \quad \textbf{(52)}$$ **Else If[1]** $\lvert F \rvert \le F_{\max}$ **Then[1]** run the *Borrowing Mode*.

where $F_{\max}$ is the upper bound of F, F^{+} is the new control law of thrust, $\ddot{L}_c^{+}$ is the new control input to eliminate L_c, k_{p3} and k_{d3} are the new control parameters that should be positive and not too large or the required F^{+} may be greater than $F_{\max}$. In this work, k_{p3} and k_{d3} are chosen as the constants, in future works, the adaption laws are desired to be considered.

The end body moves along the tether following a general PD control law of F and $\dot{L}_c$ as,

$$\begin{cases} F = F_{eq} + F_c \\ F_{eq} = -b_1^{-1} f \\ F_c = -b_1^{-1}\left(-k_{p1}\Delta\theta - k_{d1}\Delta\dot{\theta}\right) \end{cases} \tag{53}$$

$$\dot{L}_c = -b_2^{-1}\left(-k_{p2}\Delta\theta - k_{d2}\Delta\dot{\theta}\right) \tag{54}$$

where $\Delta\theta = \theta_1 - \theta_2$, the control parameters $k_{p1}, k_{p2}, k_{d1}, k_{d2} \geq 0$ are constants satisfying $k_{p1} + k_{p2} > 0$ and $k_{d1} + k_{d2} > 0$

Substituting Eqs. (**53-54**) into the equation of configuration

$$\begin{aligned} &\ddot{\theta}_1 - \ddot{\theta}_2 = f + b_1 F + b_2 \dot{L}_c \\ &f = \frac{3\omega^2\left(\sin 2\theta_2 - \sin 2\theta_1\right)}{2} - \frac{2(\omega+\dot{\theta}_1)\dot{L}_1}{L_1} - \frac{2(\omega+\dot{\theta}_2)\dot{L}_1}{L_0 - L_1 + L_c} \\ &\quad - \frac{T_2\sin(\theta_1-\theta_2)}{L_1 m_1} - \frac{T_1\sin(\theta_1-\theta_2)}{(L_0 - L_1 + L_c)m_1} \\ &b_1 = -\frac{1}{(L_0 - L_1 + L_c)m_2} \\ &b_2 = \frac{2(\omega+\dot{\theta}_2)}{L_0 - L_1 + L_c} \end{aligned} \tag{55}$$

yields a combined PD controller,

$$\Delta\ddot{\theta} = -\left(k_{p1} + k_{p2}\right)\Delta\theta - \left(k_{d1} + k_{d2}\right)\Delta\dot{\theta} \tag{56}$$

Thus, the subsystem in the Control Module II becomes a classical PD controlled system that is exponentially stable stability because of $k_{p1} + k_{p2} > 0,\ k_{d1} + k_{d2} > 0$.

Under the *Borrowing Mode*, the thruster acts as a "Borrower" who compensates the thrust limitation with the cost of the change of L_c. Under the *Repaying Mode*, the thruster pays back the borrowed L_c. The "Borrow-Repay" mechanism likes a loaning process. Thus, the objective of Control Module II can be achieved by using the Event-triggered mechanism. By the end of the cargo transfer period, the total length of the tether can be kept in its initial state.

CASE STUDY

Case 1: Configuration Keeping Using Thrusters Only on the End Body

Control Module I covers the transfer period by controlling the climber speed only, such that $u_1 = 0$. The simulation parameters are the same as in Table **1** $\theta_1(0) = 0.35\,rad$. The climber's speed changes following Eq. (**9**) $\theta_e = 0.3$ without compensating control. The results are shown in Figs. (**5** – **8**). The climber speed and the tether length evolution are shown in Figs. (**5** and **6**). The corresponding PPC input $\ddot{L}_1$, defined in Eq. (**32**), is shown as the solid line in Fig. (**6a**). The libration of the climber and the end body are shown in Fig. (**7**). Acted by the STSM, the libration angles converge in 2500*s* to the desired state smoothly. This indicates the proposed control strategy is effective to stabilize the PSE during the transfer period. Furthermore, the observing law of the tension terms is effective, as shown in Fig. (**8a**), in which the estimated disturbances match the practical disturbances well with ignorable gaps. Thus, the proposed observing method is effective for a PSE even in a long interval. Comparing the results in Fig. (**8a**) with Fig. (**4b**), shows that the gaps between the real and the estimated disturbances are smaller in the controlled case than in the uncontrolled case. This indicates that under the control, the system converges to the desired stable state, and the disturbance sections converge to zero, as shown in Eq. (**14**). Fig. (**7b**) shows the control inputs. When the libration states converge to the target state, the thrust on the climber is zero, and the thrust on the end body is kept at a nearly constant amount to hold the desired state. For the rest of the transfer period, the Event-triggered loaning control is standing by.

The total control scheme equals a PPC coordinated with an STSM. To increase the robustness of the control scheme and enhance its flexibility to deal with mission requirements, speed predictive patterns are studied in the next two cases.

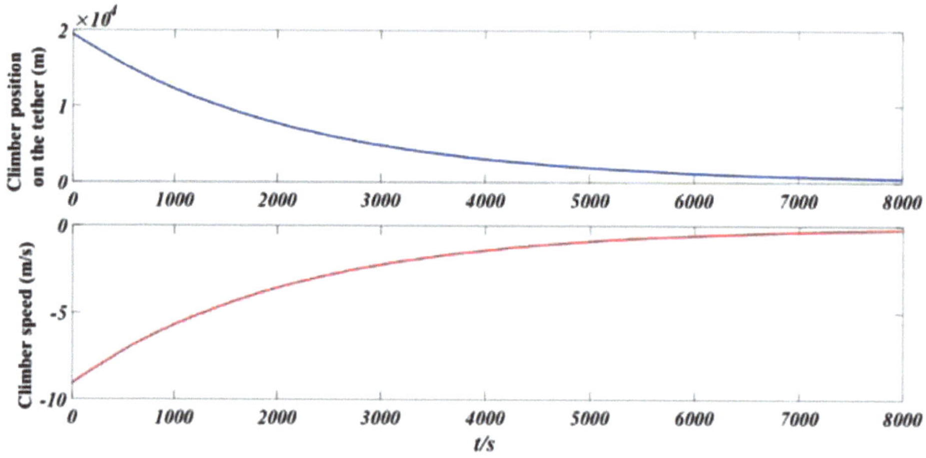

Fig. (5) Climber position and speed.

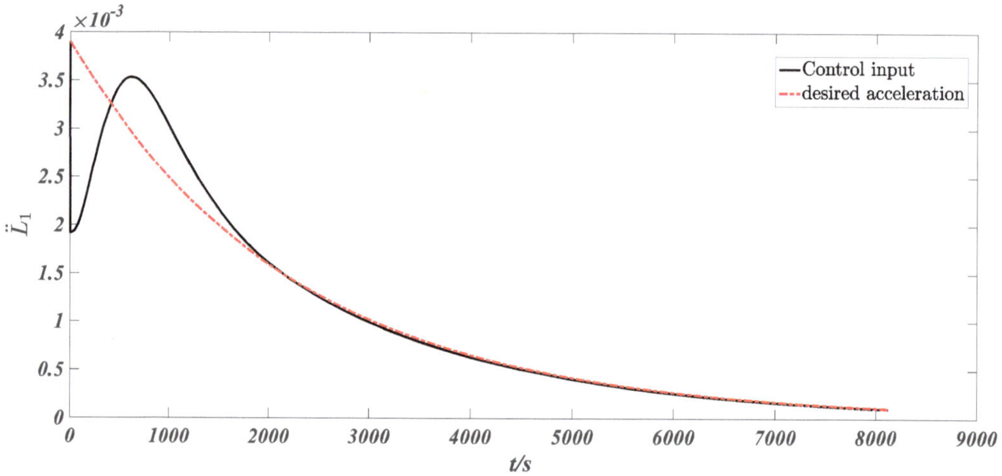

(a)

(Fig. 6) contd.....

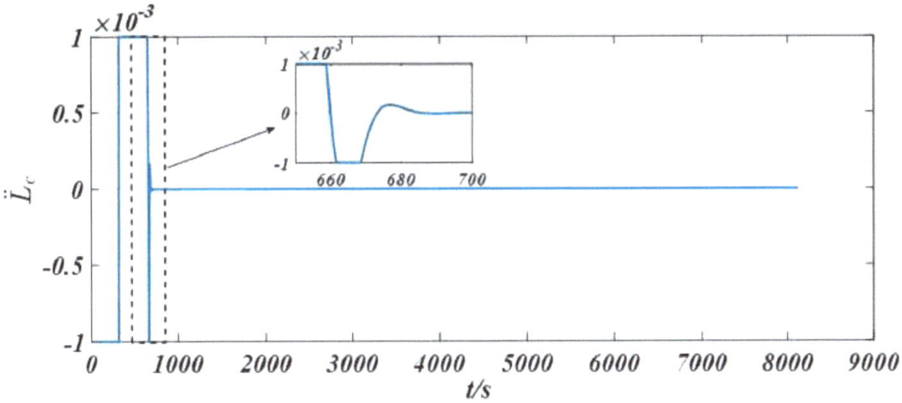

(b)

Fig. (6). Control inputs. (a) Acceleration of the climber. (b) Acceleration of the end body.

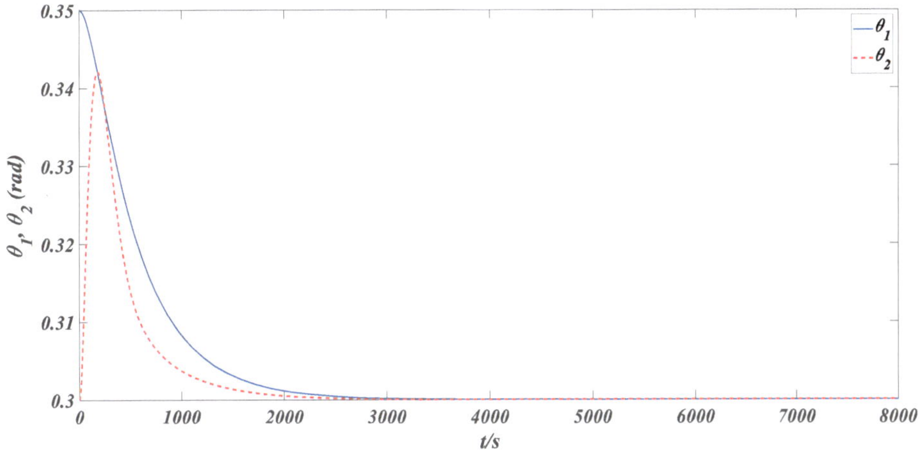

Fig. (7) Libration angles.

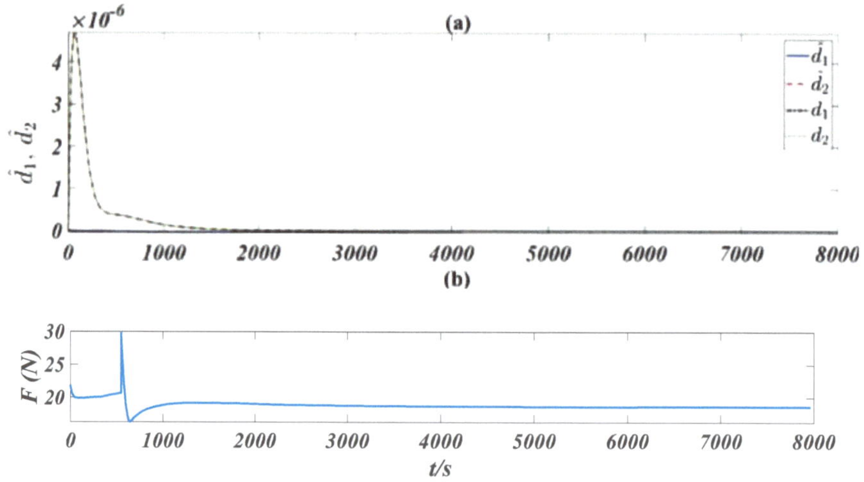

Fig. (8) (a) Disturbances. (b) Control inputs.

Case 2: Configuration Keeping Using Coordinated Control on the End Body

The event-triggered loaning control is used in Control Module II. The switching of control follows the rule in Fig. (**3**). The maximum control inputs are the same as those in *Case* 1. The control parameters of the PPC controller are the same as those in *Case* 1, and $k_{p3} = 10^{-3}$, $k_{d3} = 1$. The simulation results are shown in Figs. (**9** – **11**).

The phase plot in Fig. **9** (a) shows the PPC (for θ_1) is stable and converges to the target $\theta_e = 0.3\,rad = 0.095\pi$. The time histories of libration angles of the climber (θ_1) and the end body (θ_2) as well as the control error ($z = \theta_1 - \theta_e$) are shown in Fig. (**9b**). The θ_1 converges to the target $\theta_e = 0.3\,rad$ smoothly in 4,000*s*. Simultaneously, θ_2 is controlled to track θ_1 successfully. Although starting with different initial values, θ_2 tracks θ_1 to the peak of 0.345 *rad*, and then it decreases smoothly to catch up with θ_1 at about 4,000*s*. The corresponding control inputs $\ddot{L}_c$ and *F* are shown in Figs. (**11**). After 4,000*s*, the states of θ_1 and θ_2 agree with each other well, as shown in Fig. (**9b**) under the coordinated PD control in Eq. (30), although the curve of θ_2 fluctuates at the beginning of the control period.

The control inputs $\ddot{L}_c$ and *F* are shown in Fig. (**10**). In the first 600*s*, the control is dominated by *Borrowing Mode*, the end body is retrieved by 105m (L_c = 105 *m*). After 600*s*, *Repaying Mode* is triggered, and the control laws are switched from

Eqs. (**53**)-(**54**) to Eqs. (**49-51**). The control input $\ddot{L}_c$ deduces the magnitude of L_c. In the period of 600 - 2,000*s*, $\dot{L}_c$ is changed from negative to positive as shown in Fig. (**10b**). As a result, L_c converges from 105m to zero after 2000 *s*, see Fig. (**10a**). Simultaneously, $\theta_1 - \theta_2$ converges to and is kept at zero under the action of thrust *F* only. As shown in Fig. (**11b**), *F* fluctuates near 600*s* due to the switching of the control mode. The time histories of the tether tensions are shown in Fig. (**10c**), the tensions in this work are much smoother overall, comparing with Ref [17], without overshooting for two reasons. First, the proposed control strategy based on [29] keeps the dynamic states of a PSE at a designed steady-state, and the chattering has been eliminated. Second, the exponential characteristic of the prescribed performance function [33, 34] makes the errors of the libration states converge to zero smoothly.

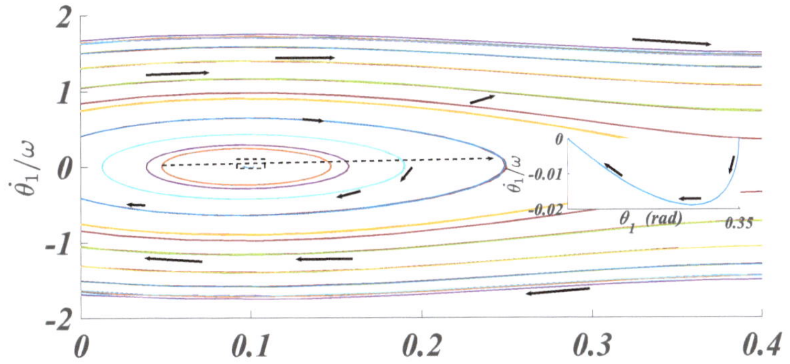

(a)

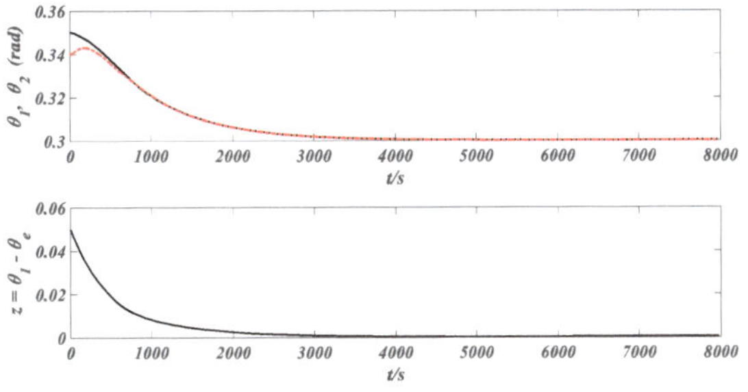

(b)

Fig. (9). Time histories of libration angles and control error *z*.

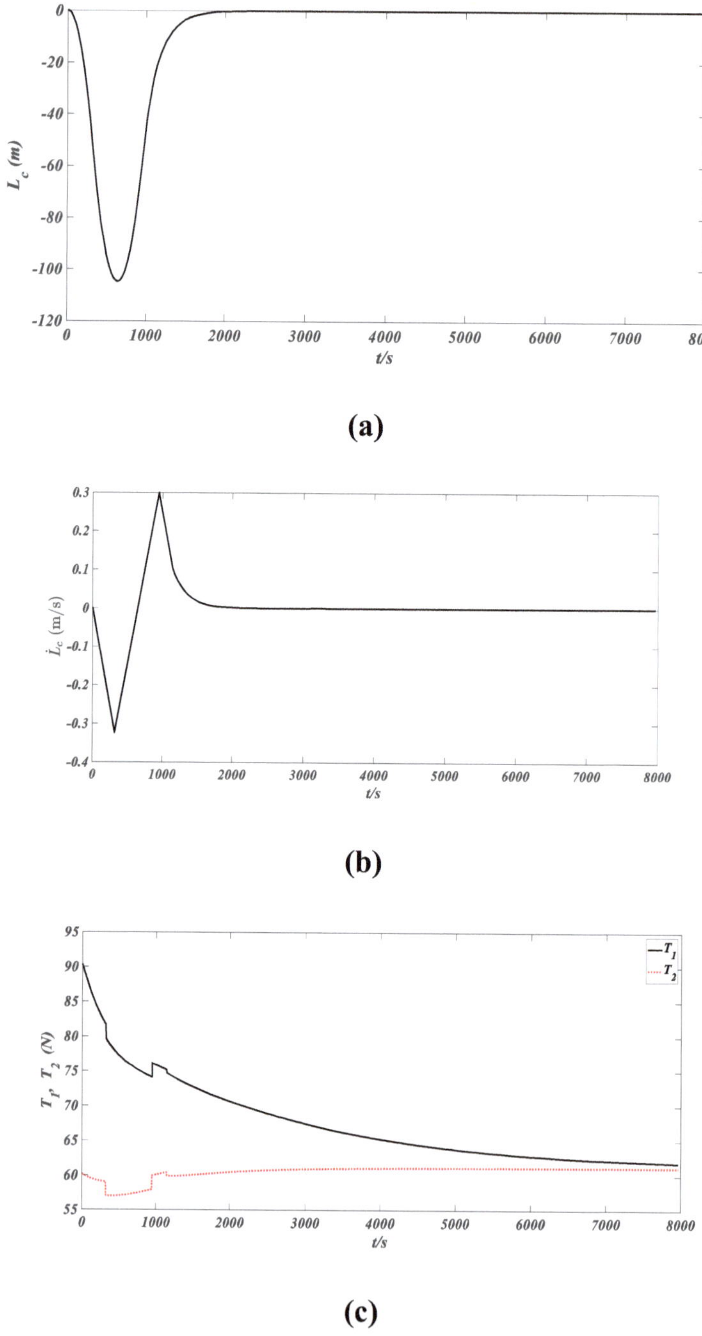

Fig. (10). System responses: (a) L_c. (b) End body speeds. (c) Tether tensions.

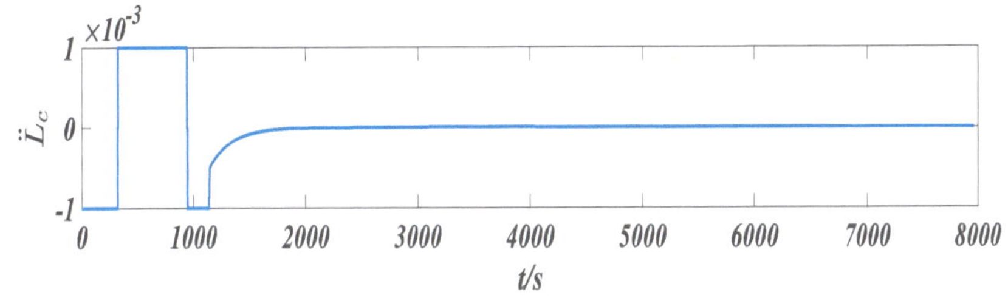

(a)

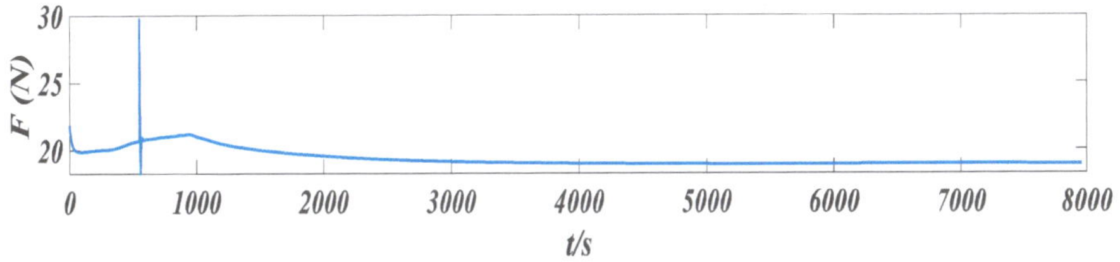

(b)

Fig. (11). Control inputs: (a) Acceleration of the end body. (b) Thrust at the end body.

In summary, the proposed event-triggered loaning control reasonably redistributes the control inputs without sacrificing the control performance, while avoiding the shortcoming of $L_c \neq 0$ at the end of cargo transportation.

CONCLUSION

This work proposes a two-module configuration keeping technology of PSE in transportation. In Control Module I, a new PPC is derived to stabilize the libration of the climber. In Control Module II, two control methods are proposed respecting two control modes for the end body. In the mode with thrust control only, an STSM is used to further stabilize the PSE with small chattering. For the mode with coordinate control inputs (thrust and tension), a new event-triggered loaning control is proposed to satisfy the configuration constraints. The total tether length remains unchanged at the end of the cargo transfer period. The simulation results reveal that the proposed control strategy can keep the configuration of PSE with a good performance implemented by multiple closed-loop control laws. Compared with the existing optimal control strategy, the calculation of the proposed control strategy is much easier and simpler which is beneficial to the implementation in the computer onboard the PSE. The proposed STSM presents good performances in eliminating the chattering with thrusters on the end body only. The proposed event-triggered loaning control is very effective in redistributing the limited control inputs in the Control Module II, while satisfying the configuration constraint simultaneously.

CONSENT FOR PUBLICATION

Not applicable.

CONFLICT OF INTEREST

The authors declare no conflict of interest, financial or otherwise.

ACKNOWLEDGEMENT

This work is supported by the Guangdong Provincial Basic and Applied Basic Research Fund (2019A1515111056).

REFERENCES

[1] Dixit U S, Dwivedy S K, Forward T W. Mechanical Sciences. Springer: Singapore, 2020.

[2] Aslanov V S, Ledkov A S. Dynamics of tethered satellite systems. Elsevier, 2012.

[3] Shi G, Zhu Z H, Li G. Stable cargo transportation of partial space elevator with multiple actuators. Advances in Space Research, 2021, 68(7): 2999-3011.

[4] Fradkov A, Andrievsky B, Guzenko P. Energy speed-gradient control of satellite oscillations. IFAC Proceedings Volumes, 2004, 37(6): 417-422.

[5] Shahov E M. Oscillations of probe satellite towed by non-stretched thread in heterogeneous atmosphere. Appl. Math. and Mech, 1988, 52(4): 567-572.

[6] Woo P, Misra A K. Dynamics of a partial space elevator with multiple climbers. Acta Astronautica, 2010, 67(7-8): 753-763.

[7] Misra A K, Amier Z, Modi V J. Attitude dynamics of three-body tethered systems. Acta Astronautica, 1988, 17(10): 1059-1068.

[8] Lorenzini E C, Cosmo M, Vetrella S, *et al.* Dynamics and control of the tether elevator/crawler system. Journal of Guidance, Control, and Dynamics, 1989, 12(3): 404-411.

[9] Shi G, Zhu Z, Zhu Z H. Libration suppression of tethered space system with a moving climber in circular orbit. Nonlinear Dynamics, 2018, 91(2): 923-937.

[10] Williams P. Dynamic multibody modeling for tethered space elevators. Acta Astronautica, 2009, 65(3-4): 399-422.

[11] Tang J L, Ren G X, Zhu W D, *et al.* Dynamics of variable-length tethers with application to tethered satellite deployment. Communications in Nonlinear Science and Numerical Simulation, 2011, 16(8): 3411-3424.

[12] Li G, Shi G, Zhu Z H. Three-dimensional high-fidelity dynamic modeling of tether

transportation system with multiple climbers. Journal of Guidance, Control, and Dynamics, 2019, 42(8): 1797-1811.

[13] Yamagiwa Y, Nohmi M, Aoki Y, *et al.* Space experiments on basic technologies for a space elevator using microsatellites. Acta Astronautica, 2017, 138: 570-578.

[14] Jung W, Mazzoleni A P, Chung J. Nonlinear dynamic analysis of a three-body tethered satellite system with deployment/retrieval. Nonlinear Dynamics, 2015, 82(3): 1127-1144.

[15] Yu B S, Wen H, Jin D P. Review of deployment technology for tethered satellite systems. Acta Mechanica Sinica, 2018, 34(4): 754-768.

[16] Kojima H, Fukatsu K, Trivailo P M. Mission-function control of tethered satellite/climber system. Acta Astronautica, 2015, 106: 24-32.

[17] Shi G, Zhu Z, Zhu Z H. Parallel optimization of trajectory planning and tracking for three-body tethered space system. IEEE/ASME Transactions on Mechatronics, 2019, 24(1): 240-247.

[18] Xu S, Sun G, Ma Z, *et al.* Fractional-order fuzzy sliding mode control for the deployment of tethered satellite system under input saturation. IEEE transactions on aerospace and electronic systems, 2018, 55(2): 747-756.

[19] Malashin A A, Smirnov N N, Bryukvina O Y, *et al.* Dynamic control of the space tethered system. Journal of Sound and Vibration, 2017, 389: 41-51.

[20] Zhao G, Sun L, Huang H, *et al.* Optimal attitude control of a tethered system for noncoplanar orbital transfer under a constant thrust. IEEE Transactions on Aerospace and Electronic Systems, 2019, 56(3): 1844-1855.

[21] Haiping R A O, Zhong R, Pengjie L I. Fuel-optimal deorbit scheme of space debris using tethered space-tug based on pseudospectral method. Chinese Journal of Aeronautics, 2021, 34(9): 210-223.

[22] Lv Y, Ren X, Na J. Online Nash-optimization tracking control of multi-motor driven load system with simplified RL scheme. ISA transactions, 2020, 98: 251-262.

[23] Rubinsztejn A, Sood R, Laipert F E. Neural network optimal control in astrodynamics: Application to the missed thrust problem. Acta astronautica, 2020, 176: 192-203.

[24] Shi G, Li G, Zhu Z, *et al.* Dynamics and operation optimization of partial space elevator with multiple climbers. Advances in Space Research, 2019, 63(10): 3213-3222.

[25] Huang P, Zhang F, Cai J, *et al.* Dexterous tethered space robot: Design, measurement, control, and experiment. IEEE Transactions on Aerospace and Electronic Systems, 2017, 53(3): 1452-1468.

[26] Shi G, Zhu Z, Zhu Z H. Stable orbital transfer of partial space elevator by tether deployment and retrieval. Acta Astronautica, 2018, 152: 624-629.

[27] Song Z, Sun K. Prescribed performance tracking control for a class of nonlinear system considering input and state constraints. ISA transactions, 2022, 119: 81-92.

[28] Liu L, Yao W, Guo Y. Prescribed performance tracking control of a free-flying flexible-joint space robot with disturbances under input saturation. Journal of the Franklin Institute, 2021, 358(9): 4571-4601.

[29] Wen X, Sun Y, Wang J, *et al.* Rejection of time-varying frequency sinusoidal disturbance using refined observer for a class of uncertain systems. ISA transactions, 2020, 100: 136-144.

[30] Zhang G, Wang Y, Wang J, *et al.* Disturbance observer–based super-twisting sliding mode control for formation tracking of multi-agent mobile robots. Measurement and Control, 2020, 53(5-6): 908-921.

[31] Shtessel Y, Taleb M, Plestan F. A novel adaptive-gain supertwisting sliding mode controller: Methodology and application. Automatica, 2012, 48(5): 759-769.

[32] Shi G, Zhu Z H. Prescribed performance based dual-loop control strategy for configuration keeping of partial space elevator in cargo transportation. Acta Astronautica, 2021, 189: 241-249.

[33] Bechlioulis C P, Rovithakis G A. Prescribed performance adaptive control for multi-input multi-output affine in the control nonlinear systems. IEEE Transactions on automatic control, 2010, 55(5): 1220-1226.

[34] Zerari N, Chemachema M. Robust adaptive neural network prescribed performance control for uncertain CSTR system with input nonlinearities and external disturbance. Neural Computing and Applications, 2020, 32(14): 10541-10554.

Advanced Control of Flight Vehicle Maneuver and Operation, 2023, 213-240

CHAPTER 8

Adaptive Fixed-time 6-DOF Coordinated Control of Spacecraft Formation Flyings

Ruixia Liu[1,*]

[1]*School of Automation, Xi'an University of Posts & Telecommunications, Xi'an 710121, China*

Abstract: In this paper, the fixed-time coordinated control problem is investigated for multiple spacecraft formation flying (SFF) system based on six-degrees-of-freedom (6-DOF) dynamic model. The system under consideration involves input quantization, external disturbance, and directed communication topology. By utilizing the neighborhood state information, a novel multi-spacecraft nonsingular fixed-time terminal sliding mode function is designed. To reduce the required communication rate, a hysteretic quantizer is employed to quantify the control torque and force. The problem addressed is the design of 6-DOF fixed-time coordinated controller such that, the controlled system is practical fixed-time stable and also ensures the relative attitude and position tracking errors can converge into the regions in fixed time. A numerical example is provided to illustrate the usefulness of the proposed control scheme.

Keywords: Coordinated control, Spacecraft formation flying, Fixed-time control, Input quantization.

INTRODUCTION

As is well known, the spacecraft formation flying (SFF) has been one of the hot research fields owing to its wild applications in various space industry such as deep space exploration, atmosphere monitoring of the Earth, and spacecraft on-orbit maintenance [1-5]. The attitude control and orbit control are arguable two equal vital technologies. For SFF missions, it is important to achieve the required attitude and position at the same time [1]. Due to the dynamic coupling between attitude motion and orbital motion, the two motions can be regarded as a whole six degree of freedom (6-DOF) motion. In recent years, the coordinated control of 6-DOF for SFF has attracted a number of researchers [2, 3]. Nevertheless, the proposed control schemes mentioned above only can ensure the asymptotic stability.

[*]**Corresponding author Ruixia Liu:** School of Automation, Xi'an University of Posts & Telecommunications, Xi'an 710121, China; E-mail: ruixialiu@xupt.edu.cn

Chuang Liu, Honghua Dai, Xiaokui Yue & Yiqing Ma (Eds.)

Fast convergence performance is a crucial requirement for coordinated control problems of SFF [4-7]. Compared with the asymptotic stabilization controller, the finite time stabilization controller has faster response speed and better anti-interference ability. Thus, the finite-time control strategies have been proposed for spacecraft formation control. Although the finite time control approach can guarantee the finite-time stability, the convergence time depends on the initial state information of the system, which brings difficulties to practical application [8]. In order to solve this constraint, the design of finite time controller is studied by using the concept of fixed-time stability, that is, the upper bound of the convergence time can be estimated regardless of the initial state of the controlled system [9, 10]. So far, the fixed-time control strategy has been utilized for a variety of industrial systems [11], but few researchers pay attention to the 6-DOF fixed time coordinated control of SFF, especially for external disturbances with unknown upper bound. Another important problem in multiple SFF missions is that the communication link between spacecraft is not always bidirectional, such as in unidirectional spacecraft laser communication system. However, the coordinated control problem is studied based on the assumption that the communication topology is undirected in some existing results.

On the other hand, networked control systems (NCSs) have received persistent research interest because they have been successfully applied to a variety of modern practical engineering systems [12, 13], such as aerospace engineering systems, nuclear power stations, unmanned vehicles etc. In the modern low-cost small spacecraft formation system, the communication between the different functional modules is connected through wireless network [14-16].

It's common knowledge that the quantization error induced by signal quantization behavior may exist in modern SFF systems, which will reduce control performance and even lead to instability [17-20]. Therefore, it is necessary to design the new control scheme for SFF, in which signal quantization is considered. Although some researches focus on the quantized control of SFF, there are no results considering the 6-DOF coordinated control of multi spacecraft formation with quantized input control signals. The complexity of the multiple spacecraft formation coordinated control task make the quantized fixed-time coordinated control a serious challenge.

In the present paper, we consider the fixed-time 6-DOF adaptive coordinated control problem for SFF involving input quantization and directed communication topology. The main contributions lie in three aspects: (1) For the coordinated controller design, the directed communication graph will bring more challenges than the undirected communication graph. (2) Based on 6-DOF dynamic model, a new multi-spacecraft nonsingular FTTSM is constructed, on which each spacecraft converges to their desired states while keeping synchronization with other formation spacecraft. (3) A fixed-time adaptive coordinate controller is designed to eliminate the effects of external disturbances and hysteretic quantizer on the control performance and ensure the practical fixed-time stability of the closed-loop system.

MODELLING AND PRELIMINARIES

6-DOF Dynamic Model

The 6-DOF dynamic model of spacecraft formation is represented as follows:

$$\begin{cases} \dot{\boldsymbol{x}}_{1i} = \boldsymbol{\Lambda}(\dot{\boldsymbol{x}}_{1i})\boldsymbol{x}_{2i} \\ \boldsymbol{G}_{fi}\dot{x}_{2i} + \boldsymbol{C}(\boldsymbol{x}_{2i}) + \boldsymbol{N}(\boldsymbol{x}_{1i}) + \boldsymbol{\tau}_i = \boldsymbol{u}_i, \quad i = 1, 2, \ldots, n \end{cases} \tag{1}$$

Where

$$\boldsymbol{x}_{1i} = \begin{bmatrix} \boldsymbol{\rho}_i \\ \boldsymbol{q}_i \end{bmatrix}$$

$$\boldsymbol{x}_{2i} = \begin{bmatrix} \dot{\boldsymbol{\rho}}_i \\ \boldsymbol{\omega}_i \end{bmatrix}, \quad \boldsymbol{\Lambda}(\dot{\boldsymbol{x}}_{1i}) = \begin{bmatrix} \boldsymbol{I}_{3\times3} & \boldsymbol{0}_{3\times3} \\ \boldsymbol{0}_{3\times3} & \dfrac{\boldsymbol{T}(\boldsymbol{q}_i)}{2} \end{bmatrix}, \quad \boldsymbol{G}_{fi} = \begin{bmatrix} m_{fi}\boldsymbol{I}_{3\times3} & \boldsymbol{0}_{3\times3} \\ \boldsymbol{0}_{3\times3} & \boldsymbol{J}_{fi} \end{bmatrix}$$

$$, \boldsymbol{C}(\boldsymbol{x}_{2i}) = \begin{bmatrix} \boldsymbol{C}(\boldsymbol{n}_0)\dot{\boldsymbol{\rho}}_i \\ \boldsymbol{\omega}_i^{\times}\boldsymbol{J}_{fi}\boldsymbol{\omega}_i \end{bmatrix}, \quad \boldsymbol{N}(\boldsymbol{x}_{1i}) = \begin{bmatrix} \boldsymbol{N}(\boldsymbol{\rho}_i, \boldsymbol{n}_0, \boldsymbol{R}) \\ \boldsymbol{0}_{3\times1} \end{bmatrix}, \quad \boldsymbol{\tau}_i = \begin{bmatrix} \boldsymbol{F}_{di} \\ \boldsymbol{z}_i \end{bmatrix}, \quad \boldsymbol{u}_i = \begin{bmatrix} \boldsymbol{u}_{fi} \\ \boldsymbol{u}_{ti} \end{bmatrix}$$

$$\boldsymbol{T}(\boldsymbol{q}_i) = \begin{bmatrix} -\boldsymbol{q}_{vi}^T \\ q_{0i}\boldsymbol{I}_{3\times3} + \boldsymbol{q}_{vi}^{\times} \end{bmatrix}, \quad C(n_0) = 2n_0\begin{bmatrix} 0 & -1 & 0 \\ 1 & 0 & 0 \\ 0 & 0 & 0 \end{bmatrix}, \quad N(\boldsymbol{\rho}_i, n_0, R) = \begin{bmatrix} -\dot{n}_0 y_i - n_0^2 \boldsymbol{x}_i - 2\dfrac{\mu}{R^3}x_i \\ \dot{n}_0 x_i - n_0^2 y_i + \dfrac{\mu}{R^3}y_i \\ \dfrac{\mu}{R^3}z_i \end{bmatrix}$$

where superscript i stands for the ith follower spacecraft; $\boldsymbol{\rho}_i = [x_i \; y_i \; z_i]^T$ represents the relative position vector from the ith follower spacecraft to the leader spacecraft; $\boldsymbol{\omega}_i \in \mathbb{R}^3$ denotes the angular velocity; $\boldsymbol{q}_i \in \mathbb{R}^4$ is the quaternion defined as

$\boldsymbol{q}_i = [q_{0i} \quad q_{vi}]^T$, where $n_0 \in \mathbb{R}$ represents angular velocity of the virtual leader spacecraft; q_{0i} is the scalar part and $\boldsymbol{q}_{vi}$ is the vector part; $m_{fi} \in \mathbb{R}$ denotes the mass; $\boldsymbol{J}_{fi} \in \mathbb{R}^{3\times3}$ is the inertia matrix; $u_{fi} \in \mathbb{R}^3$ represents the control force; $\boldsymbol{u}_{it} \in \mathbb{R}^3$ is the control torque; $\boldsymbol{F}_{di} \in \mathbb{R}^3$ is disturbance force; $\boldsymbol{z}_i = \boldsymbol{l}_i - T_{iGT}$, $\boldsymbol{\ell}_i \in \mathbb{R}^3$ is disturbance torque, $\boldsymbol{T}_{iGT} \in \mathbb{R}^3$ is the gravity gradient torque. The notation $\boldsymbol{\iota}^\times$ for the vector $\boldsymbol{\iota} = [\iota_1 \ \iota_2 \ \iota_3]^T$ represents the skew-symmetric matrix as follows:

$$\boldsymbol{\iota}^\times = \begin{bmatrix} 0 & -\iota_3 & \iota_2 \\ \iota_3 & 0 & -\iota_1 \\ -\iota_2 & \iota_1 & 0 \end{bmatrix} \tag{2}$$

It is worth to mention that the attitude and orbit are mutual coupled by $\boldsymbol{T}_{iGT} \in \mathbb{R}^3$, which is given as

$$\boldsymbol{T}_{iGT} = 3\mu \frac{\hat{\boldsymbol{R}}_{fi}^\times \boldsymbol{J}_{fi} \hat{\boldsymbol{R}}_{fi}}{(x_i^2 + (R + y_i)^2 + z_i^2)^{3/2}} \tag{3}$$

where $\hat{\boldsymbol{R}}_{fi} \in \mathbb{R}^3$ is the position unit vector. Since $\boldsymbol{T}_{iGT}$ is much small compared with control torque, $\boldsymbol{T}_{iGT}$ is always treated as disturbance.

We define the following error states

$$\boldsymbol{e}_{1i} = \begin{bmatrix} \rho_i - \rho_{di} \\ \boldsymbol{q}_{ei} \end{bmatrix}, \boldsymbol{e}_{2i} = \begin{bmatrix} \dot{\rho}_{ei} \\ \boldsymbol{\omega}_{ei} \end{bmatrix} = \begin{bmatrix} \dot{\rho}_i - \dot{\rho}_{di} \\ \boldsymbol{\omega}_i - \boldsymbol{R}(\boldsymbol{q}_{ei})\boldsymbol{\omega}_{di} \end{bmatrix} \tag{4}$$

where $\boldsymbol{q}_{ei}$ is error quaternion defined as $\boldsymbol{q}_{ei} = [q_{0ei} \ \boldsymbol{q}_{vei}^T]^T = \boldsymbol{q}_i \otimes \boldsymbol{q}_{di}$; $\boldsymbol{R}(\boldsymbol{q}_{ei})$ is the rotation matrix from the ith follower spacecraft's reference frame to its body-fixed frame; $\boldsymbol{q}_{di}$, ρ_{di}, $\boldsymbol{\omega}_{di}$, $\dot{\rho}_{di}$, denote the desired attitude, desired position, desired angular velocity, and desired velocity, respectively. Based on (4), the 6-DOF relative error dynamic model of SFF can be described as

$$\begin{cases} \dot{\boldsymbol{e}}_{1i} = \boldsymbol{\Lambda}(\dot{\boldsymbol{e}}_{1i})\boldsymbol{e}_{2i} \\ \boldsymbol{G}_{fi}\dot{\boldsymbol{e}}_{2i} + \boldsymbol{C}(\boldsymbol{e}_{2i}) + \boldsymbol{N}(\boldsymbol{e}_{1i}) + \boldsymbol{\tau}_i = \boldsymbol{u}_i \end{cases} \tag{5}$$

Where $\boldsymbol{\Lambda}(\dot{\boldsymbol{e}}_{1i})=\begin{bmatrix} \boldsymbol{I}_{3\times3} & \boldsymbol{0}_{3\times3} \\ \boldsymbol{0}_{3\times3} & \dfrac{\boldsymbol{T}(\boldsymbol{q}_{ei})}{2} \end{bmatrix}$, $N(\boldsymbol{e}_{1i})=\begin{bmatrix} N(\rho_i,n_0,R)-\ddot{\boldsymbol{\rho}}_{di} \\ \boldsymbol{0}_{3\times1} \end{bmatrix}$,

$$\boldsymbol{C}(\boldsymbol{e}_{2i})=\begin{bmatrix} \boldsymbol{C}(n_0)\boldsymbol{\rho}_i \\ -\boldsymbol{\omega}_i^{\times}\boldsymbol{J}_{fi}\boldsymbol{\omega}_i+\boldsymbol{J}_{fi}(\boldsymbol{\omega}_{ei}^{\times}\boldsymbol{\omega}_{di}-\boldsymbol{R}(\boldsymbol{q}_{ei})\dot{\boldsymbol{\omega}}_{di}) \end{bmatrix}$$

Graph Theory

It is assumed that the information flow among n follower spacecraft is described by a directed graph $\mathcal{G}=(\mathcal{V},\chi,\mathcal{A})$, where $\mathcal{V}=\{\mathcal{V}_1,\mathcal{V}_2,\ldots,\mathcal{V}_n\}$ is the set of nodes, $\chi\subseteq\mathcal{V}\times\mathcal{V}$ denotes the set of edges, and $(\mathcal{V}_i,\mathcal{V}_j)\in\chi$ denotes if and only if node $\mathcal{V}_i$ can receive the information of node $\mathcal{V}_j$. In this paper, $(\mathcal{V}_i,\mathcal{V}_j)\in\chi$ denotes only the jth spacecraft can receive the ith spacecraft's states information. $\boldsymbol{\mathcal{A}}=[a_{ij}]\in\mathbb{R}^{n\times n}$ is the weighted adjacency matrix of the graph $\mathcal{G}$ with entries

$$\begin{cases} a_{ij}>0 & \text{if}((\mathcal{V}_i,\mathcal{V}_j)\in\chi) \\ a_{ij}=0 & \text{otherwise} \end{cases} \tag{6}$$

where a_{ij} is the non-negative element of $\boldsymbol{\mathcal{A}}$, which denotes communication quality between the ith spacecraft and jth spacecraft. It is worth pointing out that self-edges are not allowed, meaning that $a_{ii}=0$.

The in-degree matrix of the graph $\mathcal{G}$ is D with entries

$$\boldsymbol{D}=\text{diag}\{d_1,d_2,\ldots,d_n\} \tag{7}$$

Where

$$d_i=\sum_{j=1}^{n}a_{ij}=\sum_{j\in\chi_i}a_{ij},(i=1,2,\ldots,n) \tag{8}$$

The Laplacian matrix $L\in\mathbb{R}^{n\times n}$ of the graph $\mathcal{G}$ is [21].

$$\boldsymbol{L}=\boldsymbol{D}-\boldsymbol{\mathcal{A}} \tag{9}$$

Hysteretic Quantizer

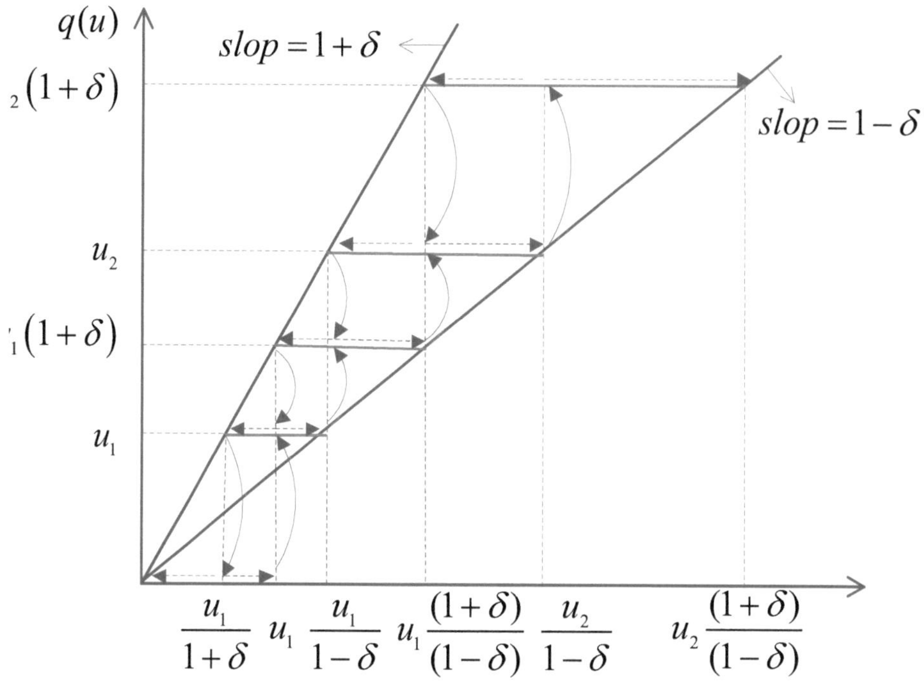

Fig. (1). Map of $q(u(t))$ for $u>0$.

To avoid chattering phenomenon caused by logarithmic quantizer, a hysteretic quantizer is employed to quantify control force and torque in this paper, which is similar as [16]. It can be described by

$$q(u(t)) \triangleq \begin{cases} u_j \operatorname{sgn}(u) & \text{if } \frac{u_j}{1+\delta} < |u| \le \frac{u_j}{1-\delta}, \dot{u}_j < 0, \text{ or } u_j < |u| \le \frac{u_j}{1-\delta}, \dot{u}_j > 0 \\ u_j(1+\delta)\operatorname{sgn}(u) & \text{if } u_j < |u| \le \frac{u_j}{1-\delta}, \dot{u}_j < 0, \text{ or } \frac{u_j}{1-\delta} \le |u| \le \frac{u_j(1+\delta)}{1-\delta}, \dot{u}_j > 0 \\ 0 & \text{if } 0 \le |u| < \frac{u_{\min}}{1+\delta}, \dot{u}_j < 0, \text{ or } \frac{u_{\min}}{1+\delta} \le |u| \le u_{\min}, \dot{u}_j > 0 \\ q(u(t^-)) & \text{othercase} \end{cases} \quad (10)$$

where $u_j = \rho^{(1-i)} u_{\min}$, $j = 1,2,\ldots n$ with $0 < \rho < 1$, $u_{\min} > 0$, and $\delta = \dfrac{1-\rho}{1+\rho}$. $q(u(t))$ is in the set $U = \{0, \pm u_i, \pm u_i(1+\delta)\}$. The map of the hysteresis quantizer $q(u(t))$ for $u > 0$ is illustrated in Fig. (**1**).

Remark 1: The parameter ρ can be termed as a measure of quantization density. From the definition of ρ, we can see that the smaller parameter ρ is, the coarser hysteretic quantizer becomes.

Preliminaries

Lemma 1: The hysteretic quantizer $q(u(t))$ is decomposed into two parts as

$$q(u(t)) = D(u)u(t) + Q(t) \tag{11}$$

where $D(u)$ and $Q(t)$ satisfy

$$1-\delta \le D(u) \le 1+\delta, \quad |Q(t)| \le u_{\min}. \tag{12}$$

The proof of lemma 1 is similar to Theorem 1 in [16].

Lemma 2: [22] If $\mathcal{G}$ is a directed graph with N nodes, then all the eigenvalues of the weighted Laplace matrix L have nonnegative real part.

Lemma 3: [22] For any matrix $\boldsymbol{M} \in \mathbb{R}^{m\times m}$, $\boldsymbol{N} \in \mathbb{R}^{n\times n}$, $\boldsymbol{X} \in \mathbb{R}^{m\times m}$ and $\boldsymbol{Y} \in \mathbb{R}^{n\times n}$, then the following equalities hold

(1) $(\boldsymbol{M} \otimes \boldsymbol{N})(\boldsymbol{X} \otimes \boldsymbol{Y}) = \boldsymbol{MX} \otimes \boldsymbol{NY}$.

(2) If the matrix $\boldsymbol{M}$ and $\boldsymbol{N}$ are invertible, then $(\boldsymbol{M} \otimes \boldsymbol{N})^{-1} = \boldsymbol{M}^{-1} \otimes \boldsymbol{N}^{-1}$.

(3) If the eigenvalues of $\boldsymbol{M}$ is $\lambda_1,\ldots,\lambda_m$ and the eigenvalues of N is $\mu_1,\ldots,\mu_n$, then the eigenvalues of $\boldsymbol{M} \otimes \boldsymbol{N}$ can be expressed by $\lambda_i \mu_j$ $(i = 1,\ldots,m; j = 1,\ldots,n)$.

Lemma 4: [26] For any $x, y \in \mathbb{R}$, if $v \in \mathbb{R}^+$ and $v > 1$, then

$$|x+y|^{\nu} \leq 2^{\nu-1} |x^{\nu}+y^{\nu}| \quad \textbf{(13)}$$

Lemma 5: [24] If $x_i \in \mathbb{R}, i=1,2,\ldots,n$, and $0 < p \leq 1$, then

$$(\sum_{i=1}^{n}|x_i|)^p \leq \sum_{i=1}^{n}|x_i|^p \leq n^{1-p}(\sum_{i=1}^{n}|x_i|)^p \quad \textbf{(14)}$$

Lemma 6: [23] Consider the nonlinear system given by

$$\dot{x} = f(x,t), \quad f(0,t)=0, \quad x \in \mathbb{R}^n \quad \textbf{(15)}$$

suppose that there exists a Lyapunov function $V(x)$ satisfies the following condition

$$\dot{V}(x) \leq -(\alpha V(x)^p + \beta V(x)^g)^k + \upsilon, \quad x \in U_o \quad \textbf{(16)}$$

where $\alpha, \beta, p, g, k \in \mathbb{R}^+, pk < 1, gk > 1,$ *and* $0 < \upsilon < \infty$. Then the origin of system (15) is practically fixed-time stable, and the residual set of the solution satisfies

$$\{\lim_{t \to T} x \mid V(x) \leq \min\{\alpha^{-\frac{1}{p}}(\frac{\upsilon}{1-\theta^k})^{\frac{1}{kp}}, \beta^{-\frac{1}{g}}(\frac{\upsilon}{1-\theta^k})^{\frac{1}{kg}}\}\} \quad \textbf{(17)}$$

where θ is a scalar satisfying $0 < \theta \leq 1$. The setting time is bounded by

$$T \leq \frac{1}{\alpha^k \theta^k (1-pk)} + \frac{1}{\beta^k \theta^k (gk-1)} \quad \textbf{(18)}$$

MULTI-SPACECRAFT NONSINGULAR FIXED-TIME TERMINAL SLIDING MODE

In this section, a multi-spacecraft nonsingular FTTSM function is designed to achieve the coordinated control for SFF. The following assumptions are made about 6-DOF dynamic model:

Assumption 1: The disturbance τ_i is supposed to be bounded, owing to gravitation, J_2 perturbations, solar radiation pressure, and magnetic forces are bounded.

Assumption 2: The desired angular velocity ω_{di} and its time derivative $\dot{\omega}_{di}$ are assumed to be bounded. The desired trajectory ρ_{di} and its time derivative $\dot{\rho}_{di}$ are assumed to be bounded.

Based on (5), the multi-spacecraft nonsingular FTTSM function is defined as

$$\boldsymbol{S} = [s_1, \ldots, s_i]^T \tag{19}$$

where $s_i = [s_{i,1}, s_{i,2}, \ldots, s_{i,k}]^T \in \mathbb{R}^{6\times 1}, i = 1,2,\ldots,n, k = 1,2,\ldots,6$ is given by

$$s_i = b_i \boldsymbol{G}_{fi}\left[\boldsymbol{e}_{2i} + \boldsymbol{\alpha}_i(\boldsymbol{e}_{1i})\right] + \sum_{j=1}^{n} a_{ij}\left[\left(\boldsymbol{G}_{fi}\boldsymbol{e}_{2i} - \boldsymbol{G}_{fj}\boldsymbol{e}_{2j}\right) + \left(\boldsymbol{G}_{fi}\boldsymbol{\alpha}_i(\boldsymbol{e}_{1i}) - \boldsymbol{G}_{fj}\boldsymbol{\alpha}_j(\boldsymbol{e}_{1j})\right)\right] \tag{20}$$

with $\boldsymbol{\alpha}_i(\boldsymbol{e}_{1i}) = [\alpha_{i,1}(e_{1i1}), \alpha_{i,2}(e_{1i2}), \ldots, \alpha_{i,k}(e_{1ik})]^T \in \mathbb{R}^{6\times 1}$, and

$$\alpha_{i,k}(e_{1ik}) = \begin{cases} \operatorname{sgn}(\sigma_{1i}\operatorname{sgn}(e_{1ik})^{p_1} + \sigma_{2i}\operatorname{sgn}(e_{1ik})^{p_2})^{k_1}, \text{ if } \bar{s}_{i,k} = 0 \text{ or } \bar{s}_{i,k} \neq 0, |e_{1ik}| > \varepsilon \\ l_{1i}e_{1ik} + l_{2i}\operatorname{sgn}(e_{1ik})^2, \text{if } \bar{s}_{i,k} \neq 0, |e_{1ik}| \leq \varepsilon \end{cases} \tag{21}$$

where

$i = 1,\ldots,n,\quad k = 1,2,\ldots,6,\quad \bar{\boldsymbol{s}}_i = [\bar{s}_{i,1}, \bar{s}_{i,2}, \ldots, \bar{s}_{i,6}]^T,\quad \bar{\boldsymbol{s}}_i = \boldsymbol{e}_{2i} + \operatorname{sgn}(\sigma_{1i}\operatorname{sgn}(\boldsymbol{e}_{1i})^{p_1} + \sigma_{2i}\operatorname{sgn}(\boldsymbol{e}_{1i})^{p_2})^{k_1}$,

$$l_{1i} = (2-k_1)(\sigma_{1i}\varepsilon^{p_1 - \frac{1}{k_1}} + \sigma_{2i}\varepsilon^{p_2 - \frac{1}{k_1}})^{k_1}, l_{2i} = (k_1 - 1)(\sigma_{1i}\varepsilon^{p_1 - \frac{1}{k_1}} + \sigma_{2i}\varepsilon^{p_2 - \frac{1}{k_1}})^{k_1},\quad b_i > 0$$

is controller gain to realize tracking control, $a_{ij} \geq 0$ is controller gain to realize coordinated control between ith and jth follower spacecraft, $\operatorname{sgn}(e_{1ik})^{p_1} = |e_{1ik}|^{p_1}\operatorname{sgn}(e_{1ik})$, $\operatorname{sgn}(e_{1ik})^{p_2} = |e_{1ik}|^{p_2}\operatorname{sgn}(e_{1ik})$, $\sigma_{1i} > 0, \sigma_{1i} > 0, p_1 > 0,$ $p_2 > 0, k_1 > 0$ are designed parameters, $\operatorname{sgn}(\sigma_{1i}\operatorname{sgn}(e_{1ik})^{p_1} + \sigma_{2i}\operatorname{sgn}(e_{1ik})^{p_2})^{k_1} = |(\sigma_{1i}\operatorname{sgn}(e_{1ik})^{p_1} +$ $\sigma_{2i}\operatorname{sgn}(e_{1ik})^{p_2})|^{k_1}\operatorname{sgn}(\sigma_{1i}\operatorname{sgn}(e_{1ik})^{p_1} + \sigma_{2i}\operatorname{sgn}(e_{1ik})^{p_2})$, $\varepsilon > 0$ is a small constant, $0 < p_1 k_1 < 1$, $p_2 k_1 > 1$, $\operatorname{sgn}(\cdot)$ is sign function.

Remark 2: The equation $a_{ij} = a_{ji}$ hold if the communication topology of SFF is undirected. This equation will simplify the analysis and design of controller. For bidirectional communication topology, this equation does not hold. Thus, it is difficult to design coordinated control under directed communication topology.

Remark 3: It is worth mentioning that the multi-spacecraft FTTSM (19) can be simplified as a modified terminal sliding mode (TSM) designed in [26] if $k_1 = 1$ and $p_2 = 0$, moreover, (19) coincides with the modified fast TSM designed in [25] for $k_1 = 1$. Note that when e_{1ik} convergence to the region $|e_{1ik}| \leq \varepsilon$, the multi-spacecraft FTTSM is converted to general sliding mode for $\bar{s}_{i,1} = 0$. Thus, the singularity problem of (19) can be effectively avoid. Moreover, by the choose of l_1 and l_2, the continuity of $\alpha_{i,k}$ and its first order time derivative is guaranteed.

By Kronecker product, the sliding mode function (19) can be rewritten as

$$\boldsymbol{S} = [(\boldsymbol{L}+\boldsymbol{B})\otimes \boldsymbol{I}_6]\overline{\boldsymbol{G}}(\boldsymbol{e}_2 + \boldsymbol{\alpha}(\boldsymbol{e}_1)) \quad (22)$$

where L is weighted Laplace matrix, which is determined by directed topology,

$$\boldsymbol{B} = \mathrm{diag}[\boldsymbol{b}_1, \boldsymbol{b}_2, \ldots, \boldsymbol{b}_n], \boldsymbol{e}_2 = [\boldsymbol{e}_{21}^T, \boldsymbol{e}_{22}^T, \ldots, \boldsymbol{e}_{2n}^T]^T,$$
$$\boldsymbol{\alpha}(\boldsymbol{e}_1) = [\boldsymbol{\alpha}_1^T(\boldsymbol{e}_{11}), \boldsymbol{\alpha}_2^T(\boldsymbol{e}_{12}), \ldots \boldsymbol{\alpha}_n^T(\boldsymbol{e}_{1n})]^T, \overline{\boldsymbol{G}} = \mathrm{diag}[\boldsymbol{G}_{f1}, \boldsymbol{G}_{f2}, \ldots \boldsymbol{G}_{fn}]$$

DESIGN OF FIXED-TIME ADAPTIVE COORDINATED CONTROL SCHEME

In this section, a fixed-time adaptive 6-DOF coordinated controller is proposed for multiple SFF with external disturbances and input quantization.

The (5) can be described by

$$\boldsymbol{G}_{fi}\left(\dot{\boldsymbol{e}}_{2i} + \dot{\boldsymbol{\alpha}}_i(\boldsymbol{e}_{1i})\right) = \boldsymbol{h}_i + \boldsymbol{D}(\boldsymbol{u}_i)\boldsymbol{u}_i + \boldsymbol{Q}_i(t) + \boldsymbol{\tau}_i \quad \textbf{(23)}$$

With

$$\dot{\boldsymbol{\alpha}}_i(\boldsymbol{e}_{1i}) = \begin{cases} \sigma_{1i} p_1 k_1 \mathrm{diag}(|\ \sigma_{1i}\mathrm{sgn}(\boldsymbol{e}_{1i})^{p_1} + \sigma_{2i}\mathrm{sgn}(\boldsymbol{e}_{1i})^{p_2}|^{k_1-1}) \\ \cdot\mathrm{diag}(|\boldsymbol{e}_{1i}|^{p_1-1})\boldsymbol{\Lambda}(\dot{\boldsymbol{e}}_{1i})\boldsymbol{e}_{2i} + \sigma_{2i} p_2 k_1 \mathrm{diag}(|\ \sigma_{1i}\mathrm{sgn}(\boldsymbol{e}_{1i})^{p_1} \\ +\sigma_{2i}\mathrm{sgn}(\boldsymbol{e}_{1i})^{p_2}|^{k_1-1})\mathrm{diag}(|\boldsymbol{e}_{1i}|^{p_2-1})\boldsymbol{\Lambda}(\dot{\boldsymbol{e}}_{1i})\boldsymbol{e}_{2i} \\ \qquad \text{if} \quad \bar{s}_{ik} = 0 \quad \text{or} \quad \bar{s}_{ik} \neq 0, \quad |e_{1ik}| > \varepsilon \\ \\ l_{1i}\boldsymbol{\Lambda}(\dot{\boldsymbol{x}}_{1i})\boldsymbol{e}_{2i} + 2l_{2i}\mathrm{diag}(|\boldsymbol{e}_{1i}|)\boldsymbol{\Lambda}(\dot{\boldsymbol{e}}_{1i})(\boldsymbol{e}_{2i}), \\ \qquad \text{if} \quad \bar{s}_{ik} \neq 0, \quad |e_{1ik}| \leq \varepsilon \end{cases} \quad \textbf{(24)}$$

$$\boldsymbol{h}_i = -\boldsymbol{C}(\boldsymbol{e}_{2i}) - \boldsymbol{N}(\boldsymbol{e}_{1i}) - \boldsymbol{G}_{fi} - \dot{\boldsymbol{e}}_{2id} + \boldsymbol{G}_{fi}\dot{\boldsymbol{\alpha}}_i(\boldsymbol{e}_{1i}), \tag{25}$$

Under Assumption1, we have

$$(\| (\boldsymbol{L}+\boldsymbol{B})\otimes \boldsymbol{I}_6 \|_1 \| \boldsymbol{\tau}_i + \boldsymbol{Q}_i(t) \|_1)^2 \le c_i, \quad i = 1,\ldots,n \tag{26}$$

where c_i are nonnegative constant numbers.

The adaptive fixed-time controller is defined as

$$\boldsymbol{u}_i = -\frac{1}{1-\delta}\boldsymbol{h}_i + \frac{1}{1-\delta}\Big(\sum_{j=1,j\neq i}^{n} a_{ij} + b_i\Big)^{-1}\Big[\sum_{j=1,j\neq i}^{n} a_{ij}((1-\delta)\boldsymbol{u}_j + \boldsymbol{h}_j) - \alpha_i \mathrm{sgn}(\boldsymbol{s}_i)^{\gamma 1} - \beta_i \mathrm{sgn}(\boldsymbol{s}_i)^{\gamma 2} - \frac{\hat{c}_i}{2\Im_i^2}\boldsymbol{s}_i\Big], \tag{27}$$

$$\dot{\hat{c}}_i = -2\kappa_i \varrho_i \hat{c}_i + \frac{\kappa_i}{2\Im_i^2}\| \boldsymbol{s}_i \|, \quad i = 1,\ldots,n \tag{28}$$

where $0<\gamma 1<1, \gamma 2>1, \Im_i>0, \kappa_i>0$ and $\varrho_i>0$ are the controller parameters.

Theorem 1: Consider 6-DOF control system (5) with the fixed-time coordinated control law (27), If the parameter uncertainty and external disturbance satisfy Assumption 1-2, then the sliding mode vector $\boldsymbol{s}_i$ will converge into

$$\| \boldsymbol{s}_i \| \le \Delta_s = \min\{\upsilon_1^{-\frac{2}{1+\gamma 1}}\Big(\frac{\upsilon}{1-\theta_o}\Big)^{\frac{2}{1+\gamma 1}}, \Big(\upsilon_2\Big(\frac{1}{2^{\frac{\upsilon+1}{2}-1}}\Big)^{n+k-1}\Big)^{-\frac{2}{1+\gamma 2}}\Big(\frac{\upsilon}{1-\theta_o}\Big)^{\frac{2}{1+\gamma 2}}\} \tag{29}$$

in fixed time, where $\beta_{\min} = \min\{\beta_i\}, \alpha_{\min} = \min\{\alpha_i\}, \iota_{\min} = \min\{\iota_i\}, \upsilon_1 = \min\{\alpha_{\min} 2^{\frac{\gamma 1+1}{2}}, \iota_{\min}^{\frac{\gamma 1+1}{2}} 2^{\frac{\gamma 1+1}{2}}\}$,

$\upsilon_2 = \min\{\beta_{\min} 2^{\frac{\gamma 2+1}{2}}, \iota_{\min}^{\frac{\gamma 2+1}{2}} 2^{\frac{\gamma 2+1}{2}}\}, \upsilon = \sum_{i=1}^{n} \varrho_i o_i c_i^2 + 1 + \sum_{i=1}^{n}\frac{\Im^2}{2} + \sum_{i=1}^{n}\Big[\Big(\frac{\iota_i}{\kappa_i}\Delta_i^2\Big)^{\frac{\gamma 2+1}{2}} - \frac{\iota_i}{\kappa_i}\Delta_i^2\Big]$,

$\iota_i = \kappa_i \dfrac{\varrho_i(2o_i-1)}{2o_i}$, $\Delta_i > 0, 0 < \theta_0 \le 1, o_i > \dfrac{1}{2}$.

Proof: we construct the following Lyapunov function candidate

$$V_1 = V_2 + V_3 \tag{30}$$

With

$$V_2 = \frac{1}{2} S^T S, \quad V_3 = \frac{1}{2} \sum_{i=1}^{n} \kappa_i^{-1} \tilde{c}_i^2 \tag{31}$$

where $\tilde{c}_i = c_i - \hat{c}_i$.

By the Kronecker product, the controller (27) can be rewritten as

$$\boldsymbol{U} = -\frac{1}{1-\delta}\boldsymbol{H} + \frac{1}{1-\delta}\left[(\boldsymbol{D}+\boldsymbol{B})^{-1} \otimes \boldsymbol{I}_6\right]\left[(\boldsymbol{A} \otimes \boldsymbol{I}_6)((1-\delta)\boldsymbol{U}+\boldsymbol{H}) - \boldsymbol{\alpha}\text{sgn}(\boldsymbol{S})^{\gamma 1} - \boldsymbol{\beta}\text{sgn}(\boldsymbol{S})^{\gamma 2} - \hat{\boldsymbol{c}}\boldsymbol{S}\right] \tag{32}$$

where
$\boldsymbol{U} = [u_1, \ldots, u_n]^T, \boldsymbol{H} = [h_1, \ldots, h_n]^T, \boldsymbol{\alpha} = \text{diag}[\alpha_1 \boldsymbol{I}_6, \ldots, \alpha_n \boldsymbol{I}_6], \boldsymbol{\beta} = \text{diag}[\beta_1 \boldsymbol{I}_6, \ldots, \beta_n \boldsymbol{I}_6]$,
$\hat{\boldsymbol{c}} = \text{diag}[\frac{\hat{c}_1}{2\Im_1^2}\boldsymbol{I}_6, \ldots, \frac{\hat{c}_n}{2\Im_n^2}\boldsymbol{I}_6]$.

Since $\boldsymbol{I}_{6n} - [(\boldsymbol{D}+\boldsymbol{B})^{-1} \otimes \boldsymbol{I}_6](\boldsymbol{A} \otimes \boldsymbol{I}_6)$ can be equivalently expressed as

$$\begin{aligned} &\boldsymbol{I}_{6n} - [(\boldsymbol{D}+\boldsymbol{B})^{-1} \otimes \boldsymbol{I}_6](\boldsymbol{A} \otimes \boldsymbol{I}_6) \\ &= [(\boldsymbol{D}+\boldsymbol{B})^{-1} \otimes \boldsymbol{I}_6]\{(\boldsymbol{D}+\boldsymbol{B}) \otimes \boldsymbol{I}_6] - \boldsymbol{A} \otimes \boldsymbol{I}_3\} \\ &= [(\boldsymbol{D}+\boldsymbol{B})^{-1} \otimes \boldsymbol{I}_6][(\boldsymbol{L}+\boldsymbol{B}) \otimes \boldsymbol{I}_6] \end{aligned} \tag{33}$$

It follows form (33) that

$$\begin{aligned} \boldsymbol{U} &= -\frac{1}{1-\delta}\boldsymbol{H} - \frac{1}{1-\delta}\{\boldsymbol{I}_{6n} - [(\boldsymbol{D}+\boldsymbol{B})^{-1} \otimes \boldsymbol{I}_6](\boldsymbol{A} \otimes \boldsymbol{I}_6)\}^{-1}[(\boldsymbol{D}+\boldsymbol{B})^{-1} \otimes \boldsymbol{I}_6] \times \left(\boldsymbol{\alpha}\text{sgn}(\boldsymbol{S})^{\gamma 1} + \boldsymbol{\beta}\text{sgn}(\boldsymbol{S})^{\gamma 2} + \hat{\boldsymbol{c}}\boldsymbol{S}\right) \\ &= -\frac{1}{1-\delta}\boldsymbol{H} - \frac{1}{1-\delta}\{[(\boldsymbol{D}+\boldsymbol{B})^{-1} \otimes \boldsymbol{I}_6][(\boldsymbol{L}+\boldsymbol{B}) \otimes \boldsymbol{I}_6]\}^{-1}[(\boldsymbol{D}+\boldsymbol{B})^{-1} \otimes \boldsymbol{I}_6] \times \left(\boldsymbol{\alpha}\text{sgn}(\boldsymbol{S})^{\gamma 1} + \boldsymbol{\beta}\text{sgn}(\boldsymbol{S})^{\gamma 2} + \hat{\boldsymbol{c}}\boldsymbol{S}\right) \\ &= -\frac{1}{1-\delta}\boldsymbol{H} - \frac{1}{1-\delta}[(\boldsymbol{L}+\boldsymbol{B}) \otimes \boldsymbol{I}_6]^{-1}\left(\boldsymbol{\alpha}\text{sgn}(\boldsymbol{S})^{\gamma 1} + \boldsymbol{\beta}\text{sgn}(\boldsymbol{S})^{\gamma 2} + \hat{\boldsymbol{c}}\boldsymbol{S}\right) \end{aligned} \tag{34}$$

Considering (22) and (23), it can be shown that

$$\begin{aligned} \dot{V}_2 &= \boldsymbol{S}^T \dot{\boldsymbol{S}} \\ &= \boldsymbol{S}^T [(\boldsymbol{L}+\boldsymbol{B}) \otimes \boldsymbol{I}_6]\left(\boldsymbol{G}_{fi}\dot{\boldsymbol{e}}_{2i} + \boldsymbol{G}_{fi}\dot{\boldsymbol{\alpha}}_i(\boldsymbol{e}_{1i})\right) \\ &= \boldsymbol{S}^T [(\boldsymbol{L}+\boldsymbol{B}) \otimes \boldsymbol{I}_6](\boldsymbol{H} + \boldsymbol{D}(\boldsymbol{U})\boldsymbol{U} + \boldsymbol{Q}(t) + \boldsymbol{\tau}) \end{aligned} \tag{35}$$

where $\boldsymbol{\tau} = [\tau_1, \ldots, \tau_n]^T, \boldsymbol{D}(\boldsymbol{U}) = \text{diag}[\boldsymbol{D}(u_1), \ldots, \boldsymbol{D}(u_n)], \boldsymbol{Q}(t) = [Q_1(t), \ldots, Q_n(t)]^T$.

Next, it can be derived that

$$\begin{aligned}
\dot{V}_2 &= \boldsymbol{S}^T[(\boldsymbol{L}+\boldsymbol{B})\otimes \boldsymbol{I}_6]([(\boldsymbol{L}+\boldsymbol{B})\otimes I_6]^{-1}\left(-\boldsymbol{\alpha}\mathrm{sgn}(\boldsymbol{S})^{\gamma 1}-\boldsymbol{\beta}\mathrm{sgn}(\boldsymbol{S})^{\gamma 2}-\boldsymbol{cS}\right)+Q(t)+\tau) \\
&= \boldsymbol{S}^T[(\boldsymbol{L}+\boldsymbol{B})\otimes \boldsymbol{I}_6](\boldsymbol{\tau}+\boldsymbol{Q}(t))-\boldsymbol{S}^T\boldsymbol{\alpha}\mathrm{sgn}(\boldsymbol{S})^{\gamma 1}-\boldsymbol{S}^T\boldsymbol{\beta}\mathrm{sgn}(\boldsymbol{S})^{\gamma 2}-\boldsymbol{S}^T\hat{\boldsymbol{c}}\boldsymbol{S} \\
&\le \sum_{i=1}^{n}\|(\boldsymbol{L}+\boldsymbol{B})\otimes \boldsymbol{I}_6\|\|\boldsymbol{\tau}_i+\boldsymbol{Q}_i(t)\|\|\boldsymbol{s}_i\|-\sum_{i=1}^{n}\frac{\hat{c}_i}{2\Im_i^2}\|\boldsymbol{s}_i\|^2-\sum_{i=1}^{n}\sum_{k=1}^{p}\beta_i|s_{ik}|^{\gamma 2+1}-\sum_{i=1}^{n}\sum_{k=1}^{p}\alpha_i|s_{ik}|^{\gamma 1+1} \\
&\le \sum_{i=1}^{n}\frac{(\|(\boldsymbol{L}+\boldsymbol{B})\otimes \boldsymbol{I}_6\|_1\|\tau_i+\boldsymbol{Q}_i(t)\|_1)^2\|\boldsymbol{s}_i\|^2}{2\Im^2}+\sum_{i=1}^{n}\frac{\Im^2}{2}-\sum_{i=1}^{n}\frac{\hat{c}_i}{2\Im_i^2}\|\boldsymbol{s}_i\|^2-\sum_{i=1}^{n}\sum_{k=1}^{p}\alpha_i(s_{ik}^2)^{\frac{\gamma 1+1}{2}}-\sum_{i=1}^{n}\sum_{k=1}^{p}\beta_i(s_{ik}^2)^{\frac{\gamma 2+1}{2}} \\
&\le \sum_{i=1}^{n}\frac{\tilde{c}_i}{2\Im_i^2}\|s_i\|^2+\sum_{i=1}^{n}\frac{\Im^2}{2}-\sum_{i=1}^{n}\sum_{k=1}^{p}\alpha_i(s_{ik}^2)^{\frac{\gamma 1+1}{2}}-\sum_{i=1}^{n}\sum_{k=1}^{p}\beta_i(s_{ik}^2)^{\frac{\gamma 2+1}{2}}
\end{aligned} \tag{36}$$

In addition, taking the derivative of V_3 yields

$$\begin{aligned}
\dot{V}_3 &= \sum_{i=1}^{n}\kappa_i^{-1}\tilde{c}_i\dot{\tilde{c}}_i \\
&= -\sum_{i=1}^{n}\frac{c_i}{2\Im_i^2}\|\boldsymbol{s}_i\|^2+\sum_{i=1}^{n}2\varrho_i\tilde{c}_i\hat{c}_i
\end{aligned} \tag{37}$$

It is noticed from (36) and (37) that

$$\dot{V}_1 \le -\sum_{i=1}^{n}\sum_{k=1}^{p}\alpha_i|s_{ik}|^{\gamma 1+1}-\sum_{i=1}^{n}\sum_{k=1}^{p}\beta_i|s_{ik}|^{\gamma 2+1}+2\sum_{i=1}^{n}\varrho_i\tilde{c}_i\hat{c}_i+\sum_{i=1}^{n}\frac{\Im^2}{2} \tag{38}$$

From

$$\varrho_i\tilde{c}_i\hat{c}_i \le -\frac{\varrho_i(2o_i-1)}{2o_i}\tilde{c}_i^{\,2}+\frac{\varrho_i o_i}{2}c_i^2 \tag{39}$$

and by substituting (39) into (38), we have

$$\begin{aligned}
\dot{V}_1 \le &-\sum_{i=1}^{n}\sum_{k=1}^{p}\alpha_i|s_{ik}|^{\gamma 1+1}-\sum_{i=1}^{n}\sum_{k=1}^{p}\beta_i|s_{ik}|^{\gamma 2+1}-\sum_{i=1}^{n}\frac{l_i}{\kappa_i}\tilde{c}_i^{\,2}+\sum_{i=1}^{n}\varrho_i o_i c_i^2 \\
&-\sum_{i=1}^{n}\frac{l_i}{\kappa_i}\tilde{c}_i^{\,2}-\sum_{i=1}^{n}\left(\frac{l_i}{\kappa_i}\tilde{c}_i^{\,2}\right)^{\frac{1+\gamma_1}{2}}+\sum_{i=1}^{n}\left(\frac{l_i}{\kappa_i}\tilde{c}_i^{\,2}\right)^{\frac{1+\gamma_1}{2}}+\sum_{i=1}^{n}\frac{\Im^2}{2}
\end{aligned} \tag{40}$$

Where $\dfrac{l_i}{\kappa_i}=\dfrac{\varrho_i(2o_i-1)}{2o_i}$.

Case 1: If $\frac{\iota_i}{\kappa_i}\tilde{c}_i^{\,2} \geq 1$, we have

$$\left(\frac{\iota_i}{\kappa_i}\tilde{c}_i^{\,2}\right)^{\frac{1+\gamma_1}{2}} \leq \frac{\iota_i}{\kappa_i}\tilde{c}_i^{\,2} \tag{41}$$

Substituting (41) into (40) yields

$$\dot{V}_1 \leq -\sum_{i=1}^{n}\sum_{k=1}^{p}\alpha_i(s_{ik}^2)^{\frac{\gamma 1+1}{2}} - \sum_{i=1}^{n}\sum_{k=1}^{p}\beta_i(s_{ik}^2)^{\frac{\gamma 2+1}{2}} - \sum_{i=1}^{n}\frac{\iota_i}{\kappa_i}\tilde{c}_i^{\,2} + \sum_{i=1}^{n}\varrho_i o_i c_i^2 - \sum_{i=1}^{n}\left(\frac{\iota_i}{\kappa_i}\tilde{c}_i^{\,2}\right)^{\frac{1+\gamma 1}{2}} + \sum_{i=1}^{n}\frac{\Im^2}{2} \tag{42}$$

According to Lemma 5, it is easily to prove that

$$\sum_{i=1}^{n}\sum_{k=1}^{p}\alpha_i(s_{ik}^2)^{\frac{\gamma 1+1}{2}} + \sum_{i=1}^{n}\left(\frac{\iota_i}{\kappa_i}\tilde{c}_i^{\,2}\right)^{\frac{1+\gamma 1}{2}} \geq \upsilon_1 V_8^{\frac{1+\gamma 1}{2}} \tag{43}$$

where $\upsilon_1 = \min\{\alpha_{\min} 2^{\frac{\gamma 1+1}{2}}, \iota_{\min} 2^{\frac{\gamma 1+1}{2}}\}, \alpha_{\min} = \min\{\alpha_i\}, \iota_{\min} = \min\{\iota_i\}$.

By applying (43), (42) can be rewritten as

$$\dot{V}_1 \leq -\upsilon_1 V_1^{\frac{1+\gamma 1}{2}} - \sum_{i=1}^{n}\sum_{k=1}^{p}\beta_i(s_{ik}^2)^{\frac{\gamma 2+1}{2}} - \sum_{i=1}^{n}\frac{\iota_i}{\kappa_i}\tilde{c}_i^{\,2} + \sum_{i=1}^{n}\varrho_i o_i c_i^2 + \sum_{i=1}^{n}\frac{\Im^2}{2} \tag{44}$$

Case 2: If $\frac{\iota_i}{\kappa_i}\tilde{c}_i^{\,2} < 1$, one can obtain

$$\left(\frac{\iota_i}{\kappa_i}\tilde{c}_i^{\,2}\right)^{\frac{1+\gamma_1}{2}} - \frac{\iota_i}{\kappa_i}\tilde{c}_i^{\,2} \leq 1 - \frac{\iota_i}{\kappa_i}\tilde{c}_i^{\,2} < 1 \tag{45}$$

Substituting (45) into (40) yields

$$\dot{V}_1 \leq -\upsilon_1 V_1^{\frac{1+\gamma 1}{2}} - \sum_{i=1}^{n}\sum_{k=1}^{p}\beta_i(s_{ik}^2)^{\frac{\gamma 2+1}{2}} - \sum_{i=1}^{n}\frac{\iota_i}{\kappa_i}\tilde{c}_i^{\,2} + \sum_{i=1}^{n}\varrho_i o_i c_i^2 + 1 + \sum_{i=1}^{n}\frac{\Im^2}{2} \tag{46}$$

Then (46) can be described by

$$\dot{V}_1 \le -\upsilon_1 V_1^{\frac{1+\gamma 1}{2}} - \sum_{i=1}^{n}\sum_{k=1}^{p}\beta_i (s_{ik}^2)^{\frac{\gamma 2+1}{2}} - \sum_{i=1}^{n}\frac{\iota_i}{\kappa_i}\tilde{c}_i^2 - \sum_{i=1}^{n}\Big(\frac{\iota_i}{\kappa_i}\tilde{c}_i^2\Big)^{\frac{\gamma 2+1}{2}} + \sum_{i=1}^{n}\Big(\frac{\iota_i}{\kappa_i}\tilde{c}_i^2\Big)^{\frac{\gamma 2+1}{2}} + \sum_{i=1}^{n}\varrho_i o_i c_i^2 + 1 + \sum_{i=1}^{n}\frac{\Im^2}{2} \quad \textbf{(47)}$$

Based on Lemma 5, one has

$$\sum_{i=1}^{n}\sum_{k=1}^{p}\beta_i (s_{ik}^2)^{\frac{\gamma 2+1}{2}} + \sum_{i=1}^{n}\Big(\frac{\iota_i}{\kappa_i}\tilde{c}_i^2\Big)^{\frac{\gamma 2+1}{2}} \ge \upsilon_2\Big(\frac{1}{2^{\frac{\gamma 2+1}{2}-1}}\Big)V_5^{\frac{1+\gamma 2}{2}} \quad \textbf{(48)}$$

Where $\upsilon_2 = \min\{\beta_{\min} 2^{\frac{\gamma 2+1}{2}}, \iota_{\min} 2^{\frac{\gamma 2+1}{2}}\}, \beta_{\min} = \min\{\beta_i\}, \iota_{\min} = \min\{\iota_i\}$

From (48), we have

$$\dot{V}_1 \le -\upsilon_1 V_1^{\frac{1+\gamma 1}{2}} - \upsilon_2\Big(\frac{1}{2^{\frac{\gamma 2+1}{2}-1}}\Big)V_1^{\frac{1+\gamma 2}{2}} - \sum_{i=1}^{n}\frac{\iota_i}{\kappa_i}\tilde{c}_i^2 + \sum_{i=1}^{n}\Big(\frac{\iota_i}{\kappa_i}\tilde{c}_i^2\Big)^{\frac{\gamma 2+1}{2}} + \sum_{i=1}^{n}\varrho_i o_i c_i^2 + \sum_{i=1}^{n}\frac{\Im^2}{2} + 1 \quad \textbf{(49)}$$

Assume that there exists a compact set

$$\Pi_i = \{(\tilde{c}_i \,||\, \tilde{c}_i \,| \le \Delta_i\} \quad \textbf{(50)}$$

where Δ_i is an unknown constant.

If $\Delta_i < \sqrt{\frac{\varsigma_i}{\iota_i}}$, we can obtain

$$\frac{\iota_i}{\kappa_i}\tilde{c}_i^2 < 1, \quad \Big(\frac{\iota_i}{\kappa_i}\tilde{c}_i^2\Big)^{\frac{\gamma 2+1}{2}} < \frac{\iota_i}{\kappa_i}\tilde{c}_i^2 \quad \textbf{(51)}$$

If $\Delta_i \ge \sqrt{\frac{\varsigma_i}{\iota_i}}$, we can obtain

$$\Big(\frac{\iota_i}{\kappa_i}\tilde{c}_i^2\Big)^{\frac{\gamma 2+1}{2}} - \frac{\iota_i}{\kappa_i}\tilde{c}_i^2 \le \Big(\frac{\iota_i}{\kappa_i}\Delta_i^2\Big)^{\frac{\gamma 2+1}{2}} - \frac{\iota_i}{\kappa_i}\Delta_i^2 \quad \textbf{(52)}$$

Denote

$$\upsilon = \sum_{i=1}^{n} \varrho_i o_i c_i^2 + 1 + \sum_{i=1}^{n} [(\frac{l_i}{\kappa_i}\Delta_i^2)^{\frac{\gamma 2+1}{2}} - \frac{l_i}{\kappa_i}\Delta_i^2] \tag{53}$$

Furthermore, we can conclude that

$$\dot{V}_1 \leq -\upsilon_1 V_1^{\frac{1+\gamma 1}{2}} - \upsilon_2 (\frac{1}{2^{\frac{\gamma 2+1}{2}-1}})^{n+k-1} V_1^{\frac{1+\gamma 2}{2}} + \upsilon \tag{54}$$

According to Lemma 6, system (5) is practically fixed-time stable. Furthermore, s_i will converge into the region

$$\Phi = \{\lim_{t \to T_s} s_i \mid V_1 \leq \min\{\upsilon_1^{-\frac{2}{1+\gamma 1}} (\frac{\upsilon}{1-\theta_o})^{\frac{2}{1+\gamma 1}}, (\upsilon_2 (\frac{1}{2^{\frac{\upsilon+1}{2}-1}})^{n+k-1})^{-\frac{2}{1+\gamma 2}} (\frac{\upsilon}{1-\theta_o})^{\frac{2}{1+\gamma 2}}\}\}$$

in a fixed time

$$T_s \leq \frac{1}{\vartheta_1 \theta_o (1 - \frac{\gamma 1+1}{2})} + \frac{1}{\vartheta_2 (\frac{1}{2^{\frac{\gamma 2+1}{2}} - 1})^n \theta_o (\frac{\gamma 2+1}{2} - 1)},$$

that is s_i will converge into the region $\| s_i \| \leq \Delta_s$.

Theorem 2: when s_i reach to the boundary Δ_s in fixed time, the tracking error e_{1i} and e_{2i} will converge to

$$\begin{aligned} |e_{1i}| &< \Delta_{e1i} = \max\{\varepsilon, (\frac{\Delta_s^{\frac{1}{k}}}{\sigma_{1i}})^{\frac{1}{p_1}}, (\frac{\Delta_s^{\frac{1}{k}}}{\sigma_{2i}})^{\frac{1}{p_2}}\} \\ |e_{2i}| &< \Delta_{e2i} = \max\{\Delta_s + l_1 \Delta_{e1ik} + l_2 \Delta_{e1ik}^2, \Delta_s + (\sigma_{1i}\Delta_{e1i}^{p_1} + \sigma_{2i}\Delta_{e1i}^{p_2})^{k_1}\} \end{aligned} \tag{55}$$

in a fixed time, where $i = 1,\ldots,n, \quad k = 1,2,\ldots,6$

Proof: Case 1: If $\bar{s}_i = 0$ we can obtain

$$\bar{s}_i = e_{2i} + \mathrm{sgn}(\sigma_{1i}\mathrm{sgn}(e_{1i})^{p_1} + \sigma_{2i}\mathrm{sgn}(e_{1i})^{p_2})^{k_1} = 0, \quad i = 1,\ldots,n \tag{56}$$

Moreover, one has

$$\begin{aligned} &\dot{\rho}_{ei} + \mathrm{sgn}(\sigma_{1i}\mathrm{sgn}(\rho_{ei})^{p_1} + \sigma_{2i}\mathrm{sgn}(\rho_{ei})^{p_2})^{k_1} = 0, \quad i = 1,\ldots,n \\ &\omega_{ei} + \mathrm{sgn}(\sigma_{1i}\mathrm{sgn}(q_{ei})^{p_1} + \sigma_{2i}\mathrm{sgn}(q_{ei})^{p_2})^{k_1} = 0, \quad i = 1,\ldots,n \end{aligned} \tag{57}$$

Construct the following Lyapunov function candidate

$$\begin{aligned} V_4 &= \frac{1}{2}\rho_{ei}^T\rho_{ei} + q_{ei}^T q_{ei} + (1 - q_{0ei})^2 \\ &= \frac{1}{2}\rho_{ei}^T\rho_{ei} + 2(1 - q_{0ei}) \end{aligned} \tag{58}$$

Denote

$$V_5 = \rho_{ei}^T\rho_{ei}, \quad V_6 = 2(1 - q_{0ei}) \tag{59}$$

Taking the derivative of V_5 and V_6 yields

$$\begin{aligned} \dot{V}_5 &= \rho_{ei}^T\dot{\rho}_{ei} \\ &= \rho_{ei}^T(-\mathrm{sgn}(\sigma_{1i}\mathrm{sgn}(\rho_{ei})^{p_1} + \sigma_{2i}\mathrm{sgn}(\rho_{ei})^{p_2})^{k_1}) \\ &= -\left|\sigma_{1i}(\rho_{ei}^T\rho_{ei})^{\frac{(\frac{1}{k_1}+p_1)}{2}} + \sigma_{2i}(\rho_{ei}^T\rho_{ei})^{\frac{(\frac{1}{k_1}+p_2)}{2}}\right|^{k_1} \\ &\le -\left(\sigma_{1i}2^{\frac{(\frac{1}{k_1}+p_1)}{2}}V_5^{\frac{(\frac{1}{k_1}+p_1)}{2}} + \sigma_{2i}2^{\frac{(\frac{1}{k_1}+p_2)}{2}}V_5^{\frac{(\frac{1}{k_1}+p_2)}{2}}\right)^{k_1} \end{aligned} \tag{60}$$

Similarly, taking the derivative of V_6 yields

$$\begin{aligned}\dot{V}_6 &= -2\dot{q}_{0ei} = \boldsymbol{q}_{iev}^T\boldsymbol{\omega}_{ei} \\ &= -\boldsymbol{q}_{iev}^T(\operatorname{sgn}(\sigma_{1i}\operatorname{sgn}(q_{ei})^{p_1} + \sigma_{2i}\operatorname{sgn}(q_{ei})^{p_2})^{k_1}) \\ &= -\left|\sigma_{1i}(\boldsymbol{q}_{iev}^T\boldsymbol{q}_{iev})^{\frac{(\frac{1}{k_1}+p_1)}{2}} + \sigma_{2i}(\boldsymbol{q}_{iev}^T\boldsymbol{q}_{iev})^{\frac{(\frac{1}{k_1}+p_2)}{2}}\right|^{k_1} \\ &\le -\left(\sigma_{1i}0.5^{\frac{(\frac{1}{k_1}+p_1)}{2}}V_6^{\frac{(\frac{1}{k_1}+p_1)}{2}} + \sigma_{2i}0.5^{\frac{(\frac{1}{k_1}+p_2)}{2}}V_6^{\frac{(\frac{1}{k_1}+p_2)}{2}}\right)^{k_1}\end{aligned} \quad \textbf{(61)}$$

Then from (60) and (61), we can obtain that system error states $(\boldsymbol{\rho}_{ei}, \boldsymbol{q}_{ei})$ converge into regions $(0,0)$, at the same time $\dot{q}_{0ei} \in (-1,1]$ converges to 1 in fixed time by using Lemma 6.

Case 2: If $\bar{\boldsymbol{s}}_i \neq 0$ and $|e_{1ik}| < \varepsilon$, which implies that $|e_{1ik}|$ has converged to the region $|e_{1ik}| < \Delta_{e1ik} = \varepsilon$ in fixed-time. Then from (21), we have

$$e_{2ik} + l_{1i}e_{1ik} + l_{2i}\operatorname{sgn}(e_{1ik})^2 = \varpi_{ik}, |\ \varpi_{ik}| \le \Delta_s \quad \textbf{(62)}$$

Moreover, it is easy to see that

$$|e_{2ik}| < \Delta_s + l_{1i}\Delta_{e1ik} + l_{2i}\Delta_{e1ik}^2 \quad \textbf{(63)}$$

which means that $|e_{2ik}|$ converges to the region $|e_{1ik}| < \Delta_{e1ik} = \varepsilon$ in fixed time.

Case 3: If $\bar{\boldsymbol{s}}_i \neq 0$ and $|e_{1ik}| > \varepsilon$ we can obtain

$$\bar{\boldsymbol{s}}_i = \boldsymbol{e}_{2i} + \operatorname{sgn}(\sigma_{1i}\operatorname{sgn}(\boldsymbol{e}_{1i})^{p_1} + \sigma_{2i}\operatorname{sgn}(\boldsymbol{e}_{1i})^{p_2})^{k_1} \neq 0, \quad i = 1,\ldots,n \quad \textbf{(64)}$$

which means

$$\bar{\boldsymbol{s}}_i = \boldsymbol{e}_{2i} + \operatorname{sgn}(\sigma_{1i}\operatorname{sgn}(\boldsymbol{e}_{1i})^{p_1} + \sigma_{2i}\operatorname{sgn}(\boldsymbol{e}_{1i})^{p_2})^{k_1} = \varpi_{ik}, |\varpi_{ik}| \le \Delta_s \quad \textbf{(65)}$$

After a simple transformation, (65) can be rewritten as

$$e_{2i} + \left(1 - \frac{\varpi_{ik}}{\text{sgn}(\sigma_{1i}\text{sgn}(e_{1i})^{\gamma 1} + \sigma_{2i}\text{sgn}(e_{1i})^{\gamma 2})^{k_1}}\right)\text{sgn}(\sigma_{1i}\text{sgn}(e_{1i})^{\gamma 1} + \sigma_{2i}\text{sgn}(e_{1i})^{\gamma 2})^{k_1} = 0 \quad \textbf{(66)}$$

When $1 - \frac{\varpi_{ik}}{\text{sgn}(\sigma_{1i}\text{sgn}(e_{1i})^{p_1} + \sigma_{2i}\text{sgn}(e_{1i})^{p_2})^{k_1}} > 0$, (66) is still kept in the form of the FTTSM as in Case 1, which implies that

$$|e_{1i}| < \left(\frac{\Delta_s^{\frac{1}{k}}}{\sigma_{1i}}\right)^{\frac{1}{p_1}}, \quad |e_{1i}| < \left(\frac{\Delta_s^{\frac{1}{k}}}{\sigma_{2i}}\right)^{\frac{1}{p_2}} \quad \textbf{(67)}$$

Thus, system state e_{1i} will converge into the region in fixed time.

$$|e_{1i}| < \Delta_{e1i} = \max\left\{\left(\frac{\Delta_s^{\frac{1}{k}}}{\sigma_{1i}}\right)^{\frac{1}{p_1}}, \left(\frac{\Delta_s^{\frac{1}{k}}}{\sigma_{2i}}\right)^{\frac{1}{p_2}}\right\} \quad \textbf{(68)}$$

On the other hand, from (65) we can obtain that system state e_{2i} will converge to the region in fixed time.

$$|e_{2i}| < \Delta_{e2i} = \Delta_s + \left(\sigma_{1i}\Delta_{e1i}^{p_1} + \sigma_{2i}\Delta_{e1i}^{p_2}\right)^{k_1} \quad \textbf{(69)}$$

Furthermore, we can conclude that

$$\begin{aligned} |e_{1i}| &< \Delta_{e1i} = \max\left\{\varepsilon, \left(\frac{\Delta_s^{\frac{1}{k}}}{\sigma_{1i}}\right)^{\frac{1}{p_1}}, \left(\frac{\Delta_s^{\frac{1}{k}}}{\sigma_{2i}}\right)^{\frac{1}{p_2}}\right\} \\ |e_{2i}| &< \Delta_{e2i} = \max\left\{\Delta_s + l_1\Delta_{e1ik} + l_2\Delta_{e1ik}^2, \Delta_s + \left(\sigma_{1i}\Delta_{e1i}^{p_1} + \sigma_{2i}\Delta_{e1i}^{p_2}\right)^{k_1}\right\} \end{aligned} \quad \textbf{(70)}$$

Thus, e_{1i} and e_{2i} $i = 1,2,3$ will converge to the regions Δ_{e1i} and Δ_{e2i} in fixed time, respectively.

Remark 4: By using the adaptive approach, the information of the parameters uncertainties and external disturbance is not required in the controller (27) design. In addition, there is no need to make additional assumption about inter-spacecraft communication topology during the fixed-time coordinated controller design procedure. Thus, the proposed control scheme is available to any communication topology.

ILLUSTRATIVE EXAMPLE

To show the effectiveness of the proposed controller, an illustrative example is given in this section. The communicate topology of SFF is described as Fig. (**2**), in which " $i(i=1,2,3)$ " is the ith follower spacecraft. The leader spacecraft is specified to a circular orbit with the radius 6878km and orbit angel velocity is $n_0 = 1.11 \times 10^{-3}$

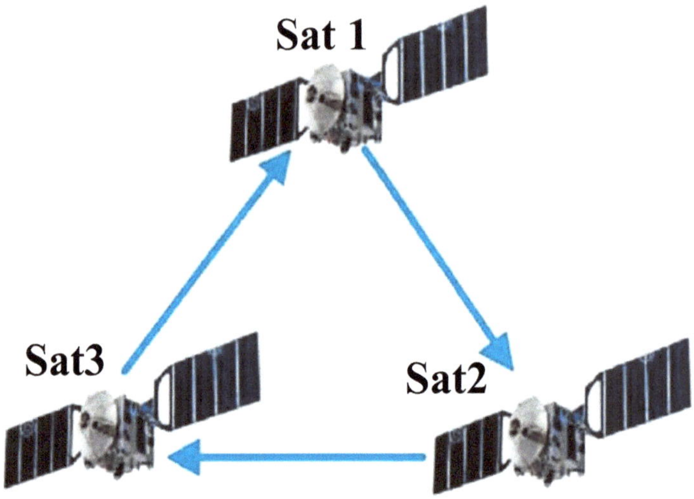

Fig. (2). The directed communication topology of three follower spacecraft.

The weighted Laplace matrix L is

$$\boldsymbol{L} = \begin{bmatrix} 1 & 0 & -1 \\ -1 & 1 & 0 \\ 0 & -1 & 1 \end{bmatrix}$$

The inertia matrix and mass are

$$\boldsymbol{J}_1 = \begin{bmatrix} 10 & 1 & 0.4 \\ 1 & 8 & 0.2 \\ 0.4 & 0.2 & 7 \end{bmatrix} \text{kg.m}^2, \boldsymbol{J}_2 = \begin{bmatrix} 11 & 0.5 & 0.4 \\ 0.5 & 9 & 0.2 \\ 0.4 & 0.2 & 7 \end{bmatrix} \text{kg.m}^2$$

$$\boldsymbol{J}_3 = \begin{bmatrix} 9 & 0.5 & 0.7 \\ 0.5 & 3.5 & 0.3 \\ 0.7 & 0.2 & 8 \end{bmatrix} \text{kg.m}^2, m_{fi} = 50\text{kg}, \quad \forall i = 1,2,3$$

The desired attitude quaternion and attitude angular velocity are

$$\boldsymbol{q}_{di} = \begin{bmatrix} 1 & 0 & 0 & 0 \end{bmatrix}^T, \boldsymbol{\omega}_{di} = \begin{bmatrix} 0 & 0 & 0 \end{bmatrix}^T \text{rad/s}, \quad \forall i = 1,2,3$$

The initial attitude quaternion and attitude angular velocity are

$$\boldsymbol{q}_1(0) = \begin{bmatrix} 0.3317 & 0.3 & 0.4 & -0.8 \end{bmatrix}^T$$
$$\boldsymbol{\omega}_1(0) = \begin{bmatrix} 0.07 & -0.05 & -0.04 \end{bmatrix}^T \text{rad / s}$$

$$\boldsymbol{q}_2(0) = \begin{bmatrix} 0.2646 & 0.2 & 0.5 & -0.8 \end{bmatrix}^T$$
$$\boldsymbol{\omega}_2(0) = \begin{bmatrix} 0.07 & -0.05 & -0.04 \end{bmatrix}^T rad / s$$

$$\boldsymbol{q}_3(0) = \begin{bmatrix} 0.5568 & 0.2 & 0.4 & -0.7 \end{bmatrix}^T$$
$$\boldsymbol{\omega}_3(0) = \begin{bmatrix} 0.06 & -0.05 & -0.05 \end{bmatrix}^T rad / s$$

The relative position and velocity are initialized

$$\boldsymbol{\rho}_i(0) = \begin{bmatrix} -10 & 160 & 50 \end{bmatrix}^T \text{m}, \quad \boldsymbol{v}_i(0) = \begin{bmatrix} 0 & 0 & 0 \end{bmatrix}^T \text{m/s}, \forall i = 1,2,3$$

To form a triangle of three follower spacecraft, the desired relative position and velocity are specified as

$$\boldsymbol{\rho}_{1d} = \begin{bmatrix} 0 & 100 & 0 \end{bmatrix}^T \text{m}, \quad \boldsymbol{\rho}_{2d} = \begin{bmatrix} 0 & 200 & 0 \end{bmatrix}^T \text{m},$$
$$\boldsymbol{\rho}_{3d} = \begin{bmatrix} 0 & 150 & 50\sqrt{3} \end{bmatrix}^T \text{m}, \quad \boldsymbol{v}_{di}(0) = \begin{bmatrix} 0 & 0 & 0 \end{bmatrix}^T \text{m/s}, \forall i = 1,2,3$$

The external disturbance of torque and force are specified as

$$z_i = \begin{bmatrix} 0.06\cos(0.3t) & 0.01\sin(0.1t) & 0.06\cos(0.2t) \end{bmatrix}^T \text{ Nm},$$
$$F_{it} = \begin{bmatrix} 0.001\cos(0.02t) & 0.002\sin(0.02t) & -0.001\sin(0.02t) \end{bmatrix}^T \text{ N}, \forall i = 1,2,3$$

The maximum of input control force and torques are limited to 5N and 1Nm, respectively.

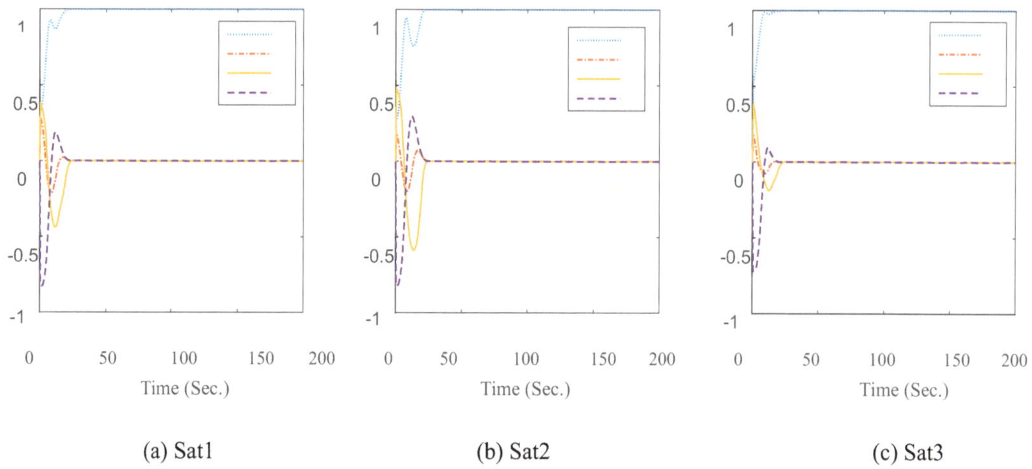

Fig. (3). Response of the relative attitude error.

To describe the synchronization accuracy and tracking accuracy of the controller (27), the performance indexes are defined as

$$\varpi_1 = \sqrt{\sum_{i=1}^{3} \|\tilde{\rho}_i\|^2}, \varpi_2 = \sqrt{\sum_{i=1}^{3}\sum_{j\in N_i} \|\tilde{\rho}_{i,j}\|^2}$$
$$\varpi_3 = \sqrt{\sum_{i=1}^{3} \boldsymbol{q}_{vei}^T \boldsymbol{q}_{vei}}, \varpi_4 = \sqrt{\sum_{i=1}^{3}\sum_{j\in N_i} \boldsymbol{q}_{i,jve}^T \boldsymbol{q}_{i,jve}}$$

where $\tilde{\rho}_i = \rho_i - \rho_{di}, \rho_j = \rho_j - \rho_{jd}, \tilde{\rho}_{i,j} = \tilde{\rho}_i - \tilde{\rho}_j$, $q_{i,jve}$ is the vector part of $\boldsymbol{q}_{i,je} = [q_{i,j0e} \quad \boldsymbol{q}_{i,jve}^T]^T = \boldsymbol{q}_{ei} \otimes \boldsymbol{q}_{ej}$. According to the definition of $\varpi_1, \varpi_2, \varpi_3$ and ϖ_4, we can conclude that smaller ϖ_1 and ϖ_3 can ensure the better formation and attitude tracking performance, respectively; smaller ϖ_2 and ϖ_4 can guarantee the better formation and attitude synchronization performance, respectively.

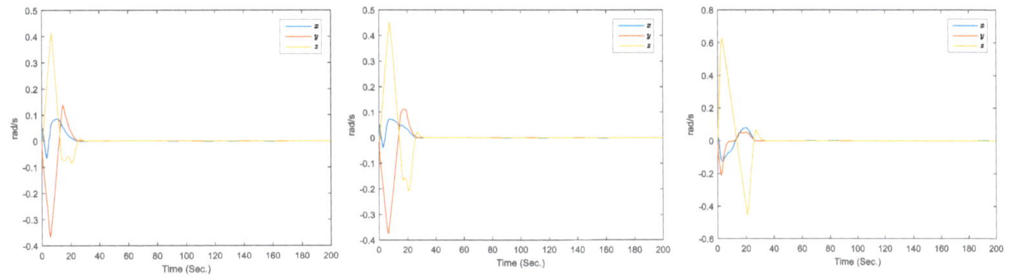

(a) Sat1 (b) Sat2 (c) Sat3

Fig. (4). Response of the relative angular velocity error.

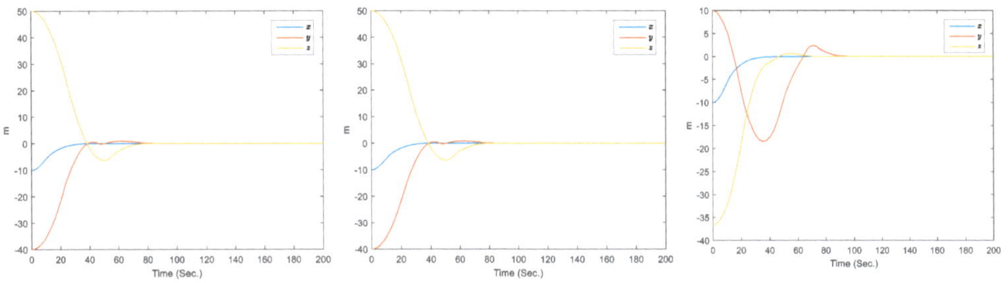

(a) Sat1 (b) Sat2 (c) Sat3

Fig. (5). Response of the relative position error.

The parameters of controller (27), adaptive law (28), and hysteretic quantizer (10) are chosen as

$$\gamma 1=0.9, \gamma 2=1.1, a_{12}=0.5, a_{23}=0.5,\ a_{31}=0.5, \alpha_i=0.1, \beta_i=1, p_1=0.4,\ p_2=1.5,$$
$$k_1=2, c_i=0.1, b_i=0.04,\ \kappa_i=0.000002, \Im_i=0.5,\ \varrho_i=0.000002,\ \forall i=1,2,3,$$
$$\delta=0.25, u_{\min}=0.00001.$$

The simulation results of the control scheme (27) are given in Figs. (**3-9**), where Figs. (**3** and **4**) plot the relative attitude error and relative angular velocity error, respectively. It can be seen that the relative attitude errors converge to near zero within 50 s, which has fast convergence rate. Figs. (**5** and **6**) show the relative position error and relative velocity error, respectively. It can be seen that the relative position errors converge to near zero about 85 s, which has fast convergence rate.

The quantized control force and torque are given in Figs. (**7** and **8**), respectively. Fig. (**4**) shows the curves of the performance indexes ϖ_1 to ϖ_4. From the simulation results, we can find that the proposed controller has good performance of tracking and synchronization.

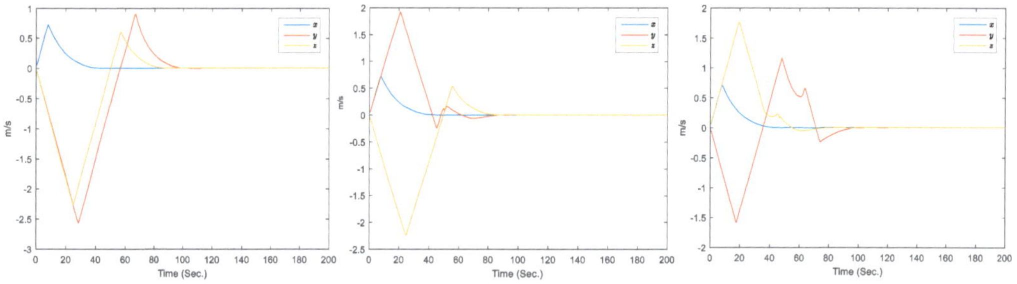

(a) Sat1 (b) Sat2(c) (c)Sat3

Fig. (6). Response of the relative velocity error.

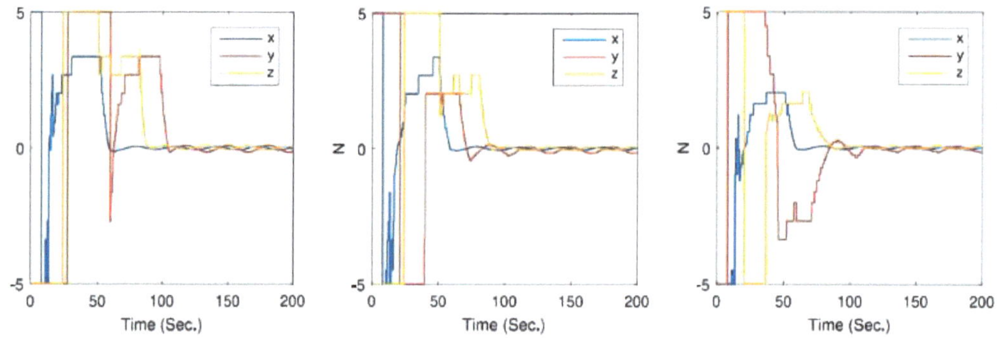

(a) Sat1 (b) Sat2(c) (c)Sat3

Fig. (7). Response of the quantized control force.

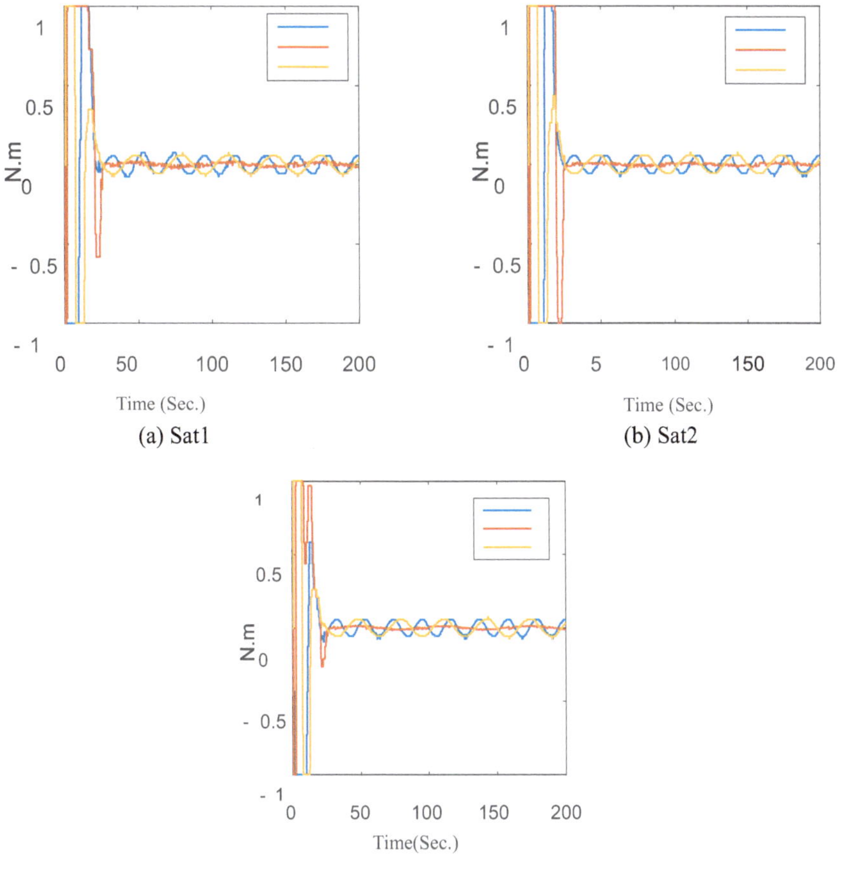

(a) Sat1

(b) Sat2

(3) Sat3

Fig. (8). Response of the quantized control torque.

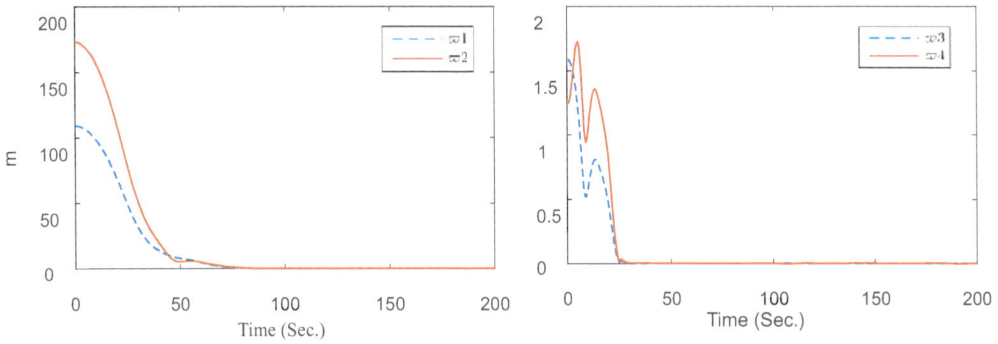

Fig. (9). Response of the performance indexes ϖ_1 to ϖ_4.

CONCLUSION

In this article, the fixed-time 6-DOF coordinated control problem has been investigated for SFF with input signal quantization and directed communication graph. An adaptive fixed-time coordinated control scheme is proposed by utilizing the designed FTTSM function, so that the closed-loop system is practically fixed-time stable, and the tracking errors converge to the desired trajectory within the fixed time. Simulation results on SFF with three follower spacecraft had clarified the effectiveness of the theoretical findings. Future study will focus on the extension of the presented control method under time-varying communication topology and communication time delay.

CONSENT FOR PUBLICATION

Not applicable.

CONFLICT OF INTEREST

The authors declare no conflict of interest, financial or otherwise.

ACKNOWLEDGEMENTS

This work is supported by Special Scientific Research Plan Project of Shaanxi Province Education Department (21JK0905) and Natural Science Foundation of Shaanxi Province of China (2022JQ-636).

REFERENCES

[1] Dong G, Cao L, Yao D, *et al.* Adaptive Attitude Control for Multi-MUAV Systems With Output Dead-Zone and Actuator Fault. IEEE/CAA Journal of Automatica Sinica, 20f21, 8(9):1567-1575.

[2] Ma H, Li H, Lu R, *et al.* Adaptive event-triggered control for a class of nonlinear systems with periodic disturbances. Science China Information Sciences, 2020, 63(5): 1-15.

[3] Wang D, Mei X, Weng R, *et al.* Hybrid filter design of fault detection for networked linear systems with variable packet dropout rate. IET control theory & applications, 2019, 13(9): 1239-1245.

[4] Chen T, Shan J, Wen H. Distributed passivity-based control for multiple flexible spacecraft with attitude-only measurements. Aerospace Science and Technology, 2019, 94: 105408.

[5] Li P, Yu X, Zhang Y. The design of quasi-optimal higher order sliding mode control via disturbance observer and switching-gain adaptation. IEEE Transactions on Systems, Man, and Cybernetics: Systems, 2018, 50(11): 4817-4827.

[6] Li Q, Deng Z. Coordinated Orbit–Attitude–Vibration Control of a Sun-Facing Solar Power Satellite. Journal of Guidance, Control, and Dynamics, 2019, 42(8): 1863-1869.

[7] Polyakov A. Nonlinear feedback design for fixed-time stabilization of linear control systems. IEEE Transactions on Automatic Control, 2011, 57(8): 2106-2110.

[8] Yang L, Yang J. Nonsingular fast terminal sliding-mode control for nonlinear dynamical systems. International Journal of Robust and Nonlinear Control, 2011, 21(16): 1865-1879.

[9] Zou A M, Fan Z. Distributed fixed-time attitude coordination control for multiple rigid spacecraft. International Journal of Robust and Nonlinear Control, 2020, 30(1): 266-281.

[10] Xia Y, Zhang J, Lu K, *et al.* Finite time and cooperative control of flight vehicles. Singapore, Springer, 2019.

[11] Qian C, Lin W. A continuous feedback approach to global strong stabilization of nonlinear systems. IEEE Transactions on Automatic Control, 2001, 46(7): 1061-1079.

[12] Liang H, Guo X, Pan Y, *et al.* Event-triggered fuzzy bipartite tracking control for network systems based on distributed reduced-order observers. IEEE Transactions on Fuzzy Systems, 2020, 29(6): 1601-1614.

[13] Liu M, Zhang L, Zheng W X. Fault reconstruction for stochastic hybrid systems with adaptive discontinuous observer and non-homogeneous differentiator. Automatica, 2017, 85: 339-348.

[14] Liu A, Zhang W A, Yu L, *et al.* Formation control of multiple mobile robots incorporating an extended state observer and distributed model predictive approach. IEEE Transactions on Systems, Man, and Cybernetics: Systems, 2018, 50(11): 4587-4597.

[15] Liu M, Zhang L, Shi P, *et al.* Fault estimation sliding-mode observer with digital communication constraints. IEEE Transactions on Automatic Control, 2018, 63(10): 3434-3441.

[16] Wang F, Liu Z, Zhang Y, *et al.* Adaptive quantized controller design via backstepping and stochastic small-gain approach. IEEE Transactions on Fuzzy Systems, 2015, 24(2): 330-343.

[17] Li H, Shi P, Yao D, *et al.* Observer-based adaptive sliding mode control for nonlinear Markovian jump systems. Automatica, 2016, 64: 133-142.

[18] Liu A, Zhang W A, Yu L. Robust predictive tracking control for mobile robots with intermittent measurement and quantization. IEEE Transactions on Industrial Electronics, 2020, 68(1): 509-518.

[19] Lv Y, Hu Q, Ma G, *et al.* 6-DOF synchronized control for spacecraft formation flying with input constraint and parameter uncertainties. ISA Transactions, 50(4):573–580, 2011.

[20] Liu C, Yue X, Shi K, *et al.* Spacecraft Attitude Control: A Linear Matrix Inequality Approach. Elsevier/ Science Press, 2022, ISBN: 978-0-323-99005-9.

[21] Na J, Huang Y, Wu X, *et al.* Active adaptive estimation and control for vehicle suspensions with prescribed performance. IEEE Transactions on Control Systems Technology, 2017, 26(6): 2063-2077.

[22] Wu B, Wang D, Poh E K. Decentralized robust adaptive control for attitude synchronization under directed communication topology. Journal of Guidance, Control, and Dynamics, 2011, 34(4): 1276-1282.

[23] Wu Y, Wang Z. Fuzzy adaptive practical fixed-time consensus for second-order nonlinear multiagent systems under actuator faults. IEEE Transactions on Cybernetics, 2020, 51(3): 1150-1162.

[24] Sun L, Huo W. 6-DOF integrated adaptive backstepping control for spacecraft proximity operations. IEEE Transactions on Aerospace and Electronic Systems, 2015, 51(3): 2433-2443.

[25] Ye D, Shi M, Sun Z. Satellite proximate interception vector guidance based on differential games. Chinese Journal of Aeronautics, 2018, 31(6): 1352-1361.

[26] Zhao L., Yu J, Lin C , Yu H, Distributed adaptive fixed-time consensus tracking for second-order multi-agent systems using modified terminal sliding mode. Applied Mathematics and Computation, 2017, 312, pp. 23–35.

CHAPTER 9

FTCESO-based Prescribed Time Control for Satellite Cluster Reconstruction

Siyuan Li[1,*], Zhaowei Sun[1], Yang Yang[2], and Fenglin Wang[3]

[1]*Research Center of Satellite Technology, Harbin Institute of Technology, Harbin 150001, China*

[2]*State Key Laboratory of Astronautics Dynamics, Xi'an Satellite Control Center, Xi'an 710043, China*

[3]*Beijing Institute of Tracking and Communication Technology, Beijing 100124, China*

Abstract: In this paper, a prescribed time control strategy for satellite cluster formation reconstruction with multiple environmental disturbances and thruster constraints is proposed. Firstly, a finite time convergent extended state observer (FTCESO) is used to eliminate the effects of external disturbances. And it has been proved to be able to accurately estimate the total disturbances of the satellite cluster in a short time. Secondly, based on the sliding mode, a prescribed time controller with piecewise control law is designed for satellite cluster reconstruction which can ensure that the satellites move to the specified configuration at a prescribed time. Then, the convergence of the controller is proved by Lyapunov stability theory. Finally, compared with a sliding mode controller, a numerical simulation is performed to demonstrate the effectiveness of the proposed method.

Keywords: Extended state observer, Formation reconstruction, Multiple disturbances, Piecewise control law, Prescribed time control, Satellite cluster.

INTRODUCTION

With the rapid development of computer, new energy and new materials technology, countries all over the world seize the opportunity to design and develop satellites with different functions. Thanks to the continuous development of aerospace industry technology, space systems are becoming more and more powerful, especially distributed satellite system [1, 2]. Compared with the traditional single large satellite, the distributed satellite system has significant

*Corresponding author Siyuan Li:** Research Center of Satellite Technology, Harbin Institute of Technology, Harbin 150001, China; Tel: +86 15776643665; E-mails: siyuan_li@hit.edu.cn; sunzhaowei@hit.edu.cn; yang842655@126.com; 13363831530@163.com

Chuang Liu, Honghua Dai, Xiaokui Yue & Yiqing Ma (Eds.)

application advantages, such as the large spatial distribution, low development and maintenance costs, high system reliability and strong flexibility [3]. Satellite formation and satellite cluster are two typical systems in distributed satellite systems. Where, satellite formation generally maintains a specific spatial configuration and the satellite cluster spends most of its operating time in a loose configuration to reduce fuel consumption [4, 5]. At the same time, satellite cluster can be reconstructed into specific spatial configurations to meet mission requirements.

When the satellite cluster orbits the earth, the disturbance force may come from the uneven distribution of the mass of earth, the gravitational attraction of other planetary bodies, the solar radiation pressure and other perturbations [6, 7]. Especially, Satellites are mainly affected by J_2 perturbation and atmospheric drag in low earth orbit. The persistent effect of the multiple disturbances causes the orbit of the satellite to change and deviate from the orbit of the two-body model [8, 9].

Disturbance observer is a very practical method for external disturbance and uncertainty which has captured considerable attention [10]. Compared with general nonlinear observer, extended state observer has the advantage of requiring less state information and faster observation speed [11]. ESO is able to estimate uncertainties, external disturbances and any unmodeled parts of the system as the so-called total disturbance [12]. Therefore, ESO is widely used in engineering applications [13-15]. Liu P constructed linear ESO with high-gain which can estimate the uncertainty items in a large spacecraft system without velocity information [16]. To deal with the problem of peaking value, a nonlinear ESO constructed from piece-wise smooth functions is investigated by Zhao Z [17]. A sliding mode controller based on ESO is designed for the attitude tracking of large flexible spacecraft by Zhang Z [18]. Li B proposed a novel continuous finite-time extended state observer which can estimate the attitude angular velocity and extended state observation errors of the satellite in a finite time [19].

On the other hand, sliding mode variable structure control is a robust control method for satellite cluster reconstruction whose main advantages are high control precision, and insensitive to parameter variation and disturbance [20-22]. Shahid K designed a sliding mode control law based on C-W equation to realize the satellite formation reconstruction which considered the effects of initial state deviation, target eccentricity and J_2 perturbation [23]. To ensue each satellite convergence to their desired state in finite time, terminal sliding mode control (TSMC) was used to solve the problem of relative motion control for spacecraft formation [24]. A nonsingular fast terminal sliding mode controller (NFTSMC) to realize the synchronization control of spacecraft formation is proposed by Liu R [25]. Although the terminal sliding mode can stabilize in finite time, the time of convergence is determined by the initial state of the system. To fix this problem, fixed time sliding mode control had been proposed whose convergence time is independent of the initial selection of the system [26-28]. On the basis of these theories, scholars have proposed prescribed time control strategy that can stabilize the system at a predetermined time [29-30]. Fixed-time control and preset-time control have been widely used in multi-agent systems [31-34], but seldom used in satellite cluster control.

However, in a space mission, satellites in cluster may be required to transform to a new specified configuration in a prescribed time. Considering the multiple disturbances, the control strategy combined with FTESO and prescribed time controller is very suitable for the requirement of satellite cluster reconstruction. To our knowledge, little research has explicitly focused on the aforementioned issues. This paper documents several key contributions as follows. Firstly, in the case of multiple environmental disturbances, a FTCESO is designed to estimate perturbations and uncertainties in a finite time. Secondly, a prescribed time control strategy is designed to ensure that the error state can converge to the equilibrium point within the user-defined time. Finally, based on a piecewise control law, we proposed a control strategy which can ensure that the satellites move to the specified configuration at a prescribed time.

The rest of this paper is organized as follows. Section 2 describes the dynamics model of the relative motion of satellite cluster and the influence of J_2 perturbation and atmospheric drag. In Section 3, a prescribed time controller based on FTCESO is developed for the satellites reconstruction and corresponding stability analysis is performed *via* a Lyapunov approach. A set of numerical simulations for comparison with normal sliding mode controller is presented in Section 4. Finally, some conclusions are drawn.

DYNAMICS MODELING

For the relative motion of satellite cluster, Earth-centered inertial (ECI) and local-vertical local-horizontal (LVLH) are the widely used coordinate systems which are shown in Fig. (**1**). Then a real or virtual satellite is regarded as the origin of the LVLH coordinate system which is called leader satellite. And the other satellites in satellite cluster are called the follower satellites. Let $\boldsymbol{r}_l$ and $\boldsymbol{r}_f$ are the position vectors of the leader and follower satellites in the ECI coordinate system, we can get that,

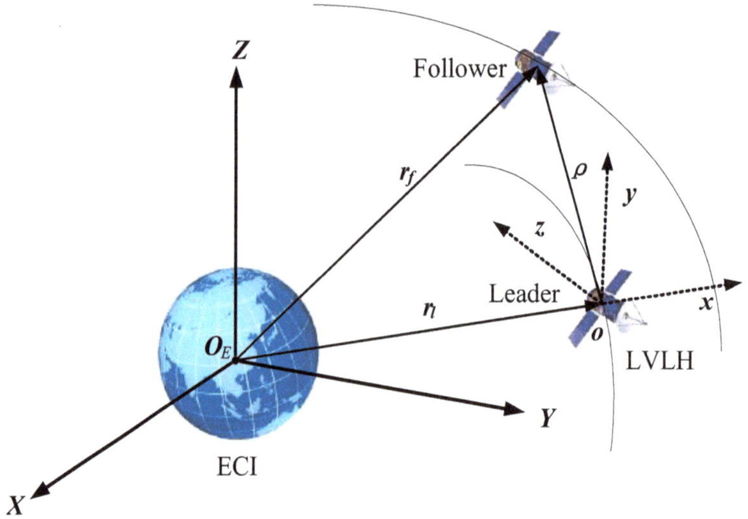

Fig. (1). Satellite coordinate system.

$$\begin{cases} \dfrac{d^2\boldsymbol{r}_l}{dt} = -\dfrac{\boldsymbol{r}_l}{r_l^3} + \dfrac{\boldsymbol{u}_l}{m_l} + \dfrac{\boldsymbol{d}_l}{m_l} \\ \dfrac{d^2\boldsymbol{r}_f}{dt} = -\dfrac{\boldsymbol{r}_f}{r_f^3} + \dfrac{\boldsymbol{u}_f}{m_f} + \dfrac{\boldsymbol{d}_f}{m_f} \end{cases} \tag{1}$$

Where $r_l = \|\boldsymbol{r}_l\|$ and $r_f = \|\boldsymbol{r}_f\|$, m_l and m_f are mass of the leader and follower satellites. $\boldsymbol{u}_l$ and $\boldsymbol{u}_f$ are the control forces, $\boldsymbol{d}_l$ and $\boldsymbol{d}_f$ are the vectors of environmental disturbances forces. According to the reference [25], we can get the satellite relative motion dynamic equations in LVLH coordinate system which can be expressed as Eq. (2).

$$\begin{cases} \ddot{x} = \dot{n}y + n^2x + 2n\dot{y} - \dfrac{\mu(r_l + x)}{[(r_l + x)^2 + y^2 + z^2]^{\frac{3}{2}}} + \dfrac{u_x}{m_i} + \dfrac{d_x}{m_i} \\ \ddot{y} = -\dot{n}x + n^2y - 2n\dot{x} - \dfrac{\mu y}{[(r_l + x)^2 + y^2 + z^2]^{\frac{3}{2}}} + \dfrac{u_y}{m_i} + \dfrac{d_y}{m_i} \\ \ddot{z} = -\dfrac{\mu z}{[(r_l + x)^2 + y^2 + z^2]^{\frac{3}{2}}} + \dfrac{u_z}{m_i} + \dfrac{d_z}{m_i} \end{cases} \tag{2}$$

Where x, y and z are the components of position vector, μ is the gravity coefficient, n is the orbital angular velocity of leader satellite, d_x, d_y, d_z are the disturbance difference between the leader and follower satellites in three coordinate axes.

When satellite cluster moves in low Earth orbit, J_2 perturbation is the main disturbance that can make the satellites to drift away. Where the gravitational potential function of the Earth is expressed by Eq. (**3**).

$$U = \frac{G_M}{R}\left\{1 - \sum_{n=2}^{\infty}\left(\frac{R_e}{R}\right)^2\left[J_nP_n\sin^2\delta - \sum_{m=1}^{n} J_{nm}P_{nm}\sin\delta\cos m(\lambda - \lambda_{nm})\right]\right\} \tag{3}$$

Where G_M is the gravitational constant; R_e is the average radius of the Earth; R is the geocentric distance; λ and δ is the longitude and latitude; P_n ad P_{nm} is the Legendre polynomials; J_n and J_{nm} are the coefficients for the zonal harmonic and tesseral and sectorial harmonics. Since the magnitude of J_2 perturbation is much larger than that of other terms, J_2 perturbation is usually used as gravitational perturbation in low Earth orbit. By converting J_2 perturbation to LVLH coordinate system, the component form can be obtained Eq. (**4**).

$$\begin{cases} \dfrac{d^2x}{dt^2} = -\dfrac{G_M x}{R^3}\left[\dfrac{3}{2}J_2\left(\dfrac{R_e}{R}\right)^2\left(1-5\dfrac{z^2}{R^2}\right)\right] \\ \dfrac{d^2y}{dt^2} = -\dfrac{G_M y}{R^3}\left[\dfrac{3}{2}J_2\left(\dfrac{R_e}{R}\right)^2\left(1-5\dfrac{z^2}{R^2}\right)\right] \\ \dfrac{d^2x}{dt^2} = -\dfrac{G_M z}{R^3}\left[\dfrac{3}{2}J_2\left(\dfrac{R_e}{R}\right)^2\left(3-5\dfrac{z^2}{R^2}\right)\right] \end{cases} \tag{4}$$

Where, J_2=0.0010826 denotes the second-order zonal harmonic coefficient. In addition to the J_2 perturbation, atmospheric drag is another important perturbation force that affects low-Earth orbit satellites and the acceleration can express as Eq. 5.

$$a = -\frac{1}{2}\frac{C_D A}{m}\rho v_{rel}\boldsymbol{v}_{rel} \tag{5}$$

Where, m is the mass of satellite, C_D is the drag coefficient, A is the cross-sectional area, m is the mass of the satellite, ρ is the atmospheric density, $\boldsymbol{v}_{rel}$ is the speed vector of the satellite relative to the atmosphere which can obtained by Eq. 6.

$$\boldsymbol{v}_{rel} = \frac{d\boldsymbol{R}}{dt} - \boldsymbol{\omega}\times\boldsymbol{R} \tag{6}$$

Where, $\boldsymbol{R}$ denotes the satellite position vector, $\boldsymbol{\omega}$ is the angular velocity of the Earth's rotation and the direction is along the positive direction of the Z axis.

CONTROLLER DESIGN

In this section, we consider a satellite cluster in low earth orbit with multiple environmental disturbances which need to change the spatial configuration to another one. According to the task requirements, the operation duration should be controlled at a certain time. Pointing at above problems, a FTCESO is designed to estimate the environmental disturbances and a prescribed time controller based on sliding mode with piecewise control law is designed for satellite cluster reconstruction.

Disturbance Estimation based on FTCESO

From what has been mentioned above we can know that satellite cluster in low earth orbit can't keep the spatial configuration all the time under disturbances. However, the disturbance model cannot completely replace the actual case, so there will exist some uncertain disturbing forces acting on the satellites inescapably. Suppose satellite cluster contains a virtual reference satellite and N real satellites which has the mass of m_i. In LVLH coordinate system, position vector of the i-th satellite relative to the leader satellite is defined as $\boldsymbol{q}_i=[x_i,y_i,z_i]^T$ and the velocity vector is $\boldsymbol{p}_i=[\dot{x}_i,\dot{y}_i,\dot{z}_i]^T$. And the desired position vector and desired velocity vector are expressed by $\boldsymbol{q}_{id}=[x_{id},y_{id},z_{id}]^T$ and $\dot{\boldsymbol{p}}_{id}=[\dot{x}_{id},\dot{y}_{id},\dot{z}_{id}]^T$.

Defining the tracking error as $\boldsymbol{e}_i=[\boldsymbol{e}_{i1},\boldsymbol{e}_{i2}]^T$ where $\boldsymbol{e}_1=\boldsymbol{q}_i-\boldsymbol{q}_{id}$ and $\boldsymbol{e}_{i2}=\boldsymbol{p}_i-\boldsymbol{p}_{id}$. Then according to Eq. (2), Eq. (4) and Eq. (5), the relative motion error model of the i-th satellite can be obtained as Eq. (7).

$$\begin{cases}\dot{\boldsymbol{e}}_{i1}=\boldsymbol{e}_{i2}\\ \dot{\boldsymbol{e}}_{i2}=f(\boldsymbol{e}_i,\dot{\boldsymbol{e}}_i,\ddot{\boldsymbol{e}}_i)+\dfrac{1}{m_i}\left(\boldsymbol{d}_{J_2}+\boldsymbol{d}_a+\Delta_i\right)+\dfrac{1}{m_i}\boldsymbol{u}_i\end{cases} \tag{7}$$

Where $\boldsymbol{d}_{J_2}$ is the J_2 perturbation and $\boldsymbol{d}_a$ is the atmospheric drag. Δ_i denotes the uncertain disturbance and $f(\boldsymbol{e}_i,\dot{\boldsymbol{e}}_i,\ddot{\boldsymbol{e}}_i)$ can be expressed as Eq. (8).

$$f(\boldsymbol{e}_i,\dot{\boldsymbol{e}}_i,\ddot{\boldsymbol{e}}_i)=\begin{bmatrix}\dot{n}_0y_i+n_0^2x_i+2n_0\dot{y}_i+2\dfrac{\mu}{R^3}x_i-\ddot{x}_{id}\\ -\dot{n}_0x_i+n_0^2y_i-2n_0\dot{x}_i-\dfrac{\mu}{R^3}y_i-\ddot{y}_{id}\\ -\dfrac{\mu}{R^3}z_i-\ddot{z}_{id}\end{bmatrix} \tag{8}$$

It is worth noting that the J_2 perturbation and atmospheric drag are calculated in the ECI coordinate system, and it should be transformed into the LVLH frame through coordinate transformation.

Then, according to reference [19], the FTCESO is designed to estimate the uncertain disturbance. According to Eq. (7), we construct an extended state $\boldsymbol{e}_{i3}$, and the relative motion error of the i-th satellite can be rewritten as

$$\begin{cases} \dot{\boldsymbol{e}}_{i1} = \boldsymbol{e}_{i2} \\ \dot{\boldsymbol{e}}_{i2} = \boldsymbol{e}_{i3} + f\left(\boldsymbol{e}_i, \dot{\boldsymbol{e}}_i, \ddot{\boldsymbol{e}}_i\right) + \frac{1}{m_i}\left(\boldsymbol{d}_{J_2} + \boldsymbol{d}_a\right) + \frac{1}{m_i}\boldsymbol{u}_i \\ \dot{\boldsymbol{e}}_{i3} = \dot{\boldsymbol{h}}_i\left(t\right) \end{cases} \tag{9}$$

Where $\boldsymbol{h}_i\left(t\right) = \frac{1}{m_i}\Delta_i$, and suppose that $\left\|\dot{\boldsymbol{h}}_i\left(t\right)\right\| \le U_i$. And the FTCESO is designed as

$$\begin{cases} \boldsymbol{\sigma}_i = \boldsymbol{e}_{i1} - \hat{\boldsymbol{e}}_{i1} \\ \dot{\hat{\boldsymbol{e}}}_{i1} = \hat{\boldsymbol{e}}_{i2} + k_1 sig^{(\kappa+1)/2}\left(\boldsymbol{\sigma}_i\right) \\ \dot{\hat{\boldsymbol{e}}}_{i2} = \hat{\boldsymbol{e}}_{i3} + k_2 sig^{(\kappa+1)/2}\left(\boldsymbol{\sigma}_i\right) + f\left(\boldsymbol{e}_i, \dot{\boldsymbol{e}}_i, \ddot{\boldsymbol{e}}_i\right) + \frac{1}{m_i}\left(\boldsymbol{d}_{J_2} + \boldsymbol{d}_a\right) + \frac{1}{m_i}\boldsymbol{u}_i \\ \dot{\hat{\boldsymbol{e}}}_{i3} = k_3 sig^{\kappa}\left(\boldsymbol{\sigma}_i\right) \end{cases} \tag{10}$$

Where $\hat{\boldsymbol{e}}_{i1}$, $\hat{\boldsymbol{e}}_{i2}$ and $\hat{\boldsymbol{e}}_{i3}$ are the observed value of $\boldsymbol{e}_{i1}$, $\boldsymbol{e}_{i2}$ and $\boldsymbol{e}_{i3}$, $\kappa \in \left(0,1\right)$, k_1, k_2 and k_3 are the observer gains, $\boldsymbol{\sigma}_i$ is the observed error, and $\text{sig}^{\kappa}\left(\cdot\right) = \text{sgn}\left(\cdot\right)\left|\cdot\right|^{\kappa}$. Further, we define the $\boldsymbol{\varepsilon}_{ij} = \boldsymbol{e}_{ij} - \hat{\boldsymbol{e}}_{ij}$, $j \in \left\{1,2,3\right\}$ as the observed error of FTCESO, and its derivative can be calculated as

$$\begin{cases} \dot{\boldsymbol{\varepsilon}}_{i1} = \boldsymbol{\varepsilon}_{i2} - k_1 sig^{(\kappa+1)/2}\left(\boldsymbol{\varepsilon}_{i1}\right) \\ \dot{\boldsymbol{\varepsilon}}_{i2} = \boldsymbol{\varepsilon}_{i3} - k_2 sig^{(\kappa+1)/2}\left(\boldsymbol{\varepsilon}_{i1}\right) \\ \dot{\boldsymbol{\varepsilon}}_{i3} = \dot{\boldsymbol{e}}_{i3} - k_3 sig^{\kappa}\left(\boldsymbol{\varepsilon}_{i1}\right) \end{cases} \tag{11}$$

For the Eq. (11), it has been proved that [19]

$$\begin{aligned} &\|\boldsymbol{\varepsilon}_i\| \le \frac{lU_i}{\upsilon_{\min}\{D\}\cdot\upsilon_{\min}\{E\}} = \varepsilon_f, \forall t > t_f \\ &t_f \le \frac{2(\alpha+\beta)}{\alpha\beta}\left(\lambda_{\max}\{E\}V(\boldsymbol{\varepsilon}_i(0))\right)^{1/2} \end{aligned} \quad (12)$$

Where $\boldsymbol{\varepsilon}_i=[\mathrm{sig}^{(\kappa+1)/2}(\varepsilon_{i1}),\ \varepsilon_{i2},\ \varepsilon_{i3}]^T$, $\boldsymbol{\varepsilon}_i(0)$ is the initial observed error. $\upsilon_{\min}\{\cdot\}$ denotes the smallest singular value of the matrix, and $\lambda_{\max}\{\cdot\}$ denotes the maximum eigenvalue of the matrix l, D and E are designed as follow

$$\begin{aligned} &l=\sqrt{k_3^2+4} \\ &D=\begin{bmatrix} k_1 & -1 & 0 \\ k_2 & 0 & -1 \\ k_3 & 0 & 0 \end{bmatrix} \\ &E=\begin{bmatrix} 2k_1/(\kappa+1)+k_2^2+k_3^2 & -k_2 & -k_3 \\ -k_2 & 2 & 0 \\ -k_3 & 0 & 2 \end{bmatrix} \end{aligned} \quad (13)$$

And the expressions of a, β, $V(\cdot)$ are

$$\begin{aligned} &\alpha=(\kappa+1)\upsilon_{\min}\{D\}\upsilon_{\min}\{E\}-2lU_i \\ &\beta=2\boldsymbol{\varepsilon}_i\upsilon_{\min}\{D\}\upsilon_{\min}\{E\}-2lU_i \\ &V(\boldsymbol{\varepsilon}_i)=\boldsymbol{\varepsilon}_i^T E\boldsymbol{\varepsilon}_i \end{aligned} \quad (14)$$

In this way, $\boldsymbol{\varepsilon}_i$ will converge to a small value within a finite time t_f by selecting the appropriate gain coefficients of k_1, k_2 and k_3.

Prescribed Time Controller Designing

In this section, we supposed that the satellite cluster need to be reconfigured within a prescribed time. So firstly, we give the definition of prescribed time stability.

Definition 1 [28]: Consider a continuous nonlinear system

$$\dot{x}(t)=f(x(t),t) \quad (15)$$

Where $x(t)\in \boldsymbol{R}^n$ is the state of the system and $f:\boldsymbol{R}^n\times \boldsymbol{R}_+\to \boldsymbol{R}^n$ denotes a continuous function which is differentiable. Let $x(t)=x(0)$ is the initial value of the system. And it is said to be prescribed time stable if it is globally asymptotically stable and there exists a user-defined time t_f satisfied that $t_f\in \boldsymbol{R}_+:\forall x_0\in \boldsymbol{R}^n:t(x(0))\le t_f$, $x(t,0)=0$ for $t\ge t_f$.

And then a time-varying function during $t\in[t_0,t_0+t_f)$ has been designed as follows

$$\varphi(t)=\exp(\eta(t_0+t_f-t))-1,\quad t\in[t_0,t_0+t_f) \tag{16}$$

Where η is a positive constant, t_0 denotes the initial time and t_f is the prescribed setting time.

Lemma 1 [30]: For the system Eq. (15), it can achieve prescribed time stable within t if there exists a positive function $V(x(t),t):\boldsymbol{R}^n\times \boldsymbol{R}_+\to \boldsymbol{R}_+$ which is continuous-differentiable such that $\dot{V}\le -bV+\frac{\dot{\varphi}}{\varphi}V+c$, where $b>0$, $c\ge 0$.

Proof: For $t\in[t_0,t_0+t_f)$, let $\bar{V}=\frac{V}{\varphi}$, and we can get the time derivative of $\bar{V}$ as

$$\dot{\bar{V}}=\frac{\dot{V}}{\varphi}+\frac{\dot{\varphi}}{\varphi^2}V \tag{17}$$

According to the lemma 1, it indicates

$$\dot{\bar{V}}\le -b\bar{V}+\frac{c}{\varphi} \tag{18}$$

And

$$\begin{aligned}\bar{V}&\le \exp(-b(t-t_0))\bar{V}(t_0)+\exp(-b(t-t_0))\int\frac{c}{\varphi}\exp(b(t-t_0))dt\\&\le \exp(-b(t-t_0))\bar{V}(t_0)-\frac{c}{\eta}\exp(-b(t-t_0-t_f))\ln\varphi\end{aligned} \tag{19}$$

Then

$$V \le \varphi \exp\left(-b\left(t-t_0\right)\right)\bar{V}\left(t_0\right)-\frac{c}{\eta}\exp\left(-b\left(t-t_0-t_f\right)\right)\varphi\ln\varphi \qquad (20)$$

It is obvious that $\varphi(t)\to 0$ at $t\to t_0+t_f$ and $\varphi\ln\varphi\to 0$ when $\varphi(t)\to 0$. Further we can get $V\left(t_0+t_f\right)=\lim_{t\to t_0+t_f} V\to 0$, so that V is globally asymptotically stable within the prescribed time t .

For system Eq. (7), we define two kinds of sliding mode function as

$$\begin{aligned} \boldsymbol{s}_{ip} &= b\boldsymbol{e}_{i1}-\frac{\dot{\varphi}}{\varphi}\boldsymbol{e}_{i1}+\boldsymbol{e}_{i2} \\ \boldsymbol{s}_{in} &= \gamma_i \boldsymbol{e}_{i1}+\boldsymbol{e}_{i2} \end{aligned} \qquad (21)$$

Where $\boldsymbol{s}_{ip}$ denotes the prescribed time sliding mode function and $\boldsymbol{s}_{in}$ is the normal linear sliding mode function. Moreover, $b>0$, $c\ge 0$ and $\gamma_i>0$. So that we adopt the piecewise control law as

$$\boldsymbol{u}_i=\begin{cases} -\boldsymbol{g}_i-\hat{\boldsymbol{\Delta}}_i-m_i\begin{pmatrix} \dfrac{\ddot{\varphi}\varphi-\dot{\varphi}^2}{\varphi^2}\boldsymbol{e}_{i1}-\left(b-\dfrac{\dot{\varphi}}{\varphi}\right)\boldsymbol{e}_{i2}+\dfrac{\boldsymbol{s}_{ip}}{2}\left(-b+\dfrac{\dot{\varphi}}{\varphi}\right) \\ +\delta_i \mathrm{sig}\left(\boldsymbol{s}_{ip}\right)\end{pmatrix} & ,t\in\left[0,t_1\right) \\ -\boldsymbol{g}_i-\hat{\boldsymbol{\Delta}}_i-m_i\begin{pmatrix} \dfrac{\ddot{\varphi}\varphi-\dot{\varphi}^2}{\varphi^2}\boldsymbol{e}_{i1}-\left(b-\dfrac{\dot{\varphi}}{\varphi}\right)\boldsymbol{e}_{i2}-k_{ip}\boldsymbol{s}_{ip} \\ -\delta_i \mathrm{sig}\left(\boldsymbol{s}_{ip}\right)\end{pmatrix} & ,t\in\left[t_1,t_1+t_f\right) \\ -\boldsymbol{g}_i-\hat{\boldsymbol{\Delta}}_i-m_i\left(-\gamma_i\boldsymbol{e}_{i2}-\delta_i\mathrm{sig}\left(\boldsymbol{s}_{in}\right)-k_{in}\boldsymbol{s}_{in}\right) & ,t\in\left[t_f,\infty\right) \end{cases} \qquad (22)$$

Where $\boldsymbol{g}_i=mf\left(\boldsymbol{e}_i,\dot{\boldsymbol{e}}_i,\ddot{\boldsymbol{e}}_i\right)+\boldsymbol{d}_{J_2}+\boldsymbol{d}_a$ is the system dynamic model, $\hat{\boldsymbol{\Delta}}_i$ is the observed value of the uncertain disturbance, k_{ip} and k_{in} are the control gain coefficients. δ_i is a positive constant satisfied $\delta_i\ge\sup\left\|\boldsymbol{\Delta}_i-\hat{\boldsymbol{\Delta}}_i\right\|$.

Remark 1: In the controller designed above, the error states can reach the sliding mode surface $\boldsymbol{s}_{ip}=0$ at t_1. Then the error states will be kept on the sliding mode surface and can asymptotically converge at time t . Finally, a normal sliding mode control law ensures system stability during $t \in [t_f, \infty)$.

Theorem 1: Considering the satellite cluster under the controller described by Eq. (22), if the observed error of the FTCESO is bounded, then the satellites in cluster will converge to the new configuration from the previous positions at a prescribed time.

Proof: During $t \in [0, t_1)$, we *choose a Lyapunov function as follows*

$$V_1 = \frac{1}{2} \boldsymbol{s}_{ip}{}^T \boldsymbol{s}_{ip} \tag{23}$$

And substituting Eq. (7), Eq. (21) and Eq. (22) the time derivative of V_1 yields the following expression

$$\begin{aligned} \dot{V}_1 &= \boldsymbol{s}_{ip}{}^T \dot{\boldsymbol{s}}_{ip} \\ &= \boldsymbol{s}_{ip}{}^T \left(-\frac{\ddot{\varphi}\varphi - \dot{\varphi}^2}{\varphi^2} \boldsymbol{e}_{i1} + \left(b - \frac{\dot{\varphi}}{\varphi} \right) \boldsymbol{e}_{i2} + \dot{\boldsymbol{e}}_{i2} \right) \\ &= \boldsymbol{s}_{ip}{}^T \begin{pmatrix} -\frac{\ddot{\varphi}\varphi - \dot{\varphi}^2}{\varphi^2} \boldsymbol{e}_{i1} + \left(b - \frac{\dot{\varphi}}{\varphi} \right) \boldsymbol{e}_{i2} + f\left(\boldsymbol{e}_i, \dot{\boldsymbol{e}}_i, \ddot{\boldsymbol{e}}_i \right) \\ + \frac{1}{m_i} \left(\boldsymbol{d}_{J_2} + \boldsymbol{d}_a + \boldsymbol{\Delta}_i \right) + \frac{1}{m_i} \boldsymbol{u}_i \end{pmatrix} \end{aligned} \tag{24}$$

Substitute the control force Eq. (22) into Eq. (24), we get

$$\begin{aligned} \dot{V}_1 &= \boldsymbol{s}_{ip}{}^T \left(\boldsymbol{\Delta}_i - \hat{\boldsymbol{\Delta}}_i \right) + \frac{\boldsymbol{s}_{ip}{}^T \boldsymbol{s}_{ip}}{2} \left(-b + \frac{\dot{\varphi}}{\varphi} \right) + \boldsymbol{s}_{ip}{}^T \delta_i sig\left(\boldsymbol{s}_{ip} \right) \\ &\le \left(-b + \frac{\dot{\varphi}}{\varphi} \right) V_1 + c \end{aligned} \tag{25}$$

For $\delta_i \ge \sup \left\| \boldsymbol{\Delta}_i - \hat{\boldsymbol{\Delta}}_i \right\|$, $c \ge 0$. So according to the lemma 1, it is obvious that the error states will reach the sliding mode surface at the prescribed time t_1. And then we still choose the Lyapunov function during $[t_1, t_1 + t_f)$ as

$$V_2 = \frac{1}{2}\boldsymbol{s}_{ip}{}^T\boldsymbol{s}_{ip} \tag{26}$$

Substituting the second part of the $\boldsymbol{u}_i$, the time derivative of V_2 is given by

$$\begin{aligned}
\dot{V}_2 &= \boldsymbol{s}_{ip}{}^T\dot{\boldsymbol{s}}_{ip} \\
&= \boldsymbol{s}_{ip}{}^T\left(-\frac{\ddot{\varphi}\varphi-\dot{\varphi}^2}{\varphi^2}\boldsymbol{e}_{i1}+\left(b-\frac{\dot{\varphi}}{\varphi}\right)\boldsymbol{e}_{i2}+\dot{\boldsymbol{e}}_{i2}\right) \\
&= \boldsymbol{s}_{ip}{}^T\begin{pmatrix} -\dfrac{\ddot{\varphi}\varphi-\dot{\varphi}^2}{\varphi^2}\boldsymbol{e}_{i1}+\left(b-\dfrac{\dot{\varphi}}{\varphi}\right)\boldsymbol{e}_{i2}+f\left(\boldsymbol{e}_i,\dot{\boldsymbol{e}}_i,\ddot{\boldsymbol{e}}_i\right) \\ +\dfrac{1}{m_i}\left(\boldsymbol{d}_{J_2}+\boldsymbol{d}_a+\boldsymbol{\Delta}_i\right)+\dfrac{1}{m_i}\boldsymbol{u}_i \end{pmatrix}
\end{aligned} \tag{27}$$

Substituting the control force Eq. (22) into Eq. (27), then we get

$$\begin{aligned}
\dot{V}_2 &= -k_{ip}\boldsymbol{s}_{ip}{}^T\boldsymbol{s}_{ip}+\boldsymbol{s}_{ip}{}^T\left(\boldsymbol{\Delta}_i-\hat{\boldsymbol{\Delta}}_i\right)-\boldsymbol{s}_{ip}{}^T\delta_i sig\left(\boldsymbol{s}_{ip}\right) \\
&\le -k_{ip}\boldsymbol{s}_{ip}{}^T\boldsymbol{s}_{ip} \\
&\le 0
\end{aligned} \tag{28}$$

It means that the error states will be kept staying at the sliding mode surface during $\left[t_1,t_1+t_f\right)$. And reference to the Eq. (21), the error states can converge to the equilibrium point at the prescribed time t_f.

Finally, we adopt a normal sliding mode function $\boldsymbol{s}_{in}$ to ensure the error system is asymptotically stable after t_f. And the proof is as follows

Defining *a* Lyapunov function as

$$V_3 = \frac{1}{2}\boldsymbol{s}_{in}{}^T\boldsymbol{s}_{in} \tag{29}$$

Then, take the derivative of V_3

$$\begin{aligned}\dot{V}_3 &= \boldsymbol{s}_{in}{}^T \dot{\boldsymbol{s}}_{in} \\ &= \boldsymbol{s}_{in}{}^T \left(\gamma_i \boldsymbol{e}_{i2} + \dot{\boldsymbol{e}}_{i2}\right) \\ &= \boldsymbol{s}_{ip}{}^T \left(\gamma_i \boldsymbol{e}_{i2} + f\left(\boldsymbol{e}_i, \dot{\boldsymbol{e}}_i, \ddot{\boldsymbol{e}}_i\right) + \frac{1}{m_i}\left(\boldsymbol{d}_{J_2} + \boldsymbol{d}_a + \boldsymbol{\Delta}_i\right) + \frac{1}{m_i}\boldsymbol{u}_i\right)\end{aligned} \tag{30}$$

Substituting the control force Eq. (22) into Eq. (30), we get

$$\begin{aligned}\dot{V}_3 &= -k_{in}\boldsymbol{s}_{in}{}^T \boldsymbol{s}_{in} + \boldsymbol{s}_{in}{}^T\left(\boldsymbol{\Delta}_i - \hat{\boldsymbol{\Delta}}_i\right) - \boldsymbol{s}_{in}{}^T \delta_i \text{sig}\left(\boldsymbol{s}_{in}\right) \\ &\le -k_{in}\boldsymbol{s}_{in}{}^T \boldsymbol{s}_{in} \\ &\le 0\end{aligned} \tag{31}$$

To this, the proof of Theorem 1 is complete. The conclusion is that the satellite cluster can rebuild a new spatial configuration within t_f which can be predefined by the requirement space mission.

NUMERICAL SIMULATIONS

For satellite cluster reconfiguration, the comparative simulation of two different control methods has been carried out in this section. The first one adopts the prescribed time controller proposed in this paper. And the other one uses a normal sliding mode controller.

Suppose that the satellite cluster contains 6 follower satellites which have the same mass of 50kg and a virtual leader satellite. Considering the constraint of thruster, the maximum thrust of each satellite in the three coordinate axes is 5N. And the satellite reconstruction mission is to transform the satellite cluster from a larger elliptical configuration to a smaller one.

Design of the initial orbit elements of the virtual leader satellite is shown in Table **1**.

Table 1. Initial orbit elements of the virtual leader satellite.

a (*km*)	e	i(*rad*)	Ω(*rad*)	ω (*rad*)	f(*rad*)
6878.14	0.01	$\pi/4$	$\pi/6$	$\pi/3$	0

The Initial states of follower satellites in the LVLH coordinate system are shown in Table **2**. And the desired states of follower satellites are shown in Table **3**.

Table 2. The initial states of the six satellites.

Sat	Initial position(*m*)	Initial velocity(*m/s*)
1	$q_1(0)=[200\ \ -692.82\ \ 346.41]^T$	$p_1(0)=[-0.383\ \ -0.442\ \ -0.664]^T$
2	$q_2(0)=[-200\ \ -692.82\ \ -346.41]^T$	$p_2(0)=[-0.383\ \ 0.442\ \ -0.664]^T$
3	$q_3(0)=[-400\ \ 0\ \ -692.82]^T$	$p_3(0)=[0\ \ 0.885\ \ 0]^T$
4	$q_4(0)=[-200\ \ 692.82\ \ -346.41]^T$	$p_4(0)=[0.383\ \ 0.442\ \ 0.664]^T$
5	$q_5(0)=[200\ \ 692.82\ \ 346.41]^T$	$p_5(0)=[0.383\ \ -0.442\ \ 0.664]^T$
6	$q_6(0)=[400\ \ 0\ \ 692.82]^T$	$p_6(0)=[0\ \ -0.885\ \ 0]^T$

Table 3. The desired states of the six satellites.

Sat	Desired position(*m*)	Desired velocity(*m/s*)
1	$q_{1d}=[-5.96\ \ -399.82\ \ -10.32]^T$	$p_{1d}=[-0.221\ \ 0.013\ \ -0.383]^T$
2	$q_{2d}=[-176.10\ \ -189.58\ \ -305.02]^T$	$p_{2d}=[-0.104\ \ 0.389\ \ 0.181]^T$
3	$q_{3d}=[-170.14\ \ 210.23\ \ -294.70]^T$	$p_{3d}=[0.116\ \ 0.376\ \ 0.201]^T$
4	$q_{4d}=[5.96\ \ 399.82\ \ 10.32]^T$	$p_{4d}=[0.221\ \ -0.013\ \ 0.383]^T$
5	$q_{5d}=[176.10\ \ 189.58\ \ 305.02]^T$	$p_{5d}=[0.104\ \ -0.389\ \ 0.181]^T$
6	$q_{6d}=[170.14\ \ -210.23\ \ 294.70]^T$	$p_{6d}=[-0.116\ \ -0.376\ \ -0.201]^T$

In simulations, J_2 perturbation and atmospheric drag are the disturbances that have been modeled. J_2 perturbation can be computed by Eq. (4) and the atmospheric density model is CIRA1961. The unmodeled perturbation is assumed to be the superposition of a constant and sinusoidal perturbation whose expression is

$\boldsymbol{\Delta}_i = 0.3 + 0.2\sin\left(0.0011t + \pi/6\right)\cdot \boldsymbol{I}_n$. For FTCESO, the parameters are designed as $k_1 = 25$, $k_2 = 840$, $k_3 = 960$. And the parameters of the prescribed time controller are designed as $t_1 = 160s$, $t_f = 180s$, $\eta = 0.01$, $b = 0.05$, $k_{ip} = 0.2$ and for the normal sliding mode function $\gamma_i = 2$, $k_{in} = 0.1$. The observation error curve of FTESO for external disturbances and uncertainty is shown in Fig. (**2**).

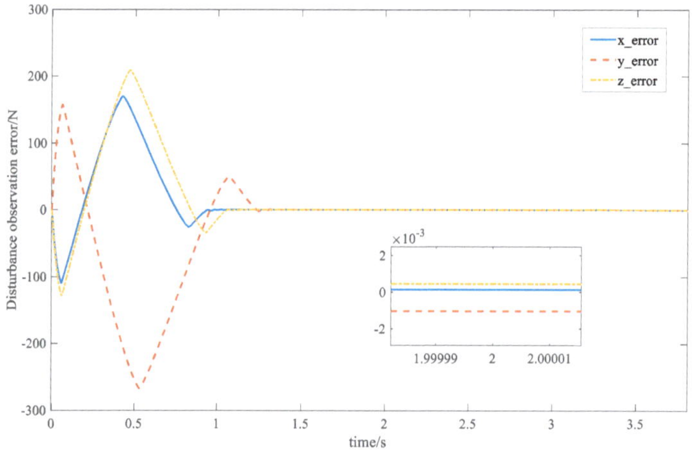

Fig. (2). Observation error of FTESO for external disturbances and uncertainty.

From Fig. (**2**), we can see that FTESO can quickly track external disturbances and uncertainties of the system within 1.5s and the observation error can be achieved 10^{-3}N. This means that the FTESO can estimate the disturbance almost without affecting the controller, so as to ensure the control effect.

The position tracking errors of the six satellites in cluster are shown in Fig. (**3**) and the velocity tracking errors of the six satellites in cluster are shown in Fig. (**4**).

It can be seen from the position errors curve that the position errors of the satellites in the three coordinate axes converge to the equilibrium point within 180s which is prescribed before simulation and the control accuracy is less than or equal to 0.02m. Moreover, for the velocity errors curve, although it converges to the equilibrium point at the time 180s, since the controller was switched at the same time, there would be some fluctuations in the velocity error, but it would converge to the equilibrium point in a short time again.

Further, the curves of control forces in the three coordinate axes are shown in Fig. (**5**).

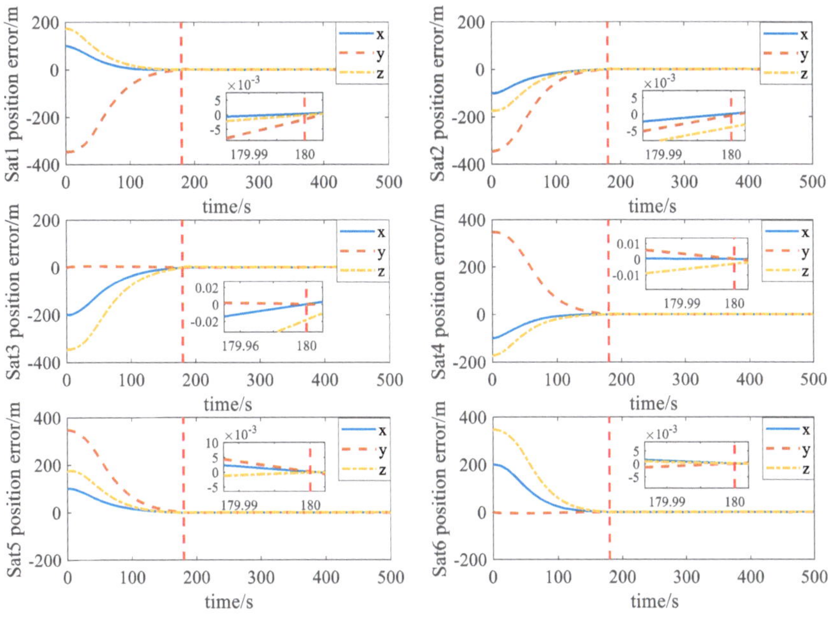

Fig. (3). Trajectories of position tracking errors of the satellites in cluster.

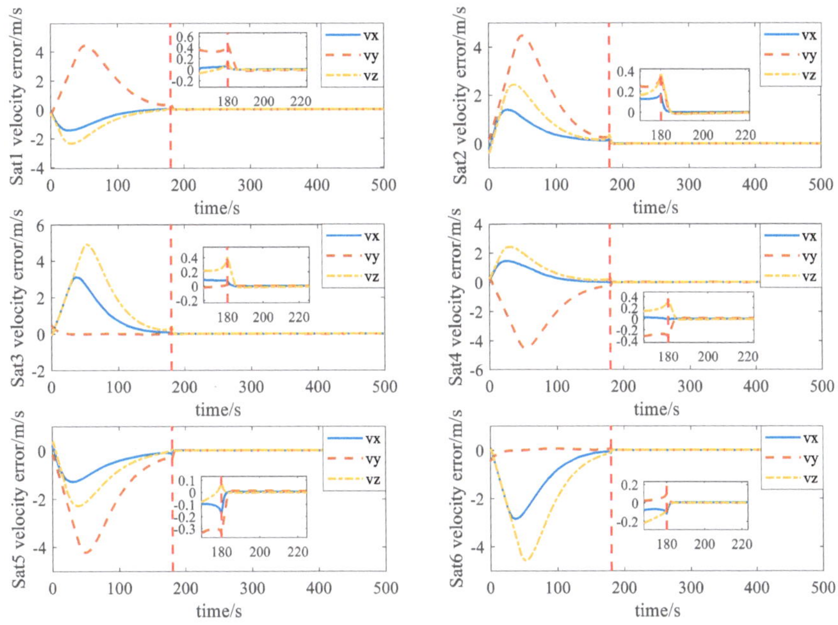

Fig. (4). Trajectories of velocity tracking errors of the satellites in cluster.

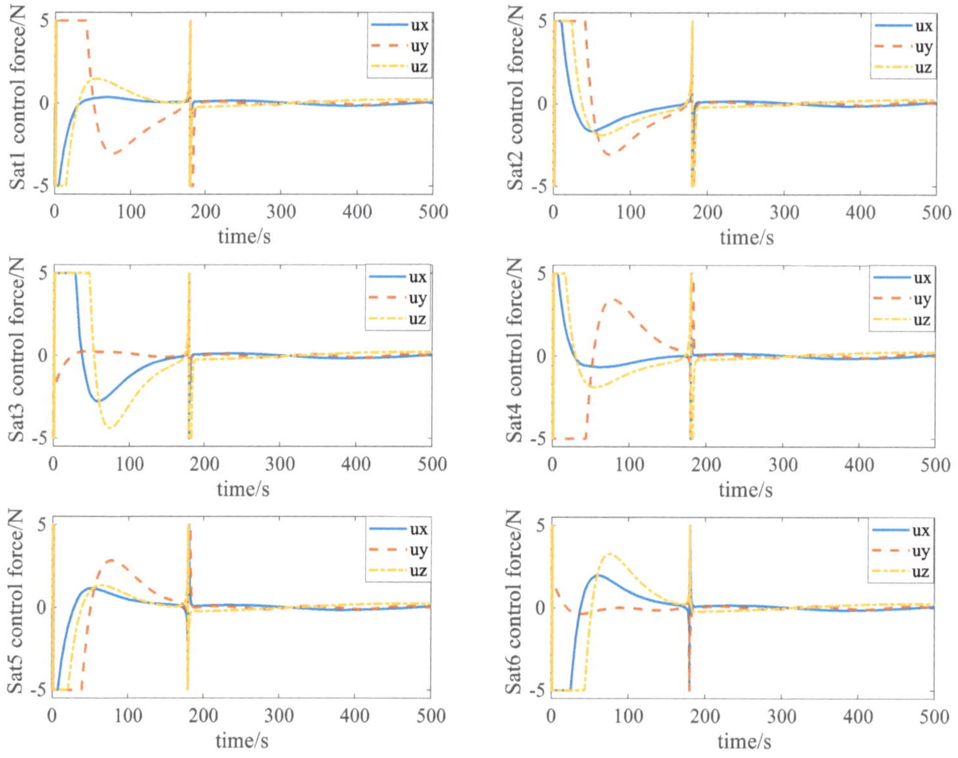

Fig. (5). The control force trajectories of the satellites in cluster.

In Fig. (**5**), it is clearly that the control force is saturated at the beginning of the simulation. This is because the initial error is large and the controller has input constraints of 5N. It is worth noting at time of 180s when the controller is switched, the error state is not on the new sliding surface. As a result, there is a sudden change in the control force at 180s which is the main reason of the fluctuation of velocity errors curve.

In order to verify the performance of the prescribed time controller, the simulation is compared with a normal sliding mode controller. And for the method of normal sliding mode, we choose the function as $\boldsymbol{s}_{in} = \gamma_i \boldsymbol{e}_{i1} + \boldsymbol{e}_{i2}$, where $\gamma_i = 2$, $k_{in} = 0.1$. And in Fig. (**6**), we can see the differences of the relative distance curve between satellite and target states using two different control methods. The simulation results are shown in Fig. (**6**).

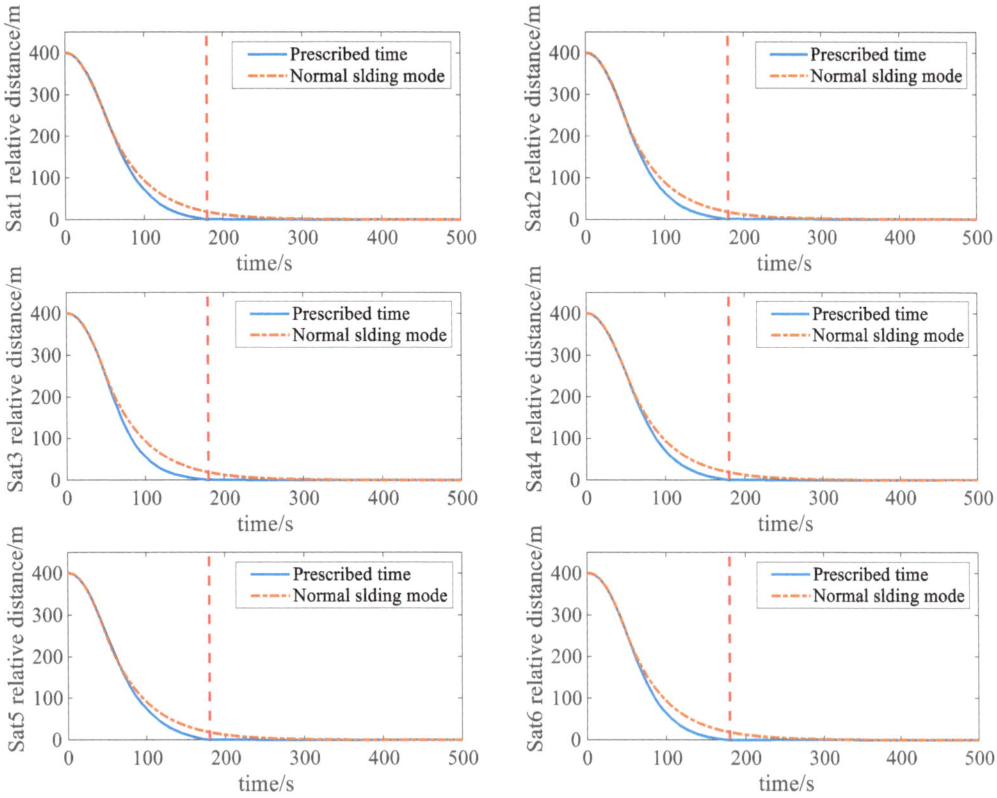

Fig. (6). Relative distance curve using two different control methods.

In Fig. (**6**), we can see the curves of relative distance to the target points of the satellites in cluster under the action of the different controllers. Because the initial configuration and the new configuration are uniformly distributed on two ellipses with different semi-major axes, the relative distance curve between each satellite and the target point is similar. And it can be seen clearly that relative distance of satellite can converge to 0 at 180s under the prescribed time controller which is faster than the normal sliding mode controller obviously. Moreover, for the prescribed time controller, the convergence time can be designed according to the task requirements and control force constraints, so it is more suitable for the reconstruction of the satellite cluster.

Finally, the reconstruction process of satellite cluster using prescribed time controller and normal sliding mode controller are shown in Fig. (**7**). We can see that both controllers can ensure the satellite cluster to complete the reconstruction.

The prescribed time controller has the faster convergence speed. Therefore, the prescribed time controller has a better control effect for the mission of satellite cluster reconstruction with time requirement.

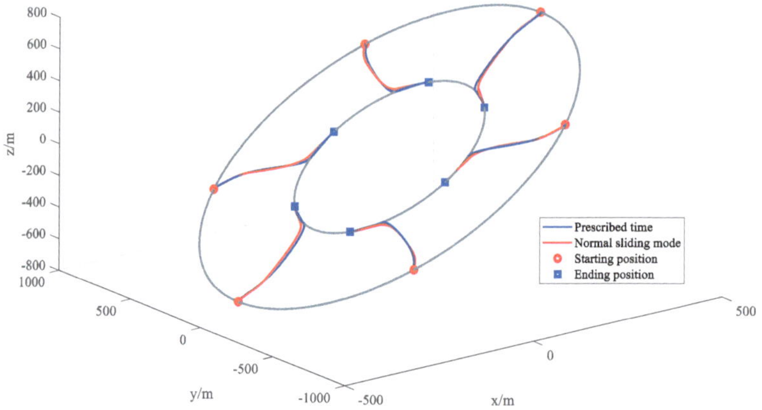

Fig. (7). Spatial motion of satellite cluster reconstruction.

CONCLUSION

This paper proposes a prescribed time control strategy for satellite cluster formation reconstruction considering multiple environmental disturbances and thruster constraints. To compensate for unmodeled external disturbances, a FTCESO is used to estimate the total disturbances of the satellite cluster in a finite time. Based on it, a prescribed time controller is designed for satellite cluster reconstruction which adopts piecewise control law. By Lyapunov stability theory, the stability of the system is proved. So that the satellite cluster can be transformed into a new specified configuration at a prescribed time. Finally, the comparative simulation has been carried out. Although the control method proposed in this paper is not smooth enough, the control effect is better and more suitable for task requirements.

CONSENT FOR PUBLICATION

Not applicable.

CONFLICT OF INTEREST

The authors declare no conflict of interest, financial or otherwise.

ACKNOWLEDGEMENT

Declared none.

REFERENCES

[1] Liu C, Yue X, Shi K, *et al.* Spacecraft Attitude Control: A Linear Matrix Inequality Approach. Elsevier/ Science Press, 2022, ISBN: 978-0-323-99005-9.
[2] Popov A M, Kostin I, Fadeeva J, *et al.* Development and Simulation of Motion Control System for Small Satellites Formation. Electronics, 2021, 10(24): 3111.
[3] Selva D, Golkar A, Korobova O, *et al.* Distributed earth satellite systems: What is needed to move forward? Journal of Aerospace Information Systems, 2017, 14(8): 412-438.
[4] Bandyopadhyay S, Subramanian G P, Foust R, *et al.* A review of impending small satellite formation flying missions. 53rd AIAA Aerospace Sciences Meeting. 2015: 1623.
[5] Zhang H, Gurfil P. Satellite cluster flight using on-off cyclic control. Acta Astronautica, 2015, 106: 1-12.
[6] L. Mazal, P. Gurfil, Closed-loop distance-keeping for long-term satellite cluster flight. Acta Astronautica, 2014,94 (1): 73–82.
[7] Zhao Y, Gurfil P, Zhang S. Long-term orbital dynamics of smart dust. Journal of Spacecraft and Rockets, 2018, 55(1): 125-142.
[8] Varma S, Kumar K D. Multiple satellite formation flying using differential aerodynamic drag. Journal of Spacecraft and Rockets, 2012, 49(2): 325-336.
[9] Morgan D, Chung S J, Blackmore L, *et al.* Swarm-keeping strategies for spacecraft under J2 and atmospheric drag perturbations. Journal of Guidance, Control, and Dynamics, 2012, 35(5): 1492-1506.
[10] Chen W H, Yang J, Guo L, *et al.* Disturbance-observer-based control and related methods—An overview. IEEE Transactions on industrial electronics, 2015, 63(2): 1083-1095.
[11] Jingqing H. The "Extended State Observer" of a Class of Uncertain Systems. CONTROL AND DECIION, 1995.
[12] Ran M, Wang Q, Dong C. Stabilization of a class of nonlinear systems with actuator saturation *via* active disturbance rejection control. Automatica, 2016, 63: 302-310.
[13] Zhang Y, Yang L, Zhu Y, *et al.* Nonlinear 6-DOF control of spacecraft docking with inter-satellite electromagnetic force. Acta Astronautica, 2012, 77: 97-108.
[14] Yang H, You X, Xia Y, *et al.* Adaptive control for attitude synchronization of spacecraft formation *via* extended state observer. IET Control Theory & Applications, 2014, 8(18): 2171-2185.
[15] Guo B Z, Zhao Z. On the convergence of an extended state observer for nonlinear systems with uncertainty. Systems & Control Letters, 2011, 60(6): 420-430.
[16] Liu P, Xue W, Chen S, *et al.* An integrated solution to ACMM problem of spacecraft with inertia uncertainty. International Journal of Robust and Nonlinear Control, 2018, 28(17): 5575-5589.
[17] Zhao Z L, Guo B Z. A nonlinear extended state observer based on fractional power functions.

Automatica, 2017, 81: 286-296.

[18] Liu C, Yue X, Shi K, *et al.* Inertia-free attitude stabilization for flexible spacecraft with active vibration suppression. International Journal of Robust and Nonlinear Control, 2019, 29(18): 6311-6336.

[19] Zhao D J, Yang D G. Model-free control of quad-rotor vehicle via finite-time convergent extended state observer. International Journal of Control, Automation and Systems, 2016, 14(1): 242-254.

[20] Massey T, Shtessel Y. Continuous traditional and high-order sliding modes for satellite formation control. Journal of Guidance, Control, and Dynamics, 2005, 28(4): 826-831.

[21] Li J, Pan Y, Kumar K D. Design of asymptotic second-order sliding mode control for satellite formation flying. Journal of guidance, control, and dynamics, 2012, 35(1): 309-316.

[22] Xu Y. Sliding mode control and optimization for six DOF satellite formation flying considering saturation. The Journal of the Astronautical Sciences, 2005, 53(4): 433-443.

[23] Shahid K, Kumar K D. Satellite formation flying using variable structure model reference adaptive control. Proceedings of the Institution of Mechanical Engineers, Part G: Journal of Aerospace Engineering, 2009, 223(3): 271-283.

[24] Shi K, Liu C, Biggs J D, *et al.* Observer-based control for spacecraft electromagnetic docking. Aerospace Science and Technology, 2020, 99: 105759.

[25] Liu R, Cao X, Liu M. Finite-time synchronization control of spacecraft formation with network-induced communication delay. IEEE Access, 2017, 5: 27242-27253.

[26] Zuo Z. Non-singular fixed-time terminal sliding mode control of non-linear systems. IET control theory & applications, 2015, 9(4): 545-552.

[27] Zhang J, Kong X, Liu C, *et al.* Agile attitude maneuver with active vibration-suppression for flexible spacecraft. Journal of the Franklin Institute, 2022, 359(3): 1172-1195.

[28] Gao Z, Guo G. Fixed-time sliding mode formation control of AUVs based on a disturbance observer. IEEE/CAA Journal of Automatica Sinica, 2020, 7(2): 539-545.

[29] Sánchez-Torres J D, Sanchez E N, Loukianov A G. Predefined-time stability of dynamical systems with sliding modes. American control conference (ACC). IEEE, 2015: 5842-5846.(definition 1)

[30] Wang Z, Liang B, Sun Y, *et al.* Adaptive fault-tolerant prescribed-time control for teleoperation systems with position error constraints. IEEE Transactions on Industrial Informatics, 2019, 16(7): 4889-4899.

[31] Cui L, Jin N. Prescribed-time ESO-based prescribed-time control and its application to partial IGC design. Nonlinear Dynamics, 2021: 1-18.

[32] Zuo Z, Han Q L, Ning B, *et al.* An overview of recent advances in fixed-time cooperative control of multiagent systems. IEEE Transactions on Industrial Informatics, 2018, 14(6): 2322-2334.

[33] Du H, Wen G, Wu D, *et al.* Distributed fixed-time consensus for nonlinear heterogeneous multi-agent systems. Automatica, 2020, 113: 108797.

[34] Ren Y, Zhou W, Li Z, *et al.* Prescribed-time cluster lag consensus control for second-order non-linear leader-following multiagent systems. ISA transactions, 2021, 109: 49-60.

SUBJECT INDEX

A

Ability 1, 2, 175
  energy convention 175
  fault-tolerant 1, 2
Adaptive local variational iteration method (ALVIM) 31
Aerospace systems 157, 182
Amplitude 164, 168, 178, 179
  forcing 178, 179
  modulation 164
Angular velocity 95, 96, 98, 111, 112, 113, 114, 115, 117, 123, 130, 149, 150, 215, 216, 245
  orbital 245
Atmospheric drag perturbations 59

B

Barbalat's lemma 80
Bifurcation analysis 160
Bilinear matrix inequalities (BMIs) 138

C

Cargo transportation 187, 209
Climber 185, 186, 187, 188, 189, 190, 191, 192, 194, 195, 198, 203, 205, 206, 209
  analytical 185, 198
  fast-transporting 186
Coefficients, equivalent damping 172
Communication 69, 127, 129
  pressure 127, 129
  wireless 69
Communication topology 214, 221, 231
  bidirectional 221
Conditions, resonance 159, 177
Consumption, fuel 73, 242
Control 1, 82, 83, 85, 112, 114, 115, 119, 121, 130, 133, 145, 146, 150, 187, 213, 216, 243
  fuzzy 1
  synchronization 243
  tension 187
  torque 82, 83, 85, 112, 114, 115, 119, 121, 130, 133, 145, 146, 150, 213, 216
Convergence 13, 16, 32, 40, 43, 54, 55, 58, 95, 96, 97, 100, 105, 123, 241, 243
  process 105
Coriolis forces 185, 195
Cycle oscillations 157

D

Damage, structural fatigue 127, 128
Damping 146, 156, 157, 158, 159, 161, 162, 169, 170, 175
  equivalent electric 175
  linear 156, 157
  truncation 162, 169
Developing broadband energy harvesting systems 157
Dexterous tethered space robot (DTSR) 187
Distance 46, 59, 245
  geocentric 59, 245
Drag force, atmospheric 60
Dynamic(s) 2, 4, 69, 71, 128, 130, 155, 158, 160, 161, 186, 187, 188, 189, 192, 199, 213
  coupling 213
  flexible 128, 130
  input-output 199
  longitudinal 2
  nonlinear 155, 160, 161

E

ECI coordinate system 243, 247
Effective vibration suppression 157, 163, 167
Effects, magnetic 135
Electromagnetic force 171, 172
Elevator angular deflection 3, 19, 22, 23, 26

Energy 68, 155, 168, 169, 170, 175, 176, 177, 181, 195
electric 170, 176, 177
harvesters 155, 181
spectrum 168, 169, 170
transfer 155
vibrational 169, 176
Energy harvesting 155, 157, 170
applications 157
system 170
techniques 155
vibration 155
Errors, sensor calibration 133
Euler-Lagrangian systems 97
Event-triggered controller 90, 129
adaptive 90
Event-triggering strategy 70, 89, 90

F

FAPI methods 60, 63
Faraday's law 172
Fast maneuvering issue 100
Finite 9, 13, 31, 100, 101, 104, 105, 108, 110, 242, 243, 249
difference methods 31
time 9, 13, 100, 101, 104, 105, 108, 110, 242, 243, 249
Finite-time 1, 95, 96, 97, 100, 103, 109, 121, 123, 214
controller, robust adaptive 95
convergence laws 1
stability 95, 96, 97, 100, 103, 109, 121, 123, 214
stabilization methods 96
Fish-scales-growing method (FSGM) 32, 33, 50, 53, 54, 55, 57, 65
Fixed-time 1, 2, 16, 27
convergence 1, 2, 16, 27
high-order regulator 2
Flexible spacecraft attitude control 90, 128
Flywheel system 157
Force evaluation 46
FTESO and prescribed time controller 243
Fuel exhaustion 127, 128
Function 74, 96, 150, 190, 220, 221, 242
desired uncompensated acceleration 190
energy index 150
nonsingular FTTSM 220, 221
smooth 74, 242
symbolic 96
Fuzzy 68, 69, 70, 75, 76, 97
logic system (FLS) 68, 69, 70, 75, 76
sensor data 97

G

Gaussian membership function 75
Gauss-Jackson method 39
Geometric tracking control design 74
Geosynchronous earth orbit (GEO) 39, 40, 44, 47, 50
Gravitational 98, 242, 245
attraction 242
gradient moments 98
perturbation 245
Gravitational constant 59, 245
geocentric 59
Gravity force 41

H

Hamiltonian system 32
Harvesting circuit 171
Harvesting efficiency 178, 179, 181
high energy 178
HEO transfer 51, 55, 58
High 39, 40, 41, 44, 46, 47, 50, 52, 59, 60, 61, 62
eccentricity 52
eccentricity transfer orbit 62
eccentric orbit (HEO) 39, 40, 41, 44, 46, 47, 50, 59, 60, 61, 62
Hopf bifurcation 161, 162
condition 161
vanish 162
Hypersonic flight vehicles 1
Hypersonic Vehicle 1, 3, 5, 7, 9, 11, 13, 15, 17, 19, 21, 23, 25, 27, 29
flexible air-breathing 1

I

Industrial systems 214
Inertia 113, 127
estimation system 113
free attitude stabilization 127
Initial 32, 36, 37, 38, 39, 51, 60, 62, 63, 64, 175, 194, 255

libration angle and angular velocity 194
slope 39
value problems (IVPs) 32, 36, 37, 38, 64
velocities 39, 51, 60, 62, 63, 175, 255
Input 2, 234
control force 234
output feedback linearization technique 2
Intervals 11, 53, 84, 151, 152, 162, 166, 167, 170
multiple isometric 53
Iteration tolerance 60
Iterative process 148

J

Jacobian matrices 32, 60

K

Keplerian solution 58
Kronecker product 222, 224

L

Lagrange multipliers 34, 35
Lambert's problem 32, 50, 51, 60, 64
Lambert transfer problems 51
Laplacian matrix 217
Law 97, 173, 187, 201
space exploration 97
switching 201
Leader satellite 243, 245, 247, 254
virtual 254
Linear 39, 157, 158, 159, 170, 172
combination 39
dampers 159
damping force 172
device 158
harvesters, traditional 157
oscillator 158, 159, 170
Lipschitz constants 14
Load 155, 156, 172, 177
dynamic 155
resistance 172, 177
Local 31, 32, 34, 36, 40, 43, 46, 47, 49, 50, 51, 63, 243, 244
variational iteration method (LVIM) 31, 32, 34, 36, 40, 43, 46, 47, 49, 50, 51, 63
vertical local-horizontal (LVLH) 243, 244
Long-duration transfer problems 55
Low earth orbit (LEO) 39, 40, 44, 46, 47, 59, 60, 61, 62, 64, 82, 83, 245, 246, 247
Lumped 1, 2, 6, 14, 27, 127, 129, 134, 188
disturbances 1, 2, 6, 14, 27, 127, 129, 134
masses 188
LVLH coordinate system 243, 244, 245, 247, 255
Lyapunov 8, 9, 11, 80, 95, 96, 97, 103, 105, 109, 220, 198, 241, 243, 252, 253, 260
and terminal sliding mode methods 96
approach 243
candidate function 198
function 8, 9, 11, 80, 103, 105, 109, 220, 252, 253
method 95, 96, 97
stability theory 241, 260

M

Magnetic forces 220
Matrix 7, 34, 71, 72, 130, 134, 135, 136, 142, 147, 200, 219, 249
coupling coefficient 130
feedback coefficient 147
transfer function 135
Methods 128, 135, 138
decoupling 138
fault-tolerant 135
robust 128
Modified Chebyshev-Picard iteration (MCPI) 32, 40, 41, 43, 44, 46, 51
method 41, 43, 44, 51
Motion, non-perturbed Keplerian 50
Motor, integrated 97
Multi-agent systems 243
Multi-revolution 51, 59
lambert's problem 59
lambert transfer problem 51

N

Negative imaginary lemma 135
NES systems 179
Networked control systems (NCSs) 214
Newton's methods 63
Nodal position finite element method (NPFEM) 186
Noise 116, 155

aerodynamic 155
Nonlinear 67, 69, 185
control techniques 67
dynamic system 69, 185
Nuclear power stations 214
Numerical methods 31, 32
traditional 32

O

Observer, nonlinear 242
Orbital 31, 33, 59
elements 33
mechanics 31
Orbits 181, 232
circular 232
periodic 181
Orbit propagation 32, 64
problem 32
Orbit transfer problems 55, 58
multi-revolution 58
Oscillators 175, 178

P

Parameters, perturbed 145
Performance 18, 20, 23, 26, 40, 46, 89, 133, 136, 140, 142, 176, 178, 181, 209, 234
attitude synchronization 234
attitude tracking 234
velocity tracking 18, 20, 23
Periodic response 169, 178, 179
linear 178
nonlinear 179
Perturbation(s) 51, 59, 95, 98, 144, 161, 242, 243, 245, 246, 247, 255
additive 144
force 246
motion 161
solar radiation pressure 59
Perturbed 50, 64
force 50
two-body system 64
Power 123, 177, 181
fractional 123
harvesting 177, 181
Problem, conservative 32
Proportion-integral-derivative (PID) 68

Q

Quasi-linearization method 34, 36

R

Reconstruction process 259
Rectangular coordinate system 33
Regime 176, 178, 179
chaotic 178, 179
linear 176
resonance vibro-impact 176
Relative motion modeling 33
Resonance frequency 162, 169, 170, 180
Response, vibrational 162

S

Satellite attitude 123
Satellite attitude stabilization 97, 100, 128
issue 100
problem 97
Satellite cluster 241, 242, 254, 260
formation reconstruction 241, 260
orbits 242
reconfiguration 254
Satellite(s) 95, 96, 98, 241, 242, 243, 244, 245, 246, 247, 254, 255, 256, 258, 259
control 96
convergence 242
coordinate system 244
formation and satellite cluster 242
formation reconstruction 242
position vector 246
reconstruction 243
virtual 243
Sensors 128, 129
airborne 129
SFF missions 213
Signorini contact law 173
Singularities, kinematic 67
Sliding mode 1, 67, 68, 69, 70, 95
control (SMC) 1, 67, 68, 69, 70, 95
controller, event-triggered adaptive 68
Sliding surface system 200
Slow invariant manifold (SIM) 164, 166
SMR 167, 179, 181
for effective vibration suppression 167
regime 179, 181

SOF controller 129, 135, 140, 143, 145, 149
Solar 83, 98
  pressure 98
  radial pressure 83
Solar radiation pressure 74, 130, 220, 242
  torque 130
Solving multi-revolution orbit transfer problems 58
Spacecraft 69, 214
  dynamics 69
  formation system 214
Spacecraft attitude 71, 74, 85
  control system 85
Spacecraft attitude tracking 70, 73, 74
  problem 70, 74
  system 73
Space transportation system 185
Spatial 242, 246, 247, 260
  configuration 242, 246, 247
  motion of satellite cluster reconstruction 260
Speed function 185, 186, 190, 191, 194
  analytical 186
Stability 70, 74, 81, 90, 129, 133, 161, 187, 200, 260
  analysis 70, 74, 129
States, sensor-measured 133
Static output feedback (SOF) 127, 129, 136, 141, 152
Strategy 32, 70, 82
  asymptotic homotopy 32
  relative threshold 70
Strongly modulated response (SMR) 157, 163, 164, 166, 167, 169, 178, 179, 181
System 68, 180, 214
  engineering 214
  response regimes 180
  robustness 68

T

Targeted energy transfer (TET) 155, 156, 157, 176, 179, 181
Technique 156, 170
  electromagnetic transduction 170
  vibrational energy harvesting 156
Technology, aerospace industry 241
Terminal sliding mode (TSM) 96, 213, 220, 222, 242, 243
Tether tensions 186, 192, 195, 198, 207, 208
TET phenomenon 155, 156
Theory 39, 67
  fuzzy logic 67
  linear differential equation 39
Tools 33
  mathematical 33
Topology 232
Torque 130, 216, 218, 233, 234, 236
  aerodynamic 130
  gravity gradient 130, 216
Total energy consumption 150, 151
  and vibration energy 151
  index 150
Traditional reaching laws 2
Trajectories 41, 54, 60, 83, 96, 101, 166, 167
  deceleration process 96
Transfer 51, 52, 60, 195, 203
  orbits 51, 52, 60
  period 195, 203
Transformation 36, 37, 111, 136
  diagonal matrix 111
  equivalent 136
Transportation 185, 190, 209
  mission 185
Triggering 69, 70, 76, 77, 79, 81, 84, 85, 86, 89, 90, 151
  instants 76, 77, 85, 89, 90, 151
  strategy 69, 70, 76, 79, 81, 84, 85, 86, 90
Tuned mass damper (TMD) 156
Two-point boundary value problems (TPBVPs) 32, 33, 36, 50, 53, 54, 55, 60, 63, 65

V

Variable estimation system 95
Variational iteration method (VIM) 34, 35, 36, 63
Velocity 19, 20, 26, 33, 46, 51, 52, 58, 60, 63, 173, 178, 216, 232, 233, 255
  boundary 52
  desired 216, 255
  orbit angel 232
Velocity vectors 60, 102, 103, 247
  angular 102, 103
Vibration(s) 105, 128, 151, 155, 156, 157, 175, 177, 181
  absorbers, effective 181
  elastic 128
  energy 151

intensity 151
mechanical 155, 177
mitigate 105
mitigation 157
transient 155
Vibration suppression 128, 157, 158, 162, 181, 182
elastic 128
Vibrational systems 155
Vibro-impact nonlinear energy sink (VINES) 158, 170, 175, 176, 177, 180, 181
VINES 157, 158, 175
damping 175
electromagnetic 157, 158
for energy harvesting 157

W

Weakly modulated response (WMR) 179
Wireless networks 69, 214

www.ingramcontent.com/pod-product-compliance
Lightning Source LLC
Chambersburg PA
CBHW050817220326
41598CB00006B/236
* 9 7 8 9 8 1 5 0 5 0 0 4 2 *